www.wadsworth.com

wadsworth.com is the World Wide Web site for Wadsworth and is your direct source to dozens of online resources.

At wadsworth.com you can find out about supplements, demonstration software, and student resources. You can also send e-mail to many of our authors and preview new publications and exciting new technologies.

wadsworth.com
Changing the way the world learns®

FOURTH EDITION

Interpersonal Process in Psychotherapy
A Relational Approach

Edward Teyber
California State University, San Bernardino

Brooks/Cole
Thomson Learning™

Australia • Canada • Denmark • Japan • Mexico • New Zealand • Philippines
Puerto Rico • Singapore • South Africa • Spain • United Kingdom • United States

Counseling Editor: Eileen Murphy
Assistant Editor: Julie Martinez
Editorial Assistant: Annie Berterretche
Marketing Manager: Caroline Concilla
Project Editor: Marlene Vasilieff
Print Buyer: Mary Noel
Permissions Editor: Bob Kauser

Production Service, Text Design, and Composition:
   Pre-Press Company, Inc.
Cover Designer: Yvo Riezebos
Cover Image: PhotoDisc
Printer/Binder: RR Donnelley & Sons/
   Crawfordsville

For permission to use material from this
text, contact us:
Web: www.thomsonrights.com
Fax: 1-800-730-2215
Phone: 1-800-730-2214

For more information contact
Wadsworth/Thomson Learning
10 Davis Drive
Belmont, CA 94002-3098
USA
www.wadsworth.com

International Headquarters
Thomson Learning
290 Harbor Drive, 2nd Floor
Stamford, CT 06902-7477
USA

UK/Europe/Middle East
Thomson Learning
Berkshire House
168-173 High Holborn
London WC1V 7AA
United Kingdom

Asia
Thomson Learning
60 Albert Street #15-01
Albert Complex
Singapore 189969

Canada
Nelson/Thomson Learning
1120 Birchmount Road
Scarborough, Ontario M1K 5G4
Canada

Library of Congress
Cataloging-in-Publication Data
Teyber, Edward.
   Interpersonal process in psychotherapy:
   a relational approach / Edward Teyber.—
   4th ed.
   p.   cm.
   Includes bibliographical references
   and index.
   ISBN 0-534-36295-8 (acid-free paper)

   1. Psychotherapist and patient. 2. Psychotherapy.
   I. Title. 1.Psychotherapist and patient. 2. Psychotherapy.
   I. Title.
   RC480.8.T38 1999
   616.89' 14--dc21   99-046368

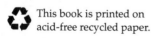
This book is printed on
acid-free recycled paper.

Dedicated to those
who are struggling
to change

ABOUT THE AUTHOR

Edward Teyber is a professor of psychology and director of the Psychology Clinic at California State University, San Bernardino. Dr. Teyber received his Ph.D. in clinical psychology from Michigan State University. He is also the author of the popular-press book *Helping Children Cope with Divorce* and coeditor with Dr. Faith McClure of the text *Child and Adolescent Therapy: A Multicultural-Relational Approach*. He has published research articles on the effects of marital and family relations on child adjustment and contributes articles on parenting and postdivorce family relations to newspapers and magazines. Dr. Teyber also enjoys supervision and clinical training, and he maintains a part-time private practice.

CONTENTS

Chapter Three
Honoring the Client's Resistance   58

Chapter Four
An Internal Focus for Change   87

Chapter Five
Responding to Conflicted Emotions 116

Part Three
Conceptualizing Client Dynamics   149

Chapter Six
Familial and Developmental Factors   151

Chapter Seven
Inflexible Interpersonal Coping Strategies   179

Chapter Eight
Current Interpersonal Factors   209

Part Four
Resolution and Change   239

Chapter Nine
An Interpersonal Solution   241

PREFACE

The relationship between the therapist and the client is the foundation of therapeutic change. Across differing theoretical approaches, and in short- or longer-term modalities, researchers find that the best predictor of treatment outcome is the therapist's ability to establish a strong working alliance early in treatment. Despite this knowledge, few texts provide specific guidelines for understanding and intervening in the therapeutic relationship. The purpose of this text is to go beyond generalizations about the importance of the therapeutic relationship and present a comprehensive treatment approach that teaches beginning therapists how to use themselves and the relationships they establish with their clients to help clients change.

Clinical training can be very stressful because beginning therapists are often painfully uncertain of how to proceed with their clients. Although trainees take courses on counseling theories and interpersonal skills development, this preparation is too abstract to give students the in-depth understanding needed for direct clinical practice. Fully cognizant of their limited experience and knowledge, these trainees are often anxiously aware that they do not really know what to do or how to help their clients. Many fell inadequate and worry about making mistakes and doing something wrong that would hurt their clients. As a result, beginning therapists are eager for practical instruction that will provide specific guidelines to help them respond effectively to their clients' problems. This text meets this need by replacing the nebulous ideas that students often have about treatment with a well-developed framework for understanding how change occurs and the role of the therapist–client relationship in the change process. Specifically, beginning counselors will learns how to conceptualize what is occurring in their real-life relationships with clients and how to use this understanding to establish a treatment focus and to guide their intervention plans.

This text elucidates each of the major issues that arise in treatment and shows how theory leads to practice. It examines actual clinical process with clarity and immediacy and captures beginning therapists' questions and concerns. The book presents a relational framework that integrates cognitive-behavioral, family systems, and interpersonal-dynamic theories. This framework fosters therapists' flexibility by allowing them to draw from a variety of theoretical perspectives while also showing trainees how to develop an effective treatment focus. Intervention guidelines are linked to a theoretical rationale and, throughout each chapter, are illustrated with numerous therapist–client dialogues, clinical vignettes, and detailed case studies. Clinically authentic and personally engaging, this text will help students understand the therapeutic process and how change occurs. It is written for graduate student therapists in practicum/internship courses who are seeing clients, and it is also suitable for students in upper-division and pre-practicum courses that provide a rigorous introduction to counseling and therapy.

Over the years, I have appreciated the opportunity to return to this work and try to make the interpersonal-process approach more helpful to beginning therapists. The goals of this revision were to further clarify certain basic concepts, such as maladaptive relational patterns and cognitive schemas; to highlight how the cognitive, affective, and interpersonal domains interact and influence each other; and to make this edition easier to read by providing more structural and organizational material. Without lengthening the book, I have also tried to foster multicultural awareness by using case studies that illustrate how the clients' cultural context effects the treatment process. Preparing trainees to practice in a managed-care milieu, I have made further links between the interpersonal-process approach and other treatment modalities: couples therapy, group therapy, family therapy, and, especially, time-limited treatment. In addition, this edition includes structured guidelines to help trainees write the case formulations, treatment plans, and process notes that are necessary in the current managed-care environment. Finally, the pervasive role of insecure attachments and attachment-related affects, especially the taboo affect of shame, has been expanded in this edition.

For the first time, this edition provides instructors and students with a complete textbook package including a student workbook, a demonstration video, and an instructor's manual. The student workbook to accompany the text is an especially important aid to help students integrate the personally challenging and evocative material presented in the text. The student workbook provides a study guide with key concepts and terms, case studies of characters from well-known films and plays, self-assessment measures of students' progress in developing process-oriented skills and awareness of their own counter transference propensities, and self-reflection questions to help students integrate and apply the material to their own lives. A demonstration video allows trainees to view two therapists applying concepts from the text in three different situations that are especially challenging for beginning therapists: (1) when a client is pressing the therapist for more self-disclosure from the therapist, (2) when an angry client is criticizing and challenging the therapist, and (3) when a client is resistant to treatment because of ethnic, religious, or gender differences. Finally, an instructor's manual accompanies the text. The manual contains an instructor's lecture outline, in-class student exercises, essay questions, and multiple-choice questions for every chapter.

Acknowledgments

Several people have continued to help me become a better therapist and to improve this book. Foremost, I wish to thank Margaret Dodds-Schumacher. I have admired her expertise both as an editor and a therapist and appreciated her generous spirit. She has made extensive contributions to the fourth edition. Thank you to the reviewers of this new edition, in particular Bob Egbert, Southern Adventist University; Dennis Myers, Baylor University; and Susan W. Grey, Barry University–Ellen Whiteside McDonnell School of Social Work. Longtime friends Faith McClure, Sam Plyler, and Keith Valone continue to help me learn more about this work and to present it more effectively. Several former graduate students, now colleagues, have also helped greatly. I would especially like to thank Meg Robertson, Linda Emerson, and Dayle Hill for all they have given. Finally, it is always a pleasure to work with Eileen Murphy, counseling editor, and my colleagues at Brooks/Cole.

Edward Teyber

FOURTH EDITION

Interpersonal Process in Psychotherapy

A Relational Approach

PART ONE

An Interpersonal Process Approach

CHAPTER ONE
Introduction and Overview

Introduction and Overview

Claire, a first-year practicum student, was about to see her first client. She had long and eagerly awaited this event. Like many of her classmates, Claire had decided to become a therapist* while working on her undergraduate degree. Counseling had always been intrinsically interesting to her. For Claire, being a therapist meant far more than having a "good job"; it was the fulfillment of a dream. How exciting, she thought, to make a living by helping people with the most important concerns in their lives.

At this moment, though, Claire felt the real test was at hand: Her first client would be arriving in a few minutes. Worries raced through her mind: What will we do for fifty minutes? How should I start? What if she doesn't show up? What if I do something wrong and she doesn't come back next time? Claire also wondered if her client—a 45-year-old Hispanic woman—would have difficulty relating to her, a white woman in her mid-twenties. Needless to say, Claire was intent upon finding a way to help this client with her problems. She had learned something about therapy in her undergraduate psychology classes, and a good deal more from her volunteer experience with callers on the local crisis hotline. But even with these experiences and a supervisor to guide her, Claire was painfully aware of her novice status and of the fact that she didn't know very much about actually doing therapy.

Although some of Claire's classmates thought she was a little too idealistic, they did share her excitement about becoming a therapist. Many of them were older than Claire and far more experienced in life. Some had raised children; others had already had careers as teachers, nurses, and businesspeople. These new therapists were often coming from life roles in which they had already been successful and felt confident. A counseling career held new hopes for these more seasoned classmates as well, but also evoked anxiety about their ability to become effective therapists. With each successive step—completing undergraduate requirements, being accepted into a graduate program, and now starting their first practicum—they were approaching their new career goals. Like Claire, however, they knew that realizing their dreams of a meaningful new career would also depend on their ability to help others. And with the arrival of their first clients, that ability was about to be tested.

*The terms "counselor" and "therapist" are used interchangeably throughout the text.

The Need for a Conceptual Framework

Usually, beginning therapists' anxieties about their ability to help are not neu-rotic insecurity or obsessive worrying. It is realistic to be concerned about one's performance in a new, complex, and ambiguous arena. Sadly, however, Claire and her classmates often lose sight of the many strengths they bring to their first counseling experiences. Most students who are selected for clinical training al-ready possess sensitivity, intelligence, and a genuine concern for others. Such personal assets, and all they have learned from their own life experiences, will prove enormously helpful to their future clients.

Personal experience, common sense, and intuition are useful, indeed. In order for a therapist to be effective with a wide range of people and problems, however, these valuable human qualities must be wed to a conceptual frame-work. Therapists must understand very specifically what they are trying to do in therapy—where they are going and why—in order to be consistently helpful to clients. Without a conceptual framework as a guide, decisions about inter-vention strategies and case management are too arbitrary to be trustworthy, and the therapist's confidence in the therapeutic enterprise lessens.

In the current managed-care milieu, there has been a profound limitation placed on psychological resources for clients. Whereas effective therapy has al-ways required thoughtful treatment planning, this now takes on additional im-portance. Practitioners will not succeed in obtaining treatment authorizations, or in negotiating additional sessions for clients from insurance caseworkers, if they cannot formulate specific treatment goals and intervention plans. As basic concepts are successively built upon throughout this text, case studies will be pro-vided that illustrate effective case formulations. Comprehensive case studies, that follow the guidelines for writing case formulations and treatment plans in Ap-pendix B, are also provided in the Student Workbook that accompanies this text.

Many therapists become painfully anxious in their early counseling work because they lack a practical conceptual framework. Uncertainty over how to conceptualize client problems and ambiguous guidelines for how to proceed in the therapy session heighten their insecurity and preoccupation with their per-formance. As a result, both their effectiveness with clients and their ability to take pleasure in this special work is diminished.

Finally, the lack of a conceptual framework hinders novice therapists as they try to find their own professional identities. Before therapists can feel se-cure in their abilities, they will need to integrate a theoretical framework that is applicable with diverse clients and congruent with their own values and life experiences. The purpose of this book is to provide a conceptual framework for understanding the therapist-client relationship and the developmental course of therapy. Throughout this text, readers will be encouraged to modify this framework to better fit their own personalities and therapeutic styles. The principles suggested here may be applied by therapists who have different per-sonality styles, have grown up in a range of different cultural contexts, and who choose to work within varying theoretical approaches. With support from supervisors and instructors, beginning therapists will be able to personalize this model, integrate it with other approaches, and modify it to make it their own.

What will this conceptual framework entail? This framework will help therapists recognize relational patterns or track repetitive themes in the concerns clients present. Finding such patterns provides a focus for treatment and gives clients structure and direction from which to proceed. As we will see, linking these relational patterns to the current interaction with the therapist is a highly effective way to intervene.

The Interpersonal Process Approach
Theoretical and Historical Context

The purpose of this applied text is to help practicum students understand their clients' problems, and use this conceptualization to formulate treatment plans and intervention strategies that guide what they say and do with clients. The therapeutic model presented here will be referred to as the *interpersonal process approach.* This is not a new theory of psychotherapy but a synthesis of ideas from three rich clinical traditions. First, the collection of interpersonally-oriented theorists who have written within a psychodynamic tradition and are known variously as Sullivanian, neo-Freudian, interpersonal, or intersubjective therapists. The second tradition is the relational branch of object relations theory (as opposed to the traditional and better known psychoanalytic school of drive/libido theory), and the attachment literature that has developed from it. The third clinical tradition is family systems theory. Although these approaches have a different focus—on interpersonal, intrapsychic, and intrafamilial domains, respectively—they share a *relational* emphasis and are fundamentally similar. This text highlights the relational dimension in each of these three approaches and links these concepts to the practice of individual psychotherapy. These theories are briefly introduced here, so as to provide a theoretical context for the clinical approach outlined in the remainder of the book. There is a wealth of practical information in these three clinical theories, but far more complexity than beginning therapists can assimilate in a short time. However, the clinical guidelines in Chapters 2–10 are derived from these sources, and readers may wish to explore these conceptual domains further as they continue their training. Discussions of the three clinical traditions follow.

Interpersonal theory   The interpersonal dimension in psychotherapy was originally highlighted by Harry Stack Sullivan and is articulated and studied today in varying forms by Hans Strupp, Michael Kahn, Donald Kiesler, Gerald Klerman, Myrna Wasserman, Lorna Benjamin, and many others. Sullivan (1968) first brought the relational focus to psychotherapy in the 1940s; he remains an enormously influential but insufficiently recognized figure. A maverick, Sullivan radically broke with Freud's biologically based libido theory and was one of the first major theorists to argue that the basic premises of Freud's drive theory (for example, sexual and aggressive instincts) were inaccurate. Also, rather than Freud's focus on fantasy and intrapsychic processes, Sullivan emphasized interpersonal relations and the child's actual experience with parents. He argued that every major tenet of Freud's theory could be understood better through interpersonal or behavioral experiences. Sullivan developed an elaborate developmental theory of psychopathology that emphasized what people do to avoid or manage anxiety in interpersonal relations. He viewed anxiety, and elaborate attempts to avoid or minimize it, as the central motivating force in human behavior. He described a painful core anxiety that was rooted in dreaded expectations of derogation and rejection by parents and others, and later by oneself. Following from this, Sullivan conceptualized personality as the collection of interpersonal strategies that the individual employs to avoid or minimize anxiety, ward off disapproval, and maintain self-esteem.

According to Sullivan, the child develops this personality, or *self-system,* through repetitive interactions with her parents. More specifically, the child organizes a self around certain basic patterns of parent-child interaction. These repetitive interactions with parents, which may be painful or anxiety-arousing, are structured into the child's personality through a collection of complementary self-other relational configurations. For example, children may develop internal images of themselves as helpless or insignificant and expectations of parents and others as demanding or critical. Alternatively, children may evolve images of themselves as special and expectations of others as admiring.

Having learned these self-other relational patterns, people systematically behave in ways that avoid or minimize the experience of anxiety. For example, suppose that particular aspects of the child, such as sadness or crying, consistently result in parental rejection or ridicule. The child learns that these parts of herself are unacceptable and represent a "bad self," and these anxiety-arousing aspects of the self are split off or disowned. The child also develops interpersonal coping styles (for example, a macho manner) that preclude being subjected to anxiety-arousing rejection again. These coping styles, or *security operations,* are interpersonal defenses that originally were necessary to protect the self in the early parent-child relationships. Unfortunately, these interpersonal defenses are overgeneralized to other relationships in adulthood and become "characterological" as the child, now grown to an adult, anticipates that new experiences with others will repeat the relational patterns of the past.

Sullivan developed his rich, if sometimes elusive, interpersonal theory through his clinical work with schizophrenic patients. He was a gifted therapist and had great respect for the personal dignity of very disturbed patients. His sensitivity to clients, compassion, and genius for understanding the intricate workings of anxiety inspired a new direction in American psychiatry. Sullivan influenced Erich Fromm (1982), Karen Horney (1970), Frieda Fromm-Reichmann (1960), and other seminal thinkers in the interpersonal field. A single integrated theory was never developed, and therapists working within this approach were never identified by a particular group name. Nevertheless, these theorists shared basic concepts and values and made enormously rich contributions to the theory and practice of counseling and psychotherapy.

Sullivan also influenced many other interpersonal/dynamic theorists such as Erik Erikson (1968), Hans Strupp (Strupp & Binder, 1984), Carl Rogers (1951), and others. He even prefigured family systems theory in his study of close relational systems. In fact, two of the founding fathers in the family therapy field, Don Jackson and Murray Bowen, trained with Sullivan. In this text, we will also be examining a host of contemporary short-term and time-limited approaches to treatment that are based on this interpersonal orientation. In sum, many of the basic concepts in the interpersonal process approach can be traced back in some form to Sullivan, Horney, Fromm-Reichmann, and the other interpersonal theorists.

In the chapters that follow, we will return repeatedly to two basic concepts from these interpersonal theorists. First, we will be exploring the *interpersonal coping strategies* that clients employ with therapists and others to avoid anxiety and maintain self-esteem at the expense of change. Second, in order to conceptualize client dynamics and better understand the interaction that transpires between the therapist and the client, we will also examine the *maladaptive relational patterns* that clients enact with therapists and others.

Object relations theory   In the present context, *objects* are people or, more precisely, internal representations of them. Object relations theory is about interpersonal relationships—especially those between parents and young children, and how those relationships are internalized. In particular, object relations theorists try to account for the meaning of attachment and the emotional need for relationships with others in the development of a self. Sullivan's theory centers on relationships, but the question that drove his work was very pragmatic: What do people do to avoid anxiety in interpersonal relations? In contrast, the question for object relations theorists is more abstract and inferential. Object relations theorists are interested in understanding how formative interactions between parents and children become internalized by the child and serve as *cognitive schemas* (relational templates, internal working models) that shape how children establish subsequent relationships with others. These cognitive models provide basic expectations or road maps for what will transpire in relationships. Although these early templates for relationships will become more complex and change over time, object relations theorists believe that these formative parent-child interaction schemas provide the basic structure for developing a sense of self, organizing the interpersonal world, and shaping subsequent relational patterns.

Object relations therapists use these schemas to conceptualize clients' problems and to understand the problematic relational patterns that clients are enacting with the therapist and others. For example, in order to understand these templates or faulty relational expectations, object relations therapists will ask themselves questions such as these:

1. What does the client want from others? (The client might want to be cared for or taken seriously, for example.)
2. What does the client expect from others? (The client might expect to be diminished or exploited, for example.)

3. What is the client's experience of self in relationship to others? (It might be burdensome or disconnected, for example.)
4. What are the emotional reactions that keep recurring? (The recurring reactions might be shame or affection, for example.)
5. As a result of these core beliefs, what are the client's interpersonal strategies for coping with his relational problems? (Possible strategies include complying with others, withdrawing from others, or trying to control others.)
6. Finally, what kind of reactions do these interpersonal styles tend to *elicit* from the therapist and others (for example, disinterest, sympathy or irritation)?

Object relations theorists also believe that the primary motive of human development is to establish and maintain emotional ties to parental caregivers (object relatedness). The greatest conflicts in life are threats to, or disruptions of, these basic ties (separation anxieties and abandonment terrors). In this regard, anxiety is a signal that emotional ties are being threatened. If parents are emotionally available and reliably responsive to a child's attachment needs, the child is secure in his ties. Through the course of development, such children can gradually internalize their parents' availability and come to hold the same constant loving affect toward themselves that their parents originally demonstrated toward them. In other words, cognitive development increasingly allows young children to internally hold a mental representation of their parents when they are away. As these children progress toward maturity, they become increasingly able to soothe themselves, function for increasingly longer periods without emotional refueling, and effectively elicit support when necessary. *Object constancy* develops as their ability to comfort themselves becomes the source of their own self-esteem and secure identity as capable, love-worthy persons. Further, they possess the internal working models necessary to establish new relationships with this same affirming affective valence.

But what if the child cannot maintain secure emotional ties to the primary caregivers? In some families, parents are preoccupied, depressed, or even abusive. Children of such parents are biologically organized to have their attachment longings fulfilled, but their emotional security is compromised by inconsistent or unresponsive parents. These children are trapped in an unsolvable dilemma. They cannot succeed in eliciting security from their parents, nor can they forsake their need for attachment. In a word, they feel anguish. What do they do? Object relations theory tries to account for the different solutions and adaptations the child must make (for example, symptoms and problems) to this poignant and all-too-human dilemma.

Object relations theorists propose complex (and competing) developmental theories to account for the intrapsychic and interpersonal mechanisms that children employ to protect against the painful separation anxieties aroused when they cannot maintain secure emotional ties. These theorists propose that all children mature through stages of development that progressively allow for a more integrated, stable view of self and others. However, if development is problematic, children remain "stuck" or else regress under stress to less integrated stages. In particular, when parents are excessively inconsistent or unresponsive, the child must resort to *splitting defenses* to maintain ties to an idealized loving parent. That is, the child internalizes the "bad" (threatening or rejecting) aspects of the parent separately from the "good" (loving or responsive) aspects of the parent. In object relations theory, in contrast to classical psychoanalytic theories, symptoms and problems do not develop from unacceptable sexual and aggressive impulses that have been repressed, but from what has been painful or threatening in children's relationships with their caretakers. In healthier parent-child interactions that provide secure attachments, however, the young child can tolerate increasing levels of frustration and disappointment as she matures. Gradually, the child will be able to increasingly integrate her ambivalent feelings toward the sometimes frustrating and sometimes responsive parent into a stable and affirming relational template of self and other, which becomes the basis of self-esteem and identity.

Under greater attachment threats, however, the child cannot do this without resorting to splitting defenses. These splitting defenses preserve the necessary image of an idealized, "all good," responsive parent with whom the child is internally connected. But the price is high: Reality is distorted; the self is fragmented; and the child becomes the one who is "bad." The frustrating parent is no longer "bad," which allows the child to view the external world as safe. The price, however, is inner conflict. The child believes that, if only he were different, parental love would be forthcoming. This situation is illustrated most poignantly by the physically abused child who is bruised and battered, but, nevertheless, continues to idealize and defend the abusive parent. Typically, the child does this by maintaining that he is bad and actually deserves the punishment. This self-negating distortion enables the child to cope, to feel that he has some control over events and is not helpless.

Why is this complex and abstract theory relevant to interpersonally oriented therapists? Realizing that the child must find some way to maintain ties with a good, loving parent helps therapists to understand why clients persist in maladaptive behavior and self-defeating relationships. When clients relinquish symptoms, succeed, or become healthier, they often feel anxious, guilty, or depressed. Loyalty and allegiance to symptoms—maladaptive behaviors originally developed to manage the "bad" or painfully frustrating aspects of parents—are not maladaptive to insecurely attached children. Such loyalty preserves object ties, or the connection to the "good" or loving aspects of the parent. Attachment fears of being left alone, helpless, or unwanted are activated if clients disengage from the symptoms that represent these internalized "bad" objects. The interpersonal process approach is designed to modify these cognitive schemas by expanding the client's relational templates and expectations of what can occur in close personal relationships. We will see how therapists can do this by providing clients with the real-life experience that, at least sometimes, some relationships can be different.

Unfortunately, much of object relations theory is presented in obtuse, inaccessible jargon. Paradoxically, object relations theorists address the most sensitive and personal dimensions of human experience in impersonal terms. However, we will be able to use certain aspects of this theory to understand clients better, relate more empathically to them, and guide our therapeutic interventions. In particular, we will be returning to the concept of *relational templates* or *internal working models* in order to understand and intervene with the maladaptive relational patterns that clients enact. We will also be exploring object ties further, in order to better understand why clients may resist change. In other words, we will see how anxiety and guilt are evoked for some clients as their attachment ties to internalized parental figures are threatened by change.

Family systems theory   Not only must children adapt to their attachment figures, they must also adapt to their familial and cultural systems. Beginning therapists can glean a wealth of information about their adult clients from family systems theory, which provides an understanding of the familial rules and roles that shaped the clients' development. We become who we are in our families of origin, and family process researchers have learned much about multigenerational family relations.

In the late 1950s, Gregory Bateson, Virginia Satir, Jay Haley, and others began studying communication patterns in families (Goldenberg & Goldenberg, 1996). They found that intrafamilial communication followed definitive but often unspoken rules that determined who spoke to whom, about what, and when. Adult clients often continue faulty communication patterns learned in their families of origin (for example, addressing conflicts through a third party rather than speaking directly to the person involved; allowing others to speak for them and say what they are thinking or feeling; never making "I" statements or saying directly what they want).

Early family researchers also learned that children are often scripted into familial roles, such as the responsible or good child; the problem or bad child; and the hero, rescuer, or invisible child. Following object relations theory, the most common role is the good child/bad child role split that occurs in so many dysfunctional families. Most adult clients are still performing these childhood roles, which distort and deny many important aspects of who they really are. In the process of adapting in their families, these children placed their families' expectations on themselves. Many clients who have played the good-child role in their families are still perfectionistically demanding of themselves, guilty about doing anything for themselves, worried about the needs of others, afraid of their anger, and confused by their sad and empty feelings. Whereas the bad child may act out personal conflicts externally (for example, through substance abuse or promiscuity), the good child often seeks treatment and enters therapy, presenting with symptoms of anxiety or depression.

Families also have unspoken rules about how, in late adolescence, children can leave home. For example, the oldest daughter may not be allowed to grow up and leave home successfully on her own. Often, she will seek emancipation through pregnancy, only to find herself even more dependent on her parents than before and forced to live at home again. A son may only be able to leave home through rejecting confrontations with parents, and then may live in another part of the country and have little or no continuing contact with the parents. Family rules, faulty communication patterns, and roles all serve to maintain family myths that, in turn, function to avoid anxiety-arousing conflicts. For example, these are some common family myths: Dad doesn't have a drinking problem; nobody is sad in our family; Mom and Dad never fight and are happily married. In sum, family myths and all of these other family characteristics are rule-bound *homeostatic mechanisms* that govern family relations and establish repetitive, predictable patterns of family interaction. They are homeostatic in the sense that they tend to maintain the stability of the family system, thus making many changes potentially uncomfortable or threatening.

Salvadore Minuchin (1974) and other family researchers have also explored the alliances, coalitions, and subgroups that make up the structure of family relations. In some families, for example, the maternal grandmother, mother, and eldest daughter are allied together, and the father is the outsider. This structural road map for reading family relations becomes even more illuminating when these family dynamics are examined in a three-generational perspective (Bowen, 1978; Boszormenyi-Nagy & Spark, 1973). It is fascinating—and sometimes disturbing—to see how these same family rules, roles, myths, and structural relationships can be reenacted across three or four generations in a highly patterned, rule-governed system. Adding to the complexity of this tapestry, many of these family rules, roles, and communication patterns operate within a broader cultural context. For example, the oldest son in a traditional Chinese family or the oldest daughter in a traditional Latino family may need to assume a responsible role. It is useful to establish the extent to which the patterns are culturally sanctioned.

The interpersonal process approach draws on these basic family systems concepts, and considers the effects of parental relations and familial experience on clients' problems. Therapists need to help clients make realistic assessments of the strengths and problems that existed in their families of origin. For many clients and therapists, however, it is culturally taboo to speak critically of parents. Alice Miller (1984) observes how this even breaks the Fourth Commandment: Honor thy parents. Herein, though, is one of the great strengths of family systems theory. The family systems therapist is trying to understand hurtful familial interactions, not to place blame or scapegoat any family member. The family therapist is concerned about the well-being of every family member, parents as well as children. In parallel, only as clients realize that their individual therapist wants to understand—rather than to blame or reject—the parent, do clients have the safety they need to explore threatening material and make significant gains in therapy. At times, however, therapists' own splitting defenses lead them to blame or reject parental figures (by saying, for example, "I can't believe how mean your mother was to you!"). It will be more difficult for these clients to progress in treatment, out of guilt and loyalty to the their caregivers. In contrast, it will be far more effective if therapists just try to understand and empathize with their client's experience (by saying, for example, "I'm sorry you were hurt so much when she did that.").

According to object relations theory, there are some good things in almost every family, even in very troubled or abusive families. There are also limitations and conflicts in the healthiest families. When clients continue to idealize parents and deny real problems that existed, as many do when they enter treatment, they are complying with binding family rules and protecting their parents at their own expense. However, clients cannot simply reject hurtful parents and emotionally cut them off, or they will tend to reenact these same problematic relations with others. This means that clients must find a way to keep parents (or healthier aspects of them) alive inside themselves as "partial identifications." How can clients accomplish this?

In order to resolve problems and make progress in counseling, clients must be able to come to terms with both the good news and the bad news in their families. As already noted, however, clients cannot achieve this integration if the therapist also employs his own splitting defenses. In other words, the therapist does not want to identify with or idealize the wounded child that exists within some clients and to reject the hurtful parent. Neither does the therapist want to support clients' continuing denial of familial problems and idealization of parents. Instead, the therapist's appropriate role is to try to understand what actually occurred in clients' development and to help them realistically come to terms with the good and the bad in their experience.

Family systems concepts inform the interpersonal process approach. In particular, we will be returning to two family systems concepts throughout this text. First, in concert with the other two theoretical approaches, family systems theory helps us clarify the problematic interaction patterns that clients are enacting with the therapist and others. Second, it helps both therapists and clients to understand the familial context that shaped many current problems. With this awareness, clients are better able to stop reenacting problematic familial roles, but without breaking off important but conflicted familial relationships.

Basic Premises

What are people's problems about? Where do they come from and how can they be resolved? Every beginning therapist grapples with such questions and will continue to do so throughout her life. One of the enduring satisfactions of a counseling career is in continuing to develop one's own ideas about the nature of people, their strengths and their problems.

This section introduces some ideas and assumptions that are central to the interpersonal process approach:

1. Problems are interpersonal in nature.
2. Familial experience is the first and most significant source of learning about ourselves and others.
3. The therapeutic relationship is the most important avenue for helping clients resolve problems.

In other words, this overview examines how problems are expressed, how they have developed, and how they can be resolved.

Problems are interpersonal in nature: A separateness-relatedness dialectic
Most client conflicts arise in close relationships, and clients often enter therapy when they are beginning or ending a primary relationship. One useful way to

conceptualize many of these interpersonal problems is along a continuum of separateness and relatedness. This will now be examined from a family systems perspective.

Most family systems theorists agree that the basic dimension of family life is a developmental continuum of separateness-relatedness. Although there will be great variation in the specific ways these basic family functions are carried out in different cultural contexts, all families must nurture young children and provide them with a secure sense of acceptance and belonging (relatedness). As children grow older, the developmental task of the family shifts to training them for independence and preparing them for a successful emancipation (separateness) that encompasses continuing ties and familial responsibilities. In this context, separateness does not mean disconnectedness. It means that the child has internalized aspects of the parents, can function more independently, and is centered in his authentic self. Thus, in order to function well, the family must have the flexibility and interpersonal range to meet offspring's changing needs for both attachment in early childhood and for greater autonomy in later years. As we will see, healthier families achieve the interdependent balance between the poles of separateness and relatedness and encompass aspects of both dimensions at the same time.

If there are problems in forming an attachment bond or in meeting the dependency needs of young children, family members will be disconnected from one another. Children from disengaged families will not develop a shared sense of family identity or loyalty to each other. Most importantly, these children will not be motivated to comply with parents' socialization demands and will not internalize social rules and self-controls adequately unless an affectional bond exists. They do not feel responsibility toward others or the culture at large and are likely to develop acting-out symptoms and to externalize blame. These offspring are more likely to come in contact with police and judges than with mental-health professionals.

Problems also arise at the other end of the continuum. When there is too much closeness and family members are enmeshed, parents do not support the individuation and emancipation of their offspring. For example, enmeshed families stifle individuation by making anger or differences of opinion unacceptable, and by avoiding conflict rather than addressing and resolving it. Individuation is also blurred by ineffective communication patterns, as when family members express what another is thinking or speak to each other through a third person rather than directly. Enmeshed families also inhibit emancipation by discouraging developmentally appropriate steps toward autonomy. For example, these families do not give offspring in late adolescence permission to leave home psychologically, succeed on their own, and form new attachments. Such offspring often feel a binding guilt as they try to become mature adults who expand and shift commitments to love relationships and career choices (Teyber, 1983).

In contrast to offspring in disengaged and enmeshed families, offspring in families that function well develop a balance between these two extremes and integrate the relatedness and separateness dimensions. These offspring develop a sense of self that allows them to feel emotionally connected to others while still remaining separate from them or individuated (Jordan, Kaplan, Miller, Stiver, & Surrey, 1991; Stierlin, 1972.

The same dialectic of separateness and relatedness continues into adulthood. Here again, the challenge of successful relationships is to balance intimacy and autonomy. On the one hand, we need to have a sense of belonging and being cared about. It is important that others acknowledge our experience and respond to us in caring and affirming ways. At the same time, however, we must also be separate. That is, all of us have the right to our own feelings and perceptions, even if they differ from those of significant others. We need to be able to express our own preferences and beliefs and to pursue our own interests and goals. Healthier individuals can accommodate the polarities of separateness and relatedness. They can set limits, have personal boundaries, and pursue their own interests while still being responsible and committed.

Most clients present problems that reflect an inability to achieve an integrated sense of self as both an independent and competent person and as an emotionally available and committed person. Thus, several unifying themes may be identified in clients' varied problems. Client conflicts will often embody separateness issues of autonomy, initiative, and power, on the one hand, or relatedness issues of intimacy, trust, and commitment, on the other.

Familial experience: Source of affective learning   Social learning in the family of origin is the most significant source of long-standing personality strengths and problems. Family homeostasis requires that many patterns of interaction and communication become repetitive and rule-bound. These transactional patterns are strongly maintained in the family of origin and can be highly resistant to change in adulthood.

There are two reasons why familial experience has such a powerful, long-term impact on the individual. The first is the sheer repetition of family transactional patterns. The same types of affect-laden interchanges are reenacted thousands of times in daily family life. Suppose, for example, that a parent has difficulty responding positively to a child's successes. When the young child enthusiastically seeks the parent's approval for an accomplishment, the parent might ignore the child and change the topic, compare the child unfavorably with a sibling's greater accomplishment, turn away and look vaguely sad or hurt, or take the success away from the child by making it his own. The same type of parental response usually occurs when the child shows the parent a favorite drawing, makes a new friend, wins a race at school, or earns a star from her teacher. This transactional pattern becomes a powerful source of learning when it continues over a period of years and even decades. As an adult, the child is likely to feel conflicted about completing her educational degree or about taking pleasure in a promotion earned at work. Thus, the most important problems in people's lives develop from these habitual response patterns. Contrary to popular belief and portrayals in Hollywood's movies, long-standing personality problems are shaped more by repetitive family transactional patterns (*strain* trauma) than by isolated traumatic events (*shock* trauma) (Wenar, 2000). Thus, enduring problems do not arise so much from the wounding events themselves. They come instead from the lack of validation or empathy from caregivers for the painful events that occurred.

Second, the impact of these repetitive transactions is magnified because of the intensity of the affect involved. Parents are the pillars of the child's universe, and children depend on them with a life-and-death intensity. These repetitive transactional patterns have been reenacted in highly charged affective relationships with the most important people in one's life. The child in the preceding example may well feel a desperate need somehow to win the parent's approval, yet simultaneously feel increasing anxiety about trying to succeed or about approaching the rejecting parent. Thus, our sense of self in relation to others is learned in the family of origin and, in many ways, will carry over to adulthood.

In summary, family interaction patterns may be hurtful and frustrating for the individual, validating and encouraging, or most commonly, a mixture of the two. For better or worse, these repetitive patterns of family interaction, roles, and relationships are internalized and become the foundation of our sense of self and the social world. Of course, other factors are influential as well. Familial relationships are molded within a cultural context that can affirm or repudiate parental behavior, and familial experience will not be central to many problems that clients present (Pipher, 1994). Familial experience does, however, provide our first and most long-lasting model for what goes on in close relationships. It will figure significantly in the individual's choice of marital partner and career, in how adult offspring will, in turn, parent their own children, and in many of the other enduring problems and satisfactions of adult life.

Although the influence of the family of origin is profound, problematic transactional patterns can be relearned. Change is indeed possible, and occurs in part through a relational process of experiential relearning, which we consider next.

The therapist-client relationship: The key to resolving problems   The relationship between the therapist and the client is the foundation of the therapeutic enterprise. The nature of this relationship is the therapist's most important means of effecting client change; it determines the success or failure of therapy. In order to utilize the therapeutic relationship as a vehicle for change, however, therapists must understand very specifically the meaning of their interactions with their clients. Thus, the interpersonal process approach focuses on understanding what goes on between the therapist and client in terms of their interaction or *process.*

The therapist-client relationship is complex and multifaceted; different levels of communication occur simultaneously. For example, there is a subtle but important distinction between the overtly spoken *content* of what is discussed and the *process* dimension of how the therapist and client interact. In order to work with the process dimension, the therapist must make a perceptual shift away from the overt content of what is discussed and begin to track the relational process of how two people are interacting as well. This means that, at times, therapists need to step beyond the usual social norms and talk with the client about their current interaction. Because this may conflict with familial or cultural prescriptions, it may feel awkward at first for some beginning therapists. For others, it may seem impolite to speak forthrightly about "you and me." However, we will see how effective it can be when the therapist can link the problem that the client is talking about with others to the therapist-client relationship, or is willing to explore with the client what may be going on between them. We will also see how the therapist can use this metacommunication in a tactful way that does not make the client feel uncomfortable. Consider the following example.

During the first session, the client tells the therapist that he resents his wife because she is always telling him what to do. He explains that he has always had trouble making decisions on his own and, as a result, his wife has often simply made decisions for him. Even though he felt that she was just trying to help him with his indecisiveness, he resented her "bossiness" and "know-it-all attitude." After describing the presenting problem in this way, the client asked what the therapist thought he should do.

Let us look at the process dimension of their interaction. Suppose the therapist complied with the client's request and said, "I think that the next time your wife tells you what to do, you should. . . ." If therapy continues in this prescriptive vein, the therapist and client will begin to reenact in their relationship the same type of conflict that led the client to seek treatment; the therapist will be telling the client what to do, just as his wife has done. The client has certainly invited this response, and will probably welcome the therapist's suggestions at first. In the long run, however, he will probably come to resent the therapist's directives just as much as he resents his wife's, and will ultimately find the therapist's suggestions to be of as little help as hers.

Alternatively, the therapist might respond to the other side of the client's conflict and say, "I don't think I would really be helping you if I just told you what to do. I believe that clients need to find their own solutions to their problems." Frustrated, the client responds, "But I told you I don't know what to do! It's hard for me to make decisions. Aren't you the expert who is supposed to know what to do about these things?"

This response throws the client back on the other side of his conflict and leaves him stuck in his own inability to make decisions. If this mode of interaction continues and comes to characterize their relationship, their process will recapitulate the other side of the client's problem. His inability to initiate, make his own decisions, and be responsible for his own actions will immobilize him in therapy, just as it has in other areas of his life.

What we see here is that clients do not just talk with therapists about their problems in an abstract manner. Rather, they actually recreate in their relationship with the therapist the same conflicts that have led them to seek treatment. *This recapitulation of the client's problem is a regular and predictable phenomenon that will occur in most therapeutic relationships.* We will return to this scenario later in the chapter and explore better therapeutic responses.

To resolve problems, clients must experience in their real-life relationship with the therapist a new and more satisfying response to their problems than they have found with others. When clients terminate prematurely or therapy reaches an impasse, the therapist and client are usually reenacting in their relationship subtle aspects of the same conflict that the client has been struggling with in other relationships, although neither of them may be aware of this reenactment. For example, clients may feel that: they are being controlled by the therapist and have to do everything his way, they have to take care of the therapist and meet her needs, or they must please the therapist and win his approval, just as they are doing with others in their lives. When this occurs, *the therapeutic process has metaphorically repeated the conflicted interaction that clients have not been able to resolve in other current relationships, and that they have often experienced in earlier formative relationships.*

As clients begin to play out with the therapist the same relational patterns that originally brought them to therapy, the therapist must be able to respond in a new and more effective way that allows clients to resolve the conflict within their relationship. This concept may seem easy to understand but it is often challenging to put into practice in affect-laden relationships with clients. Accordingly, this basic theme will be repeated in different ways and illustrated with case examples throughout the text.

In successful therapy, clients experience a new kind of relationship. They participate in a relationship in which their old conflicts or relational patterns are intensely aroused, but the therapist does not respond in the same problematic way that others have in the past. Thus, throughout each session, the guiding question for therapists is: Am I co-creating a new and healing relationship, or am I being drawn into a repetitive pattern that is familiar but problematic for the client? When the therapist and client are able to resolve these expectable reenactments that occur in most therapeutic relationships, change has begun to occur.

In this process, clients begin to more successfully distinguish between past and present reality. They also learn that they no longer have to respond in their old ways or always receive the same unwanted responses from others. Thus, clients begin to develop a wider range of expectations and to become more flexible in their relationship with the therapist. As we will see, it is often relatively easy to take the next step and help the client generalize this experience of change with the therapist and adopt more adaptive responses with others outside of the therapy setting. This *corrective emotional experience* is the fundamental premise of interpersonal process psychotherapy and the basic mechanism of therapeutic change. Let's examine more closely what occurs in these pivotal interactions.

Clients believe actions, not words. Accurate interpretations, relevant educational inputs, and effective cognitive restructuring will be useful in almost every therapy, but they are not the primary seat of action in the interpersonal process approach. This is a performance-based or experiential learning model. Clients change when they live through emotionally painful and ingrained relational scenarios with the therapist, and the therapeutic relationship gives rise to outcomes different from those anticipated and feared (Strupp, 1980). When the client reenacts important aspects of her conflict with the therapist, and the therapist's response does not fit the old relational templates or schemas, the client has the experience that relationships can be another way. It is powerful and enlivening to find that, at least this time, the same unwanted relational pattern did not occur. Among other important effects, the new or corrective response from the therapist creates *greater interpersonal safety* for the client. This is not the general safety that comes from being taken seriously and treated with respect, but a more significant safety that results when the client is not hurt again by receiving certain interpersonal responses that are familiar and expected, but unwanted or even dreaded. As we have noted, this pivotal moment is not the end point in treatment, but a window of opportunity for a range of important new behaviors to emerge. Clients may be empowered to:

- Feel better about themselves or forgive themselves for behavior they have felt unrealistically shameful or guilty about
- Have the safety to experience more fully the pain of how much this has hurt them in other relationships
- Remember more specifically when and with whom this particular relational conflict originally was learned
- Remember relationships in the past that were more affirming
- Make meaningful connections between current behavior and developmental relationships
- Feel bolder and risk addressing, or making overt, conflicts that have been occurring with the therapist
- Make decisions or risk trying out new ways of responding

In these ways, corrective emotional experiences with the therapist begin to break old relational templates and repetitive relational patterns, and a wider range of interpersonal responses becomes available to the client. We will see later how the reparative response from the counselor can also change clients' self-esteem and sense of themselves.

Corrective emotional experience has been the cornerstone of many dynamic and interpersonally oriented therapies (Alexander & French, 1980). Although originally unappreciated in the 1940s by the psychoanalytic community and disdained as merely superficial role-playing, their more active, direct approach provided clients with a real-life experience of change in the here-and-now relationship with the therapist. In large part, Alexander and French originally advocated the corrective emotional experience in order to shorten therapy.

A related term in the literature for the corrective emotional experience is "passing transference tests" (Weiss & Sampson, 1986). They elucidate how the therapist must respond to clients' conflicts in more helpful ways than caregivers did originally. If therapists do this consistently throughout treatment, they will systematically disconfirm the client's "grim, unconscious pathogenic beliefs" that developed in problematic parent-child interactions (for example, "If I get stronger and do what I want, I will hurt my parent," or "I must always be resented because my needs are too demanding"). Focusing more specifically on the process dimension, Kell and Mueller (1966) also illuminated how the client's central conflicts tend to get reenacted with the counselor. These conflicts can then be either resolved or merely recapitulated again, depending on the nature of the relationship that the therapist and client can work out. These and other clinical theories we will be exploring (especially the contemporary, short-term therapies of Strupp, Sifneos, Luborsky, and others) use different terms to describe the same direct, immediate experience of change in the real-life relationship between the therapist and client. Perhaps Frieda Fromm-Reichmann (1960) captured this central tenet best by saying that the therapist must provide the client with an *experience* rather than an *explanation.*

In order to provide this corrective emotional experience, however, the therapist must be able to work with the process dimension. Let's return to our example of the client who complains that his wife tells him what to do. When the client asked his therapist what he should do, one option would have been for the therapist to offer a *process comment* that described their current interaction and made it overt as a topic for discussion. For example, the therapist might have said, "It seems to me that you're asking me to tell you what to do, which will bring up the same problem for us in therapy that you are having at home with your wife. Let's see if you and I can do something different in our relationship. Rather than having me tell you what to do, let's try and work together to understand what is going on for you when you are feeling indecisive. Where do you think is the best place to begin?"

In this instance, the therapist is being supportive by offering the client a new and different *collaborative* relationship in which they can work together on the client's problems. If this collaborative effort continues to develop over the course of treatment, the client soon will begin to find that he can make decisions and initiate with the therapist. Then, with the counselor's support, he will next begin to try out these new behaviors with his wife and others. Of course, the therapist's initial attempt to offer the client a corrective response will need to be repeated in many different ways for change to occur. If the therapist can maintain this type of collaborative relationship, however, it will facilitate other types of therapeutic interventions as well. Educational inputs, behavioral alternatives, dynamic interpretations, and interpersonal feedback will all be more effective when they occur in this type of interpersonal context.

As we have seen, the guiding principle in the interpersonal process approach is to provide clients with an experience of change. With this in-vivo relearning, clients live-out a new relationship with the therapist that disconfirms their faulty expectations and expands their templates of what can occur for them in other relationships. For example, they learn that sometimes they can ask for help, set limits and say "no," give priority to their own needs, succeed without evoking envy, and so forth. This work demands personal involvement on the part of the therapist, however, who must be willing to engage in an authentic relationship and risk being effected by the client. In order to have the emotional impact necessary to provide a corrective emotional experience and produce change, the relationship must hold real meaning for both participants. In this simple human way, it is the relationship that heals. If the therapist is merely an objective technician, psychologically removed and safely distant from the client, the relationship will be too insignificant to effect change.

On the other hand, therapy will also falter if the therapist becomes inappropriately close. If therapists overidentify with the client or become too invested in the client's changing, perhaps in order to shore up their own feelings of adequacy, therapy will not progress. In this situation, therapists lose sight of the process they are enacting with the client and typically begin to respond in problematic ways that recapitulate the client's maladaptive relational patterns.

One of the most common problems for beginning therapists is to experience the client's situation or concerns as identical to their own. The solution to this overidentification is a supervisory relationship that helps therapists see the ways in which their own experience actually differs from the client's. There may be important similarities but, in reality, the client's problems are never the same as the therapist's. Once able to differentiate their own issues from those of the client, therapists will often be able to see how the therapeutic process has been reenacting aspects of the client's conflict. This understanding often enables therapists to alter the way they have been interacting and establish a more productive interpersonal process.

Beginning therapists should expect to make many "mistakes" with their clients—to become overinvolved or underinvolved with certain kinds of clients, to reenact the client's maladaptive relational patterns at times, and so on. Such mistakes are an inevitable part of the counseling process for beginning and experienced therapists alike. Few clients are fragile, however, and therapeutic relationships are often remarkably resilient. Unfortunately, most beginning therapists do not know this, and concern about making mistakes remains one of their biggest anxieties. Beginning therapists will do better therapy—and enjoy it more—once they discover that *mistakes can be undone*. In fact, mistakes provide important therapeutic opportunities when therapists are willing to work with the process dimension—to talk with clients about problems that may be occurring between them. (For example, the therapist may say, "I think I might have misunderstood what you just said, and you seem to have become more distant from me in the last few minutes. What do you see happening between us here?") The therapist breaks the old mold and helps to make this relationship different from past problematic scenarios by clarifying the miscommunication or responding to whatever is problematic. (The client might say, for example, "When you are being quiet like that and don't say very much, I feel like you are judging me. Would you talk more?") As their performance anxiety declines, beginning therapists will find that they are better able to identify and work with the process dimension. By doing so, therapists can readily undo mistakes and realign the therapeutic relationship when it has gone awry.

In closing, remember the maxim that clients will not hold it against you for making a mistake, only for not owning it. Sometimes we need to apologize.

Client Diversity and Response Specificity

We have already begun to see that beginning therapists need a theoretical framework to guide their therapeutic interventions. What must such a theory provide if it is to be of help to therapists and their clients?

One feature of an effective clinical theory is that it must have the flexibility and breadth to encompass the diversity of clients who seek treatment. Although unifying patterns in personality certainly exist, every client is different. Each client has been genetically endowed with a unique set of features and each has been raised differently in her family. Socialization is different for women and men; members of different cultures have different experiences; and economic class shapes opportunity and expectations. To be helpful, a clinical theory must be able to help therapists work effectively with all of these highly diverse clients.

Moreover, each therapist is a different person. Like clients, therapists differ in age, gender, ethnicity, sexual orientation, and developmental background. Therapists also bring diverse values, worldviews, and personal styles to their clinical work. How can any theory help such a diversity of therapists respond to the extraordinary range of human experience that clients present?

Client response specificity will be one of our best tools. This approach requires therapists to be *flexible,* so that they can respond on the basis of each client's personal history and ways of viewing the world. It is helpful to be familiar with the experiences of particular groups (for example, to know that African Americans' history includes slavery and racial discrimination). However, this approach attempts to understand and respond to each client as a unique individual. Thus, we will not seek treatment rules or intervention guidelines that apply to all clients, or even to specific diagnostic categories or groups, such as men, Latinos, or Christians. Rather, we will be trying to find case-specific recommendations in which we explore what it means to this unique individual to be depressed, lesbian, or bi-racial. Let's examine further what this means.

Interpersonal process psychotherapy is a highly idiographic approach; it emphasizes the personal experience or subjective worldview of each individual client. Diagnostic categories are certainly useful, but they are only the map and not the territory. Personality typologies also highlight potential directions that therapy might take, but the actual therapeutic process and issues covered will be client-specific. Throughout each session, therapists are encouraged to join the client in a process of *mutual exploration* to try to find the subjective meaning that this particular experience holds for this client (saying, for example, "Okay, when I hear you say that, it sounds to me more like sadness or longing than depression. What would you say?"). It is challenging for therapists to enter the client's *subjective worldview.* The meaning that this particular experience holds for the client often differs greatly from the meaning it holds for the therapist or for other clients that the therapist might see. Therapists are well-advised to listen carefully to clients' language and choice of words, and to collaboratively explore clients' own metaphors (saying, for example, "Help me understand what you're saying about our relationship when you tell me that you just want to 'sail' right out of here?"). When someone who cares about you listens seriously and tries hard to understand, meaning is created.

With client response specificity, there is no standardized treatment protocol. Each individual client's developmental history, cultural context, and current life circumstances serve to guide the therapist's treatment plans and intervention strategy. Thus, the interpersonal process approach does not advocate a particular therapeutic stance toward clients in general such as prescriptive, nondirective, active, neutral, supportive, and so forth. Each of these modalities might be helpful with a particular client and ineffective with another. This approach asks something different and far more demanding of therapists: (1) to conceptualize the specific relational experiences that this client needs in order to change; and (2) to be flexible enough to modify their interventions and re-spond in the ways that this particular client utilizes most productively. In fact, Strupp (1980) found that many therapists in his study did not demonstrate flexibility in their technical approach to the client or have the interpersonal range to modify their interventions and provide the responses that each client could utilize best!

To illustrate client response specificity more concretely, consider a much debated question: Should therapists self-disclose to their clients? Client re-sponse specificity emphasizes that self-disclosure (or any other response) will hold very different meanings for different clients. In fact, the same response may even have the opposite effect on two different clients with contrasting de-velopmental histories and cultural contexts. For example, if a client's parent was distant or aloof, the therapist's judicious self-disclosure may be helpful for the client. In contrast, the same type of response may be anxiety-arousing for a client who had had an emotionally seductive caregiver or who had been the confidant of a depressed parent.

Greater sharing with the therapist may help the first client learn that, contrary to his deeply held beliefs, he does matter and can be of interest to other people. For the second client, self-disclosure may inadvertently impose the unwanted needs of others. Therapists will be most effective if they first consider how the client's relational templates may shape the impact of their interventions.

To illustrate further, let's examine another question: Should therapists give opinions when clients ask for advice? As before, the therapist's response is in-formed by each client's particular circumstances. For example, offering advice may be counterproductive for a compliant client who is trying to become more assertive with a dominating, dependency-fostering spouse. For such a client, it may be better for the therapist to ask, "What do you think would be best to do?" In contrast, giving advice and directives may work well with clients whose care-givers were unable to help them solve their problems or were uninterested in preparing them for aspects of adult life. To withhold advice from these clients may only reenact developmental conflicts and impede their progress. Once again, the same response from the therapist will have very different effects on different clients. From this metaperspective, techniques or interventions from any theoretical approach may help or hinder the client; it depends on whether the response serves to reenact or resolve problematic relational patterns for this particular client. Thus, in order to conceptualize client dynamics and guide treatment interventions, counselors will be given specific guidelines:

1. To clarify the specific relational patterns that originally created problems for the client
2. To anticipate how these problematic patterns could be expressed or played out in the therapeutic relationship
3. On the basis of this awareness, to track the process dimension and offer different responses that provide a more reparative relationship than the client has found in the past

It is time to examine another aspect of client response specificity. Therapists can *assess* the effectiveness of their interventions by observing how the client utilizes or responds to what they have just done. For example, if the therapist observes that the client becomes anxious or distant in response to the therapist's self-disclosure, the therapist is learning how to respond most effectively to this particular client. When the therapist provides a corrective response, many clients will make progress right away by feeling safer, acting more boldly, bringing forth relevant new material, becoming more engaged with the therapist, and so forth. In contrast, when the therapist responds in a way that recapitulates problematic old patterns, clients will behaviorally inform the therapist of this by acting "weaker" in the next few minutes—for example, by being more distant, confused, or compliant (Weiss, 1993). If therapists track clients' immediate responses to the varying interventions they employ (providing interpersonal feedback, making process comments, cognitive re-framing, and so forth), clients will behaviorally show therapists how best to respond. As emphasized earlier, however, therapists then must be flexible enough to adjust their interventions to provide clients with the responses they can utilize best.

To sum up, working in this highly idiographic way adds complexity to the therapeutic process and places more demands on the therapist. However, such an approach gives therapists the flexibility to respond to the specific needs and unique experiences of the diversity of clients that seek help. Client response specificity is a core concept in the interpersonal process approach. In subsequent chapters, we will utilize guidelines for keeping process notes (Appendix A) and writing case conceptualizations (Appendix B) to help therapists formulate the specific relational experiences that their clients need.

Model of Therapy

Drawing on some of the concepts introduced in this chapter, we now provide a brief overview of the therapeutic model developed in Chapters 2–10. For most clients with Axis I and Axis II disorders, certain basic developmental tasks were left unfinished in their family of origin. Although enduring problems are often complex and have multiple causes, this unfinished business usually includes the faulty beliefs and ineffective coping strategies that clients have adopted to cope with insecure attachments and other developmental conflicts. The purpose of therapy is to resolve the interpersonal problems that have resulted by providing a corrective emotional experience with the therapist. The therapist does this by providing a relationship that enacts a resolution of the client's maladaptive relational patterns, rather than going down the same problematic lines again. Providing such reparative experiences is often easier in theory than in practice. As we will explore, clients are often adept at engaging the therapist in the same scenarios that are causing problems in other relationships.

Simultaneously, the client seeks to avoid reexperiencing painful conflicts, and to find a new and more satisfying response to them. To this end, the client tries to assess safety and/or danger in the therapeutic relationship. The client's

relational problems with others are activated with the therapist and are especially likely to be played out along the process dimension. Throughout treatment, the client continues to assess whether the therapist responds in new and
safer ways than others have in the past or whether the therapist responds in
similarly problematic ways. For example, suppose the client risks becoming
stronger and begins successfully pursuing her own interests and goals. The
client will be intensely interested in determining if the therapist sincerely takes
pleasure in the client's success or feels threatened and competitive, as the client's
parent used to feel. Or, suppose the client reveals vulnerabilities and has emotional needs for the therapist. Will the therapist respond with genuine acceptance or will the therapist, instead, feel overwhelmed by these needs, as the
client's parent used to feel?

Clients resolve their conflicts when the therapist's response repeatedly disconfirms such problematic expectations. Through this corrective emotional experience, clients find that it is safe to act in new and more adaptive ways—at
least with some people, sometimes. Clients can then begin exploring new aspects of themselves and better ways to establish meaningful relationships. This
begins the next phase of treatment: *generalizing their experience of change with
the therapist to other relationships in their lives*. We will see that clients can do
this, in part, by learning how they can change their own responses to others in
current relationships—even when significant others continue to respond in familiar but unwanted ways.

Therapists cannot simply tell clients that relationships can be different from
those that the clients have experienced in the past; they must show them.
Deeply ingrained relational templates, that may involve interpersonal themes of
mistrust or compliance, have fundamentally organized the client's relational
world for decades. Such over-learned responses will change as the therapist and
client address how similar relational themes are being activated in their relationship and, together, they work out a different type of relationship that does
not fit the old pattern. Finding this interpersonal solution often entails conflict
and anxiety for both the therapist and the client. Thus, the power to effect enduring change does not come primarily from insight or new behavioral alternatives. Rather, the impetus for change occurs when clients find that they are experiencing the same types of issues with the therapist (for example, control,
power, mistrust) that cause problems in other relationships. This time, however,
they are able to resolve the old problem and find that, at least sometimes, relationships can develop in better ways. This experiential relearning is the basis of
interpersonal process psychotherapy. It places real demands on the personhood
of the therapist but holds the great reward of helping people change.

Limitations and Aims

Limitations   In this text, beginning therapists are introduced to the practice
of interpersonally oriented psychotherapy with individual adult clients. However, it is important for beginning therapists to learn how to work in different

treatment modalities and in varying treatment lengths. Sadly, it has become more challenging for practitioners to earn a living in the contemporary managed care milieu. Clinicians today must market themselves strategically (business courses and special training will be necessary for some). Trainees are encouraged to develop skills in a variety of additional treatment modalities, including child, marital, group, and family. While the treatment model presented here emphasizes individual adult treatment, the process orientation provides basic skills that are readily applied to each of these modalites. (For instance, the first journal published in the new field of family therapy was titled *Family Process*.)

The therapeutic model presented here is designed for an intermediate treatment length of 6–12 months. This treatment length is not necessary for many clients. Unfortunately, it is needed for some, but is no longer available in many health maintenance organization (HMO) and managed care settings. However, the treatment fundamentals presented here also provide many of the basic skills necessary for brief therapy, including initial intakes, crisis intervention, and short-term therapy. Specific guidelines for linking the interpersonal process approach to time-limited, marital, group, and family therapy will be provided at various points throughout the text.

In particular, the interpersonal process approach is readily adapted to short-term treatment modalities of 10–16 sessions. In recent years, interpersonal/dynamic theorists have developed effective, time-limited treatment approaches that are grounded in a relational or process approach (Davanloo, 1980; Levenson, 1995; Luborsky & Marks, 1991; Malan, 1976; Mann & Goldman, 1982; Safran & Muran, 1998; Sifneos, 1987). Before they can succeed in these short-term models, however, beginning therapists first need to acquire certain basic clinical skills. For example, short-term work with this model requires therapists to "bring the conflict into the relationship" and to make overt how the therapeutic process may be reenacting aspects of the client's conflicts with others. Most beginning therapists are not ready to be so forthright and to address these relational reenactments with the client as quickly as time-limited approaches demand. Also, short-term treatment approaches require therapists to establish a treatment focus by identifying maladaptive relational patterns more quickly and accurately than most beginning therapists can do. Further, in the closing session(s) of time-limited treatments, beginning therapists cannot be expected to manage the client's unresolved grief reactions and strong feelings of loss that are regularly evoked. Furthermore, it is not realistic for beginning therapists to be able to work effectively with clients' negative transference reactions, particularly angry and blaming accusations that the therapist is betraying or abandoning them (just as others have done before), which are commonly evoked in time-limited therapy during termination.

Beginning therapists will need more experience and specific training in brief psychotherapy before they can succeed in this essential treatment mode. Relevant guidelines to working in this important modality will be introduced at many points. As they progress in their training, readers are encouraged to examine some of the excellent, short-term treatment approaches available, such as the interpersonal treatment of depression (Klerman, Weissman, Rounsaville, & Chevron, 1984), personality disorders (Benjamin, 1995; Horowitz et al., 1984), and psychotherapy (Strupp & Binder, 1984).

Aims   The interpersonal process approach presented here emphasizes three broad aims that are essential to helping clients change. First, therapists must learn how to establish a strong working alliance with a wide diversity of clients. Change is predicated on the nature of the therapist-client relationship, and specific guidelines to provide a collaborative or working alliance are provided in the next chapter and throughout.

Second, therapists must be able to respond effectively to the client's emotions and inner life. Therapy is a personal and private sharing in which therapists often respond to clients' vulnerability and help them with their painful feelings. Making contact with clients in these very personal ways is a critical aspect of therapy, yet it is the dimension along which therapists are most likely to stumble. Most therapists, like other people reared in our culture, have been socialized to avoid rather than approach strong feelings such as anger, fear, and shame. One of the most effective ways for therapists to facilitate change, however, is to help clients come to terms with emotional reactions that they have not been able to resolve on their own. Specific guidelines will be provided to help beginning therapists understand and respond to their clients' conflicted emotions.

The third component of effective therapy is to conceptualize the client's problems and formulate what needs to occur in treatment for this client to change. This conceptualization should enable the therapist to identify the client's faulty beliefs and expectations, maladaptive relational patterns, and problematic coping strategies. Focusing on these three issues will provide direction for the ongoing course of treatment. Because the most effective therapists are as adept conceptually as in the affective and relational domains, this text aims to facilitate the therapist's effectiveness in all three arenas.

This text will also provide an overview of the nature and course of a therapeutic relationship. Therapeutic relationships usually have a coherent life course and follow a predictable developmental pattern from beginning to end. Although variations occur with every client, therapists can often identify an ordered sequence of stages. At each successive stage of therapy, the therapist must negotiate certain therapeutic tasks with the client. If the therapist achieves this, the next set of issues and concerns will often emerge from the client. An awareness of this developmental structure will help the therapist to formulate intervention goals and strategies for each successive stage. The chapters that follow are organized to reflect this developmental schema and to parallel the course of therapy from beginning to end.

Readers of this book must also have realistic expectations for themselves. Beginning therapists who are seeing their first clients will find much to help them in their initial work with clients. However, more information is presented here than a novice clinician can fully integrate and apply. The concepts embodied in the interpersonal process approach are complex and challenging. Beginning therapists are advised to incorporate these concepts into their practice gradually, at their own pace. Especially during their first year of seeing clients, some may begin to feel that the more they learn, the less they know. With more experience, however, second-year students will often begin to find that they are effectively employing many of these concepts with their clients. Typically, trainees are able to make these concepts their own and utilize them easily and effectively in about three years. Since it will take some time to learn how to work in this way, the best approach is to be patient and enjoy the learning.

Finally, throughout this book, beginning therapists will be encouraged to be themselves with clients rather than trying to fulfill the role of a counselor. Perhaps Kahn says it best, "When all is said and done, nothing in our work may be more important than our willingness to bring as much of ourselves as possible to the therapeutic session. . . . One of the great satisfactions of this work comes at the moment students realize that when they enter the consulting room, they don't need to don a therapist mask, a therapist voice, a therapist posture, and a therapist vocabulary. They can discard those accouterments because they have much, much more than that to give their clients" (1997, p. 163).

Let's turn now to the first stage of therapy: establishing a working alliance.

Suggestions for Further Reading

1. In the beginning, it may not be easy to differentiate the process dimension from the overt content of what clients are discussing. The case study of Theresa in Chapter 1 of the Student Workbook to accompany this text illustrates how therapists can distinguish process from content. It also shows how therapists can use the process dimension to intervene more effectively.

2. One useful book for trainees interested in the interpersonal process approach is Michael Kahn's *Between Therapist and Client: The New Relationship* (New York: Freeman, 1997). This engaging book provides a historical overview of the major concepts and theorists of the therapist-client relationship. In particular, Kahn articulates the cardinal issue of *nondefensiveness* in the face of clients' attempts to pull the therapist into their own relational reenactments. This book deserves to be read and reread by beginning and experienced therapists alike.

3. Chapters 3 and 5 of Salvadore Minuchin's *Families and Family Therapy* (Cambridge, MA: Harvard University Press, 1974) provide a succinct presentation of structural family relations and family developmental processes. This discussion will help therapists who work with individual clients to understand the genesis and familial context of their clients' problems.

4. A highly readable application of object relations and attachment theory to psychotherapy can be found in Chapters 6–9 of John Bowlby's book *A Secure Base* (New York: Basic Books, 1988). An excellent conceptual overview of object relations theory may be found in *Freud and Beyond* by S. Mitchell and M. Black (New York: Basic Books, 1995).

5. The great novelists best bring to life the profound impact of childhood and familial experience on adult life. See, for example, John Steinbeck's *East of Eden* (New York: Viking, 1952) and Franz Kafka's *Letter to His Father* (New York: Shocken, 1966).

PART TWO

Responding to Clients

CHAPTER TWO

Establishing a Collaborative Relationship

Conceptual Overview

Psychotherapy is a profession based on trust. Clients enter therapy with a need: They are often in pain and asking for help with something they have not been able to resolve on their own. A trustworthy response is to respect the clients' requests for help and to respond compassionately to their concerns. Yet to be most effective, the therapist must also respond in a way that helps clients achieve a greater sense of their own *self-efficacy*. The goal is not only to resolve specific situational problems, but to do so in a way that leaves clients empowered with a greater sense of their own competence and mastery. In this way, the interpersonal process approach strives to foster personal growth as well as to provide symptom relief.

Clients cannot resolve problems and achieve a greater sense of their own personal power in a hierarchical or one-up/one-down therapeutic relationship. Clients need to share ownership of the change process and must be active, egalitarian participants when working with the therapist, rather than being passively "cured" or told what to do. Thus, this chapter presents a model for a *collaborative relationship* or working alliance that accepts the client's need for understanding and guidance, but equally encourages the client's own initiative and autonomy.

Although they have tried long and hard, researchers have not been able to find consistent empirical support for the superiority of any one treatment approach over another. However, they have found strong evidence of great *variability* in the effectiveness of individual therapists within each treatment approach (Teyber & McClure, 2000). Perhaps the strongest finding in the psychotherapy outcome literature is that the most common feature of effective therapists, across different theoretical approaches, is the therapist's ability to establish a strong *working alliance* early in treatment. This chapter explores how therapists can establish a strong working alliance, and use process comments to restore it when the misunderstandings and problems that occur in every therapeutic relationship temporarily rupture the therapeutic alliance.

31

Chapter Organization

One of the most important concerns for beginning counselors is whether their clients will return after the first session and remain in treatment. Many clients do indeed drop out after the first session; such experiences painfully exacerbate beginning counselors' concerns about their own adequacy and performance. This chapter, and the one to follow, provide basic guidelines for conducting a successful initial session.

This chapter is divided into two sections. The first introduces the concept of a collaborative relationship and then elaborates this cardinal concept in terms of a balance between directiveness and nondirectiveness in the therapeutic relationship. Sample therapist-client dialogues will be used to illustrate how the therapist can begin to establish a collaborative working alliance in the initial session.

The second section presents ways to foster the collaborative alliance and further engage the client as a full working partner. These goals are achieved by listening with presence, empathically entering the client's subjective worldview, and understanding the core meaning in what the client is saying. The therapist can better understand the client's experience by (1) clarifying the repetitive relational themes that recur throughout the different narratives the client presents; (2) identifying the pathogenic beliefs and problematic expectations that provide unifying links for the client's problems; and (3) approaching the client's conflicted feelings in a direct and caring way. Following the central theme of this book, both sections will focus on interpersonal process characteristics of the therapist-client relationship, including how these three themes are replayed in the therapist-client relationship and how they can be used to facilitate change.

A Collaborative Relationship

At each successive stage of treatment, therapists have a different overarching goal to guide their interventions. In the first stage, the therapist's principal goal is to establish a collaborative relationship or working alliance with the client. A collaborative alliance is established when clients perceive the therapist as a capable and trustworthy ally in their personal struggles. In order to achieve this credibility and become someone who matters to clients, the therapist successfully communicates that she:

- Sees their predicament and recognizes their distress
- Feels with them and is moved or touched by their pain
- Is on their side and has their best interests at heart
- Has an abiding commitment to help them through this predicament

The concept of the therapeutic alliance was originally developed by psychoanalytic theorists. In particular, Ralph Greenson (1967) clarified three interrelated components in all therapeutic relationships: the real relationship; the transference configuration; and, most importantly, the working (collaborative, therapeutic) alliance. Found in many other theoretical contexts, the working alliance is akin to Bowlby's *holding environment* in attachment terms, where the client's distress is emotionally "held" or contained in the safe relational envelope of the therapist's understanding (Karen, 1998). In existential therapy, Rollo May (1977) elucidates the therapist's *presence* with the client. He emphasizes that the therapist must be fully present with the client in an authentic manner throughout each session. The collaborative alliance is also closely linked to Carl Rogers's core conditions of *empathy, genuineness,* and *warmth* (Rogers, 1981). Similarly, within a cognitive-behavioral framework, Borns and Auerbach (1996) underline the importance of empathy and the therapeutic alliance.

Researchers have defined the working alliance as a collaborative process whereby both client and therapist agree on therapeuticals; collaborate on tasks designed to bring about successful outcomes; and establish a relationship based on trust, acceptance, and competence (Gaston, 1990). The therapist's ability to establish a working alliance in the initial sessions strongly predicts successful outcomes in both short-term and longer-term treatment (Gelso & Carter, 1994; Hovarth and Greenberg, 1994b). One of the therapist's primary aims in the initial sessions is to articulate clear expectations for working in this collaborative manner and, more importantly, to behaviorally enact these spoken expectations by giving the client the *experience* of working together on her problems. If a collaborative relationship is maintained throughout treatment, this interpersonal process will go a long way toward helping clients resolve their presenting problems and will meet our broader goal of helping clients achieve a greater sense of their own personal power as well.

Balancing Directive and Nondirective Initiatives

At the outset, the therapist should ask clients what they know about counseling, explore their expectations of the therapist and the treatment process, and ask how they think the therapist and client could best work together. Perhaps the most widely held misconception among clients is that the therapist is a doctor who will prescribe their route to mental health. Often, the therapist is perceived as the sole agent responsible for change—via advice, explanations, interpretations, and simply telling the client what to do. Therapists need to explain to some clients that the therapeutic process is different than seeing their medical doctor. Although the therapist is trained to work with problems such as those the client is presenting, therapy will be more effective if the client can be an active participant in his treatment—for example, by orienting the therapist to the issues and concerns that are most important right now, identifying what's been helpful and what has not, and sorting through with the therapist shared treatment goals. Therapists may also explain that, even though the client has not been able to change on his own and the therapist will be an active partner, the client's willingness to work with the therapist to understand his issues will be important. Unless the therapist works with this process dimension, a hierarchical doctor-patient or teacher-student relationship is likely to develop.

Such a one-up/one-down relationship brings with it a number of problems. As we would expect from client response specificity, a doctor-patient relationship will work well with some clients in the short run. Many clients who enter treatment believe they have to be compliant in close relationships. Regardless of what the therapist says about collaborative relationships, these clients still believe that the therapist, like others, truly wants them to merely follow his lead. Such clients will passively follow the therapist early in treatment, and even actively seek direction from the therapist. However, this compliance will ultimately evoke shame and anger toward the therapist, just as it does in other relationships. Although most clients will not recognize or understand what is transpiring, this faulty interpersonal process will prevent them from utilizing the therapist's help, making progress, or even remaining in treatment. Even if clients don't have compliance issues, clients remain dependent on the therapist if this hierarchical mode is enacted. They will not be able to gain a greater sense of their own personal power as long as they hold the dysfunctional belief that the source of potency resides in the therapist, rather than in themselves. Let's examine this further.

Is the interpersonal process approach a nondirective approach to treatment? No. Some clients bring relational templates to the treatment setting that do not encompass collaborative relationships or mutuality, and these clients may insist that the therapist assume a directive or leadership role. If clients bring to therapy an authoritarian upbringing and hierarchical internal working models, then it would most likely be fruitless to insist that they go along with a more egalitarian process. They have not experienced this in the past and cannot yet encompass it. Instead, the therapist can be flexible, meet the client on his terms, and accept the client's request to provide more direction or advice. At the same time, however, the therapist can start the change process by giving the client permission to initiate more whenever he might like, and by tracking the process that results from this hierarchical interaction. That is, the therapist can watch for opportunities when the client seems to be rejecting the directives he has just elicited from the therapist (by saying, for example, "Yes, I'll try that, but . . .").

Whereas a directive stance may work in the short run, a purely nondirective approach will often sputter right from the start. Understandably, most clients feel frustrated when their requests for help or direction are merely reflected back. A negative cycle may ensue: The client becomes increasingly angry and demands direction from the therapist, who further eschews this role and talks about inner direction and finding one's own answers. This, in turn, further frustrates the client, who may not feel that she has the answer to anything at that moment, and only sees the therapist's attempts to be nondirective as evasive. Many clients do indeed drop out of therapy prematurely because the therapist was too quiet or unwilling to accept their appropriate requests to be more actively involved and responsive. Therapists need to have a genuine presence and be alive in the relationship. They need to provide feedback about the relational and cognitive patterns they observe, empathic responsiveness to the client's feelings, and process comments about what may be occurring in their current interaction. Beginning therapists who lose clients because of their own lack of responsiveness often report feeling afraid of "making a mistake" or feeling insecure about "what to do." More subtly, beginning therapists' inhibition or constriction may be prompted by their own concerns about acting more strongly and expressing their own thoughts and observations, or by apprehension at the responsibility implied in becoming important to the client and having an impact on her life.

Thus, clients do need direction from an expert, and the therapist can provide guidelines for how to proceed without falling back into a hierarchical relationship. How can this be achieved? In the initial interview, the therapist can make an overt bid to establish a collaborative relationship in which the therapist and client work together to resolve the client's problems. For example, this attitude may be conveyed by the following overture.[1]

Therapist:
> Let's work together to figure out what's been wrong. Tell me your ideas about what the problem has been. I'd also like to know what you'd most like to change or get help with, and where you think is the best place to begin.

The therapist is trying to communicate in words and actions that she will be an active and responsive ally. It is clear from this invitation, however, that the client is not going to be a passive recipient, but an active participant. The therapist must follow through on this invitation, however, by enacting a collaborative enterprise in which the client is encouraged to contribute his ideas and finds that his input is valued. This establishes a new, middle ground of shared control, between the polarizations of prescriptively taking charge or nondirectively following the client's lead. For many clients, this collaborative alliance will provide a corrective interpersonal process that they may not have experienced in other close relationships, and will be an essential aspect of helping them change. *Regardless of the therapist's theoretical orientation or techniques, one of the most important determinants of treatment outcome is whether the therapist and client can continue this process of mutual collaboration.* Let's look more specifically at ways to establish a collaborative relationship in the initial session.

Beginning the Initial Interview

Therapists can begin to establish a collaborative relationship in their initial contact with the client. The therapist structures the session by providing the client with guidelines and direction for what is going to occur in the interview. However, it is essential that the treatment focus reflect the client's own wishes and that the treatment goals be experienced by the client as her own. In other words, the therapist is trying to help clients articulate their own goals and agenda, and then collaboratively join clients in addressing their concerns. This can be done

---

[1]Throughout this book, sample dialogues are provided to illustrate the concepts presented. Beginning therapists should not use these exact words (or try to say what a supervisor might say in a similar situation), but should find their own ways to express these concepts. Therapists lose their creativity and own sense of self efficacy when they try to imitate someone else.

simply and briefly by offering clients an open-ended invitation to talk about what brings them to therapy.

Therapist:
> I'm looking forward to working together with you on the problems that have brought you in. I'd like to begin by learning more about what you see as the difficulty.

OR

Therapist:
> Why don't you help me understand what's wrong in your life or troubling you right now. It might be difficult to describe all the issues at first, but I think we can work through this together. Tell me what you think we should start with—or what concerns you feel are most pressing—and what you think needs to change.

This type of inquiry communicates several important things to the client. At the simplest level, it ends the opening phase of social interaction that occurs as the therapist and client are introduced and walk to the interview room. More importantly, it tells clients that the therapist is someone who is willing to talk directly about their personal problems—as they experience or perceive them—and is ready to respond to their need for help by listening. It does so, however, in a way that still gives clients the freedom to choose where they want to start and leaves them in charge of how much they want to disclose. From the outset, clients are sharing control of the interview by choosing what they want to talk about, but the therapist is an active participant who has offered some direction for where they are heading. This type of collaborative alliance does not occur if the therapist gives the client a more specific cue.

Therapist:
> When we talked on the telephone, you said you were having trouble with your boss. What is the problem there?

OR

Therapist:
> You've been having anxiety symptoms. How long has this been going on, and how severe have they been?

The difference between these two openings may seem insignificant. However, it is important to communicate that clients should talk about what they want to talk about and not feel that they have to follow the therapist's agenda. From the start, we want the client to take an active role in directing the course of therapy, while still feeling that the therapist is participating as a supportive ally. *Enacting this process dimension in the initial session is more important than the content of what is discussed.* Following the process while simultaneously attending to the content is a challenging balancing act for beginning counselors, but becomes much easier with practice.

In most cases, clients will readily accept the therapist's open-ended invitation and begin to share their concerns. The therapist can then follow the client's lead and begin to learn more about this person and his difficulties. Therapy is underway when this occurs. Before we go on to the next step, however, we must examine two exceptions in which the client does not accept the therapist's offer to begin: when another therapist has conducted an initial screening interview and when the client has conflicts over initiating.

Previous screening interview   If another therapist has previously conducted an intake interview, the client may not be so ready to begin.

Client: *(impatiently)*
> I've already been through all of this in the intake with Dr. Smith. Do I have to go over it all again just for you?

OR

Client:
> I don't know how much you already know about me. What has Dr. Smith told you about me?

Our initial goal is to develop a working alliance between the therapist and the client. In these two examples, the previous intake therapist is a third party who is psychologically still in the room with them. This is especially problematic for those clients who have come from families in which a third person was *triangulated* into every two-person relationship (Bowen, 1978). That is, whenever two people were close, or in conflict, a third family member would be drawn in and would disrupt the dyad (Haley, 1980). To keep therapy from recapitulating this dysfunctional family interaction, the therapist and client must begin their own relationship as a stable dyad. It is important to maintain a dyadic therapeutic relationship that does not allow others to disrupt the therapeutic alliance.

Therapist:
> I know that you have already spoken with Dr. Smith, and I have learned a little bit about you from his intake notes. But just you and I are going to work together from now on, and I'd like to hear about you in your own words. It may be a little repetitious for you, but this way we can begin together at the same point.

Third-party interference or triangulation in the therapist-client dyad may also occur in the initial therapy sessions because of the unseen presence of the clinical supervisor. Through insecurity, compliance issues, or other factors, beginning therapists may triangulate their supervisors into the therapeutic dyad. For example, the therapist might tell the client, "I'll have to ask my supervisor about that." More commonly, beginning therapists may silently invoke their supervisors—by wondering what they would say or do at a particular point in the session or how they would evaluate the therapist at that moment. This self-critical monitoring does not facilitate therapy. It diminishes the therapist's self-confidence and ability to engage with the client. Further, it disempowers therapists by inhibiting them from being themselves in the session, from finding their own words and acting on their own perceptions, and from developing their own therapeutic styles. Beginning therapists are encouraged to trust themselves enough to simply be a person sharing the story of another person. Supervisory input and evaluation, while necessary, usually will be utilized more productively when it is processed outside of the therapy hour. The therapeutic process is awry when the dyadic relationship between the therapist and the client is broken by third-party interference.

At other times, the client may be curious about what the supervisor thinks of her or the progress she is making. This may happen especially when a videotape or audiotape is used during sessions. It is important for the therapist to explore the reasons for the client's concern. Some clients may feel insecure working with a trainee. Others may feel protective of the therapist and may be overly compliant so that the therapist looks good to the supervisor. Whenever the therapist senses the presence of a third party in the room, it is recommended that he make this presence overt and explore the impact this third-party has on the therapeutic relationship.

Conflicts over initiating   There is another common circumstance in which the client does not respond to the therapist's initial request to begin. The problems that some clients bring to treatment involve conflicts over initiating. This type of client cannot begin at the therapist's request. By asking the client to begin, the therapist has inadvertently presented the client with his central conflict. It is difficult for this type of client to decide what to talk about or to assume any responsibility for the course of therapy. Thus, a therapist who nondirectively waits for the client to lead places an impossible demand on this client.

On the other hand, a directive therapist who takes charge and begins the session by telling the client what to talk about often reenacts the same unwanted responses the client has found with others. Thus, therapy stalls right from the start if the client has problems with initiative and the therapist responds in either a directive or nondirective manner. What can the therapist do in this case? One way to find the more effective middle ground between directive and nondirective extremes is to make a process comment. An effective intervention is simply to ask or wonder aloud about what may be occurring between them at that moment.

Therapist:

> It seems to be hard for you to get started. Maybe we can begin right there. Is it often hard for you to begin, or is there something about this situation in particular that is difficult for you?

By first identifying this as a problem, and then encouraging the client to explore it, the therapist has offered the client a focus and assisted the client in moving forward. However, the therapist has provided this focus without taking over and telling the client what to do, which would only recapitulate old relational patterns. The therapist's open-ended inquiry is supportive, in that it responds to the client's immediate concern. Yet the client can take this issue of initiating wherever he wants, and can share responsibility for the direction of treatment. This type of response provides a new opportunity for clients to explore their problems in a supportive environment. Our first goal is met as the client experiences a collaborative interaction with the therapist, rather than merely having a conversation about the need to work together. Such a collaborative experience sets important expectations for the future course of therapy.

Finally, age, class, gender, culture, religion, and other issues of diversity are involved in initiating therapy. For example, the beginning therapist needs to be sensitive to the fact that, in some cultures, it may be seen as disrespectful to presume to lead with one's elders, an educated person, or an authority figure such as a therapist. A person of color working with a white therapist has probably experienced racism and may be hesitant to begin this relationship. Therapists can best respond to such concerns by acknowledging differences and inviting clients to express or explore the concerns together.

Understanding the Client

The therapist has begun the session by giving the client an open-ended invitation to talk about whatever feels most important to her. Now imagine that the client has begun to share her concerns with the therapist and is clarifying the background and context of her problems. The therapist is trying to find the subjective meaning that each successive vignette holds for the client and grasp what is most significant to the client *from the client's point of view* (Kelly, 1963). In other words, therapists try to have the cognitive flexibility to de-center, enter into the client's subjective experience, and appreciate the meaning that this particular issue holds for the client (McClure & Teyber, 2000). The client will feel understood by the therapist when the therapist can find a common meaning that links together the client's concerns and/or distills the central feeling in her experience. When this occurs, it facilitates the working alliance: Clients become more engaged with the therapist and are encouraged to explore their problems more fully. Let's examine this pivotal sequence more closely.

As used here, *understanding* connotes warmth and a feeling of concern for clients. It signals to clients that their situation is seen and their distress matters to the therapist. When therapists can grasp their dilemma and express this understanding, they are providing Bowlby's secure base or holding environment to contain the client's distress. Although this is not sufficient to resolve clients' problems, such empathic understanding will often ease clients' initial distress and, for some, their presenting symptoms may even abate. A secure attachment configuration is established, and the therapist's attuned responsiveness may temporarily relieve the entire family of attachment affects:

- Anxiety may diminish from this reassuring emotional contact.
- Depression may lessen; the experience of being seen and accepted, rather than dismissed or judged, engenders hope in the client.
- Anger from frustration over not being seen or feeling invalidated by others may be assuaged by the therapist's affirmation.
- Shame over having needs revealed by entering therapy and asking for help may be lessened by the therapist's respect.

The sections that follow examine further what it means to *understand* the client's experience. We will see why it is important to acknowledge and affirm the client's subjective experiences, explore what therapists can do to better understand the client, examine how process comments can be used to clarify misunderstandings, and see how beginning therapists' own performance anxieties may make it difficult for them to listen with presence.

Clients Do Not Feel Understood or Affirmed

Most clients are concerned that others do not really listen to them, take them seriously, or understand what they are saying. Clients often describe themselves as feeling invisible, alone, strange, or unimportant. Many clients feel this way because their subjective experience was not *validated* in their family of origin. While growing up, most clients repeatedly received messages that denied their feelings and invalidated their experience.

> You shouldn't feel that way.
> Why would a silly thing like that make you mad?
> You can't possibly be hungry now.
> I'm cold. Put your sweater on.
> How can you be tired? You've hardly done anything.
> You shouldn't be upset at your mother. She loves you very much.
> You don't really want to do that.
> We don't talk about those things.
> What's the matter with you—how could you feel that way!

Further, family members often changed the topic or simply did not respond when the client expressed a feeling or concern. One of the most effective ways therapists can help their clients change is to affirm their subjective experience. R. D. Laing goes so far as to suggest that people stop feeling "crazy" when their subjective experience is validated (Laing & Esterson, 1970).

Consistent invalidation or mystification in the client's family of origin has profound, long-lasting consequences. In its severe forms, some authors describe it as "soul murder" because clients lose themselves when they lose the validity of their own experience (Schatzman, 1973). Whereas disconfirmation has occurred to some extent in most clients' families, victims of abuse have experienced pervasive invalidation. One of the most serious consequences of such systematic invalidation is disempowerment. When reality-based feelings and perceptions are repeatedly denied, offspring become incapable of setting limits, saying "no," and refusing to go along with what does not feel right to them.

Consistent denial of their experience leaves clients unsure of what has actually happened to them and of the subjective meaning that events hold. They no longer trust their own perceptions of what may be occurring or making them uncomfortable. Denying the validity of their own experience, these clients characteristically say to themselves, "Oh, that didn't really happen" or "I'm just exaggerating this; it wasn't that bad," when it actually was significant. More basically, they often cannot find words to communicate their experience and, even if they can, do not expect others to respect or understand what they say.

When their subjective experience has been denied repeatedly, as commonly occurs in alcoholic, abusive, or highly authoritarian families, clients do not know what they are feeling, what they like or value, or what they want to do. In place of clear feelings and confident perceptions, a vague, painful feeling of internal dissonance results. Fortunately, this undifferentiated feeling state can be replaced with emotional clarity and an accompanying sense of greater personal power if the therapist consistently listens to clients with respect and validates their experience. To affirm clients and support their efficacy, therapists listen intently, take seriously whatever matters to the clients, and reflect their understanding of what the clients have said. Although such simple empathy and validation may sound like common human responses, they are not. Many clients have not had their most important feelings and perceptions validated in their most significant relationships. Thus, therapists must validate the client's experience by identifying and articulating the central meaning that this particular experience seems to hold for the client. For example, in the following questions and comments, the therapist is not necessarily agreeing or disagreeing with the client, but trying to capture the central meaning or issue in what the client has just said.

> It didn't seem fair to you.
> I'm wondering if you were frightened when he did that?
> It's been too much for you, more than you can stand.
> It felt great to be so effective and in charge!
> It was disappointing; you wanted more than that.
> Here again, are you having to take care of everyone else?

The therapist does not have to be an exceptionally perceptive person to understand the client's experience, and therapists need to know they will not be accurate all the time. They just have to be right often enough to form an empathic bond with their client; Winnicott (1965) referred to this as "good enough mothering." When the therapist does not understand what clients are saying, however, the therapist should not feign understanding by saying "yes" or "OK." Instead, the therapist should acknowledge his confusion and/or ask the client to say something in a different way, so that the therapist can understand and they can get back on track. Further, if therapists feel they have been inaccurate, they can simply check out their perceptions and invite the client to clarify them. For example, the therapist replies: "As I listen to you, it seems as if you feel so discouraged that you just want to give up. Is that how it is for you?" This invitation for a dialogue takes the performance pressure of having to be right off the therapist, diminishes unwanted concerns about mistakes, and furthers the collaborative alliance. The therapist's perceptions certainly will be inaccurate at times, but his sincere efforts to understand clients' experience by engaging in this process of mutual clarification will often make clients feel cared for.

In most instances, a primary working goal for the therapist is to provide validation throughout each session by grasping the client's core messages and affirming the central meaning in what the client relays. To appreciate how important validation is, beginning therapists may remember what others have done to help them during their own crisis periods. Almost universally, helpful responses include validation of one's feelings and perceptions.

Providing validation is particularly important when working with people of color, gay men and lesbians, economically disadvantaged clients, and others who feel "different." These clients will bring issues of oppression, prejudice, and injustice into the therapeutic process, and their personal experiences have often been invalidated by the dominant culture. These clients, in particular, will not expect to be heard or understood by the therapist. The first step in working with all people is to listen empathically and hear what is important to them. Therapists respect the personal meaning that experiences hold for different people when they enter into and affirm the client's subjective experience. The most effective therapists, of every theoretical orientation, offer these basic human responses especially well. Pioneering clinicians as varied as Beck, Bowlby, Kohut, Rogers, Satir, and Yalom all share this respect for the client's subjective experience, and emphasize the need for therapists to respond empathically and enter the client's worldview.

Demonstrate Understanding

To engage a client in a working alliance, the therapist listens to the client's experience, finds the feeling and meaning that each successive vignette holds for the client, and accurately reflects back what is most significant in the client's experience. An accurately empathic reflection that captures the meaning of the client's experience is a complex intervention; it is never just a rote parroting of what the client has said. An effective reflection is more akin to an interpretation. It goes beyond what the client has said and communicates that the therapist understands the core message, registers the emotional meaning, or distills what is most important in the client's experience. Rogers (1975) believed that communicating such understanding helps provide a deep acceptance that is a prerequisite for any meaningful change. Beginning therapists will find that they achieve credibility with their clients, strengthen the working alliance, and become important to their clients when they *demonstrate* their understanding in this tangible way.

According to one common stereotype, a therapist is someone who says, "I hear you. I know just what you mean. I understand completely." However, this type of global, undifferentiated response is not effective and, paradoxically, often furthers the client's sense of not being seen or heard. The therapist does not simply say, "I understand," but demonstrates that understanding by *articulating* the central meaning of what the client has said. Therapists show the client that they can be helpful when they capture and express the specific meaning in the client's experience, rather than offering well-intended but vague reassurances. And by checking with the client to see if that understanding is accurate, the counselor further strengthens the collaborative alliance.

How can the beginning therapist put these ideas into practice? As an illustration, we now examine a brief case study of a client who initially had the experience of not being heard by her therapist but later felt understood with a second therapist.[2]

While growing up, Marsha did not feel heard or understood by her parents. Her father was distant. He was not comfortable talking with his adolescent daughter and believed his wife should handle the children. Her mother was critical, demanding, and intrusive. Whenever her mother felt or believed something, she demanded that her children see it the same way. For example, if Marsha felt something that her mother did not feel, her mother would chide her, "That's ridiculous. What's wrong with you?" As an adolescent, Marsha was painfully self-conscious and insecure. She didn't really understand why she felt so unhappy, however, and often thought, "There's just something wrong with me." Marsha frequently cried alone in her room.

Marsha had always thought that everything would get better when she went away to college. To her dismay, though, she found herself becoming even more depressed during her first semester. She could not stop crying and started gaining weight. More confused about herself than ever, Marsha began seeing a counselor at the Student Counseling Center. Although she did not really have words for what was wrong, Marsha tried to help her counselor understand her problems:

Marsha:
I feel empty—kind of lonely.
Therapist:
Do you have any friends? What are your classmates like?
Marsha:
I guess I have friends. I have a roommate in the dorm.
Therapist:
What do you do with your friends—go to the movies, shopping?
Marsha:
Yeah, I do those things. I belong to a foreign language club, too.
Therapist:
Do you like your friends? You're new to the university; maybe you need some new friends here at school.
Marsha:
Well, I've had friends, but I just feel empty.
Therapist:
But you've just left your family and come to college. You must miss your family and feel lonely. It's natural to feel lonely when you move away from home. Most of the other kids in the dorm feel that way, too. I know I sure did when I moved away to college.
Marsha:
Oh.
Therapist:
This is not unusual at all. You're going to be just fine once you get through this transition.

---

[2]The clinical examples in this text are based on actual cases but identifying information, including the gender of the client or therapist, has been altered to ensure confidentiality.

Marsha:
I hope so. Maybe my family is different, though. In high school, I always thought that my family had more problems than my friends' did.
Therapist:
Yeah, but, like I said, it's real natural for you to be kind of emotional at this time in your life.
Marsha:
It is? But I still feel different from everybody else.
Therapist:
Sure, it's real natural for you to feel different from everybody else. Late adolescence, moving away to college—it's a tough time in life. You're going to be just fine. Do you have a boyfriend?
Marsha:
Yes. I try to tell him what's wrong, but it doesn't work. And then he gets angry at me because nothing helps. I just don't know why I get depressed a lot. My mother used to yell at me a lot, though.
Therapist:
Do you have trouble eating or sleeping?
Marsha:
No trouble eating; I've gained ten pounds. I wake up at night some-times—worrying about things that don't really matter, feeling like crying.
Therapist:
Are you eating alone? Maybe you should be eating with friends.
Marsha: *(with resignation)*
Maybe that would help.

Marsha could not say exactly why, but she did not like seeing the counselor and she did not go back. She became more depressed as the semester went on, continued to wonder what was wrong with her, and struggled in her classes. During advising for second semester, a concerned professor saw her distress and talked her into going back and trying another counselor. Reluctantly, Marsha agreed.

Therapy began more slowly this time. Marsha found herself impatient with the therapist and reluctant to share much. She alternately acted blasé and then distressed, but would never allow the therapist to stay with her experience for very long. The therapist was not frustrated, however, and was effective at communicating his continuing interest in her. After a few weeks, Marsha sensed that this relationship might be safe enough to risk disclosing her feelings again.

Marsha:
I feel empty.
Therapist:
What's it like to feel empty?
Marsha:
I just feel empty.
Therapist:
Show me in some way what that emptiness feels like.

Marsha: *(long pause; nothing is said)*
Therapist:
> Maybe you can use just adjectives, or make a sound or a gesture that will describe the emptiness.

Marsha:
> There's just an emptiness inside. The wind blows right through me. I'm always hungry. It never gets filled up.

Therapist:
> That's right. You just feel empty inside. Sex, alcohol, food—nothing fills it up. The hole just stays there.

Marsha: *(doesn't speak; nods and becomes teary; looks at therapist)*
Therapist: *(holding her gaze kindly)*
> I'm feeling for you right now and can see how much you are hurting. It's sad that things have happened to make you feel this empty and alone, but I'm so honored that you are willing to risk sharing this with me.

Marsha: *(cries harder)*
> There's just something wrong with me; there always has been.

Therapist:
> Tell me about it; help me understand what's always been wrong.

Marsha was heard and felt understood this time. Breaking the old relational pattern, she was not crying alone in her room anymore. In subsequent sessions, the therapist consistently listened and responded to what was most important for Marsha. Although there were ups and downs, this empathic connection forged a collaborative alliance that ultimately allowed Marsha to resolve her life-long dysthymia.

The concept of understanding the client sounds simple enough. All the therapist has to do is listen carefully, grasp the underlying feelings and core messages, and find a way to reflect this understanding to the client. In actual practice, however, this is not so easy to do. Most of us have been strongly socialized to "hear" in a limited, superficial way that denies the emotional messages or interpersonal conflicts embedded in clients' remarks. This was why Marsha's first counselor could not hear what she was truly saying and avoided the difficult emotions involved. The client kept coming back to her core message, but the therapist neither heard the content of what she was saying nor did he follow the process. By reassuring her and trying to talk her out of her feelings, he metaphorically reenacted her developmental conflict of not being seen or heard.

Most beginning therapists are aware that they possess a "third ear"—that they already have a highly developed ability to hear the basic meaning in what people say. However, they often feel they must avoid acknowledging the true content of these underlying, and often nonverbal, messages because of their emotional content. Therapists want to avoid or downplay these emotional responses for many reasons: feeling concerned that they might embarrass the client; feeling unsure of how to respond; feeling reluctant to violate cultural or familial proscriptions about forthright communication; or feeling a need to take care of others by shielding them from their own pain or discomfort. As a result, many beginning therapists have a habit of switching automatically to a more superficial level to avoid the conflict and pain that may emerge from empathic listening.

Marsha's second counselor broke the social rules, however, and responded directly to the core affective message, "I feel empty." As a result, Marsha felt understood for the first time in therapy. In a small but significant way, this was a corrective emotional experience. The therapist joined her in her experience—which significant others had not done before. Disconfirming her expectations, he did not move away from her feelings or try to talk her out of them; he actually welcomed them and encouraged her to enter them further. When the therapist is able to provide such experiences consistently, over time trust builds, and the client becomes more confident in the expectation to be seen and heard. For many beginning therapists, stepping out of the social norms they grew up with is an exciting but stressful component of clinical training. Therapists are encouraged to try responding in this more direct, empathic way; use the ability they already possess to hear what is most important to the client; and take the risk of saying what they see.

To summarize, therapy offers clients an opportunity to be understood more fully than they have been in other relationships. When this understanding is offered in the initial session, clients feel that they have been seen and are no longer invisible, alone, strange, or unimportant. At that moment, the client begins to perceive the therapist as someone who is different from most other people and possibly as someone who can help. In other words, hope is engendered when the therapist understands and articulates the personal meaning that each successive vignette holds for the client. With this in mind, let us look more specifically at how the therapist can find the central meaning in what the client presents.

Identify Recurrent Themes

As we have seen, the therapist's goals at the beginning of the session are to:

- Encourage the client's initiative
- Actively join the client in exploring her concerns
- Identify the core messages or central meaning in what the client says
- Articulate this empathic understanding to the client

Although this, too, may sound simple, it is not. It requires the therapist to give up a great deal of control over the direction of therapy and over the timing with which issues are brought up for discussion. The "letting go" of what the client discusses in a given session may be particularly discomforting for the novice therapist who is accustomed to being in control, albeit subtly, and who is characteristically adept in social interactions. However, the same social adroitness that facilitates interpersonal relations outside of therapy may limit a client's progress within the therapeutic relationship. Here, the therapist is placed in the far more demanding position of responding to the diverse and unpredictable material that the client produces, rather than subtly directing what the client will cover. This approach achieves our overarching goal of enacting an interpersonal process that fosters self-efficacy by sharing more responsibility with clients for resolving their problems. However, it also puts significant personal demands on the therapist. Specifically, *the therapist must be able to relinquish control over what will occur next in the therapeutic relationship, tolerate the ambiguity of not knowing what clients will produce or where the current topic will lead, and make sense of the varied material that clients present.*

What assistance is available to help therapists meet the demands of working in this way? If therapists can identify *recurrent themes* in what the client relays, it will help them make sense of the client's subjective experience and better understand what is most important out of the wide-ranging material the client presents. This empathic understanding, in turn, provides therapists with the only legitimate control they can have in therapeutic relationships—over their own responding—as opposed to ineffective attempts to direct or control the material that clients produce. Additionally, the appropriate inner controls that come from this understanding will enable therapists to tolerate the ambiguity inherent in this work.

Suppose the client tells the therapist about his reasons for coming to therapy. As the therapist follows the different recollections and descriptions the client chooses to relate, the therapist is trying to discern an *integrating focus* for the wide diversity of material that the client is presenting. The therapist can do this by identifying three types of common factors that recur throughout the different material the client presents:

1. Repetitive relational themes or interpersonal patterns
2. Pathogenic beliefs, automatic thoughts, or faulty expectations
3. Recurrent affective themes or central feelings

As we will see, identifying these superordinate patterns provides unifying themes that help counselors conceptualize their clients' dynamics and formulate treatment plans. To beginning therapists, the disparate material that clients present often seems unrelated and disconnected. As they progress in their training, however, beginning therapists will increasingly recognize that synthesizing themes are present in all three domains if they listen for them. In order to help therapists develop this essential skill, guidelines for writing process notes within an interpersonal process framework are provided in Appendix A. Keeping these process notes after each session is one of the best ways for counselors to begin recognizing the organizing patterns that exist in their clients' functioning. Let's examine each of the three types of common factors in turn.

Repetitive relational patterns   Although terminology varies and different facets are emphasized, the bedrock of all interpersonal/ dynamic treatment approaches is identifying the repetitive relational patterns that are central to the client's problems. The therapist listens for the interpersonal themes that keep recurring throughout the different narratives the client relates. For example, "Anyone I trust betrays me by leaving"; "They expect or need so much from me, and I feel overwhelmed by their demands"; "I have to give up what I want in order to be close. If I ever try to meet my own needs or say what I want, my partner will leave me." As beginning therapists review their process notes, they will often observe that most of the varying problems that each client talks about consist of only two or three basic themes or patterns.

Interpersonally oriented psychotherapy has always been focused on clients' repetitive, self-defeating relationship patterns. Since the 1980s, however, this approach has been developed more systematically as an effective focus in short-term or time-limited treatments (Ecker & Hulley, 1996; Levenson, 1995; Luborsky & DeRubeis, 1984; Sifneos, 1987; Strupp & Binder, 1984). These effective, programmatic treatments all focus on the client's maladaptive relational patterns, which are variously labeled as the core conflictual relationship theme, cyclical maladaptive pattern, maladaptive transaction cycle, or repetitive transactional pattern. In each approach, the therapist tries to identify the relational themes that recur in the different interpersonal conflicts the client is experiencing. The therapist highlights these relational patterns to clients so that they can learn to recognize these scenarios by themselves. The client's pattern, for example, might be to feel controlled by others, let down, or abandoned. Perhaps the client always feels criticized, or idealized, or disdained. It is compelling for clients when the therapist can highlight the same themes occurring in three different spheres: in current interactions with significant others; in formative relationships with family members; and, most importantly, in the current, here-and-now interaction with the therapist. Let's restate and extend this basic approach.

The therapist points out repetitive relational patterns as clients are relaying vignettes from current or past relationships. More importantly, the therapist actively looks for opportunities to *link* the relational patterns that are occurring with others to the current interaction between the therapist and the client. In order to facilitate change, the therapist must have the flexibility and willingness to work through this reenactment with the client and help create a better solution in their real-life relationship than the client has found with others. Throughout this text, we will focus on how therapists can help clients rework this repetitive relational pattern that is causing so many problems with others.

Pathogenic beliefs   Clients' repetitive relational patterns involve problematic beliefs about themselves and others and faulty expectations for what is going to occur in relationships and in the future. Weiss and Sampson (1986) and Weiss (1993) have clarified the pathogenic beliefs that help to create and sustain such maladaptive relational scenarios. For example, regarding guilt/compliance issues, the client's pathogenic belief might be: "I am being selfish when I do what I want; others are hurt by my independence; I must sabotage or not enjoy my own success or I will be disloyal." A pathogenic belief regarding rejection/shame dynamics might be: "I am unimportant and do not matter; others will see me as childish and needy; they would ignore me if they knew what I needed." Beck (1976), Beck and Freeman (1990), and others have illuminated the automatic thoughts and dysfunctional interpretations that are central to clients' problems. They write compellingly about the cognitive schemas people develop—that is, ways of viewing themselves, others, the world, and the future. In particular, these authors have identified a *cognitive triad* in which people come to view themselves as defective or unlovable; the world as unmanageable or overwhelming; and the future as bleak and hopeless. As a result of formative developmental experiences, *clients begin to exercise a selective bias in processing information. As a result of this cognitive filter or selective attention process, clients synthesize the same repetitive themes from diverse interpersonal experiences.* They continuously construe events to arrive at the same conclusions, such as disinterest or unfairness, which come to characterize their experiences.

These dysfunctional beliefs and expectations are often learned in repetitive interactions with caregivers and may operate consciously or without much awareness. Once again, therapists try to identify these core beliefs that provide integrating threads throughout the client's relational problems; make them overt to clients or name them—especially as they are occurring in the therapist-client relationship; and question the truthfulness or utility of these beliefs in current relationships. As Beck emphasizes, automatic thoughts and dysfunctional beliefs are not so evident when clients are functioning well. It is when clients are upset or distressed (as Beck would say, in their *hot cognitions*) that pathogenic beliefs and relational templates will be revealed. At these crisis points, they are expressed as the client's reality and experienced as the only way that relationships have ever been and ever will be! In other words, when clients are not upset, they can imagine a wide range of interpersonal scenarios and outcomes, entertain a more realistic set of beliefs about themselves and others, and have flexible expectations about what may occur in relationships. In contrast, clients' beliefs and relational expectations often become rigid and unidimensional when life circumstances have activated their conflicts. Thus, it is when clients are in distress that therapists can best assess their orienting schemas and relational templates. In closing, the interpersonal process therapist is just as likely to ask: "How did you explain that to yourself" as "How did you feel about that?"

Recurrent affective themes   Therapists can also identify common themes in clients' emotional reactions. Recognizing and responding to these *recurrent affects* is a meaningful gift therapists can offer their clients. (The therapist might say, for example, "This is so hard for you. Here again, when _____ happens, you are always left feeling _____.") Often, a single, core affect comes up again and again for the client (Mueller & Kell, 1972). As the therapist listens to the client, an overriding feeling such as sorrow, bitterness, or shame may pervade the client's mood or characterize the different experiences the client relates. (The therapist might then say, for instance, "As I listen to you, it sounds as if you have felt hopelessly burdened all of your life.") When the therapist can identify and accurately name this core affect—the feeling that the client experiences as the central or defining aspect of her existence—it has a profound impact. Clients often feel that the therapist understands who they really are and sees what their existence is like in a way that others have not been able to do. Perhaps nothing goes further toward establishing a collaborative alliance than the therapist's ability to identify a core affect and accurately reflect the profound meaning it holds for the client.

In most cases, responding to the recurrent affects or repetitive emotional themes that result from clients' maladaptive relational patterns is one of the most significant interventions therapists can provide. Unfortunately, beginning therapists often feel insecure about responding to clients' feelings, and, to complicate matters, this topic often receives too little attention in clinical training. We will explore these interventions in depth in Chapter 5. The process notes provided in Appendix A will also help therapists identify the recurrent patterns in all three arenas.

To sum up, if the therapist can identify the repetitive interpersonal themes, pathogenic beliefs, and primary affects that link together the clients' experience and problems, clients will feel that the therapist understands them in a way that others do not. Providing this highly differentiated empathy is a gift to clients and, therefore, an orienting guideline for therapists throughout the course of treatment. It is not an isolated intervention, but a consistent attitude that characterizes the therapist's way of being with the client. When the therapist strives both to understand what a client is experiencing in this moment, and to understand the integrating themes that unfold and reveal the coherence of the clients' life, she helps the client find himself in his own reflection from the therapist.

The basic model described here also applies to short-term therapies. Although therapists must be more focused and limited in their goals, their principal aim in both modalities is to provide a focus on maladaptive relational patterns. The primary challenge of successful time-limited therapy is that, far more quickly and accurately than in longer-term treatment, therapists must be able to (1) establish a collaborative alliance; (2) identify the recurrent themes in all three arenas; and (3) make these patterns overt by highlighting how they are occurring right now in the current interaction with the therapist. Therapists must learn how to enact these processes with clients in longer-term counseling first, then they will be successful in adapting these skills to time-limited treatment.

Process Comments Facilitate A Collaborative Alliance

Carl Rogers legitimized concern about the quality of the therapist-client relationship. His contribution to the field of counseling cannot be overstated: He "changed the game" (Kahn, 1997). His conviction was that the therapist does not cure clients but, rather, offers them understanding and caring (agape). Rogers's core conditions for the success of the therapeutic enterprise are empathy, genuineness, and warmth (Rogers, 1961). In practice, however, researchers find that many therapists cannot remain empathic, or even neutral, when clients repeatedly diminish or compete with them or when clients are critical, demanding, or controlling toward the therapist. In a series of studies known as Vanderbilt I, Strupp and Hadley (1979) found that clients' maladaptive relational patterns routinely elicited hostility and control from seasoned therapists, and when these complementary responses occurred, it led to poor therapeutic outcomes. That is, *clients with angry, distrustful, and rigid relational styles tended to evoke countertherapeutic hostility and control—even in a carefully selected sample of highly trained and experienced therapists* (Henry, Schacht, & Strupp, 1990)! Rogers's core conditions are invaluable and provide a good working definition of a collaborative alliance, but are not easily achieved with some clients. Process comments provide therapists with a method of intervening to help resolve, rather than merely reenact, the client's maladaptive relational patterns.

Process comments make the interaction between the therapist and client overt and put the relationship "on the table" as a topic for discussion. Process comments are not confrontations, demands, or judgments. They are simply observations about what may be occurring between the therapist and client at that moment. They are offered tentatively, as an invitation for further dialogue and mutual sharing of perceptions. Often they lead to greater closeness. (For example, the therapist might say, "As you're talking about this, I'm wondering if you might be trying to tell me how you are feeling about me and our relationship as well?") Process comments enliven the relationship and are especially effective when the interaction has become repetitive or has lost its direction. They bring *immediacy* to the therapeutic interchange as the therapist and client stop talking about others out there and begin talking instead about "you and me"(Robitschek & McCarthy, 1991). These metacommunications also give therapists a way to alter the client's relational patterns and to change reenactments when they occur.

Therapist:

I think that something important might be going on right now. Can we talk together about what just happened between us? I know people don't usually talk together this way, but I think it could help us understand what's been going wrong with your wife and others as well.

In addition to activating stalled relationships and altering maladaptive relational patterns, process comments also do the following:

- Invite a fuller sharing between therapists and clients
- Promote further self-exploration and help clients see discrepancies between what they are saying and doing
- Provide interpersonal feedback and help clients see themselves from others' eyes and learn about the impact they have on others

Other terms besides immediacy have been used to describe interventions similar to process comments. For example, Kiesler discusses *therapeutic impact disclosure* —interpersonal feedback from the therapist to make overt how the client's maladaptive relational patterns are affecting the therapist at that moment and are being played out in their current interaction (Kiesler & Van Denberg, 1993). Similarly, object relations theorists provide *metacommunicative feedback* to register the unspoken emotional quality of a relationship (or, in object relations' strained terminology, to make the client's *projective identifications* overt) (Cashdan, 1988).

Therapist:

Although you've never actually said anything like this, I'm afraid that if I disagree with you, you'll be angry and leave our relationship. What do you think about this feeling I'm having?

Closely akin to process comments are *self-involving statements,* which may be especially useful for beginning therapists. Whereas *self-disclosing statements* refer to the therapist's own past or personal experiences, self-involving statements express the counselor's current reactions to what the client has just said or done (McCarthy, 1982; Tyron & Kane, 1993). (For example, the therapist might say, "Right now, I'm feeling _____ as you're telling me about this.") Contrast these two responses:

Therapist:
> I have a temper too, sometimes.

VERSUS

Therapist:
> As I listen to you speak about this, I find myself feeling angry too.

As we would expect on the basis of client response specificity, self-disclosing comments may be useful at times. Too often, however, clients' focus is shifted away from themselves when therapists reveal personal information. More productively, self-involving statements keep the focus on the client and reveal information about what is happening in the relationship and/or how it makes the counselor feel.

Therapist:
> I can feel my stomach tighten right now as you are reading that letter to me.

Sharing personal reactions to what clients have just said or done conveys personal involvement with clients and acknowledges emotional resonance. It also gives clients feedback about the impact they are having on the therapist. When sincere, this type of self-involving statement or process comment goes a long way toward facilitating a collaborative alliance.

Depending on the therapist's ability to communicate in a sensitive manner, of course, any of these metacommunicative interventions that focus on the current, here-and-now interaction can be used ineffectively or effectively.

Therapist:
> I'm feeling a little bit bored by this. Maybe we should talk about something else.

VERSUS

Therapist:
> I think I'm missing you right now, and not grasping the real meaning this holds for you. Can you help me get closer to what is most important for you here?

In particular, there are two situations when process comments will not work. First, therapists should not jump in with process comments without first considering the possibility that their reaction reflects more about their own *countertransference* issues than it does about the client. As a rule of thumb, therapists may want to wait until they have seen an interaction occur two or three times and had time to reflect upon it—or discuss it with a supervisor—before venturing this observation with the client. Differentiating the client's concerns from the therapist's own issues is an important topic; we will return to it in subsequent chapters.

Second, process comments create problems for clients when therapists promise something they cannot provide. The process comments and related concepts introduced here all encourage open and direct communication, which is also an invitation for further closeness. However, if the therapist then responds in impersonal, distancing, or judgmental ways that do not honor the honesty invited, clients are set back by this mixed message. This double-binding situation will be especially problematic for clients whose caretakers gave promises of authentic emotional involvement but failed to follow through.

To sum up, process comments are powerful interventions that facilitate a collaborative alliance, reveal important new issues for further exploration, and provide the most effective way to alter problematic relational patterns that are being reenacted with the therapist. However, beginning therapists should not utilize process comments until they feel comfortable communicating in this forthright way and are ready to meet the client in the authentic manner that process comments invite. For helpful illustrations of how process comments facilitate group interaction, see the classic text by Yalom (1975).

Performance Anxieties

Let's let therapists be human, too. They cannot always hear or fully assimilate what they have just been told or be accurately empathic. Understandably, they will miss the feeling and meaning in the client's experience at times. This section examines performance anxieties that make therapists less empathic and cause therapists to push clients to change prematurely.

Therapists will not be as effective with clients when they are burdened by their own excessive performance demands. Commonly, beginning therapists are trying too hard (to be helpful, to win approval from a supervisor, to prove their own adequacy to themselves, to be liked by the client). When therapists are trying too hard, it is almost impossible to de-center, empathically enter into the client's subjective worldview, and be emotionally present.

Too often, novice therapists are thinking about where the interview should be going next, wondering how best to phrase what they are going to say next, preparing advice, or worrying about what their supervisors would expect them to do at this point in the interview. Such self-critical monitoring often immobilizes student therapists, blocks their own creativity, and, sadly, keeps them from enjoying this rewarding work. Further, it prevents therapists from listening and understanding as fully as they could. Thus, beginning therapists are encouraged to slow down their own inner monitoring process and to focus more on the moment, on their current interaction with the client, and on the meaning that this particular experience seems to hold for the client right now.

Identifying recurrent themes, as described in this chapter, can itself become a disabling performance demand. If therapists try too hard to find the essential meaning in the narratives that the client relays, they may lose the broader context of the client's life that clarifies the patterns therapists are trying to find. It is better for therapists to relax, be more patient, and help clients unfold their story as they experience it. Undistracted by performance demands, therapists will be better able to see the repetitive themes and basic meanings in clients' experiences.

When they feel anxious about their helping abilities, beginning therapists may develop a need to *do* something to make the client change. Novice therapists often believe they are supposed to make the client think, feel, or act differently within the initial therapy session. As a result, beginning counselors often change the topic abruptly (at least from the client's point of view) and pull clients away from their own concerns into what the therapist thinks is important. Although therapists will certainly want to help clients make bridges to new topics they haven't considered, these attempts will usually fail when they are driven by the therapist's own internal pressure to fix the problem. Paradoxically, when the therapist acts on these unrealistic expectations, clients are more likely to drop out of therapy because they feel that the therapist does not understand them or is just not helpful.

To sum up, therapists with excessive performance anxieties are less effective because they respond more to their own internal needs (to be helpful, competent, or liked) than to the client's need to be understood. The most appropriate way for therapists to manage their realistic concerns about their performance is to discuss them openly with a respected supervisor. Therapists do not need to labor under the burden of having to know what is going on with the client. What is asked, instead, is only to keep trying to form a collaborative alliance with the client in order to learn together what is occurring.

Care and Understanding as Preconditions of Change

The therapist does need to help the client change, of course, but change is most likely to occur if the client first experiences the therapist as someone who understands and cares. In the initial sessions, the therapist's primary goal is to begin a working alliance. The therapist does this by establishing an emotional connection with the client through empathy and respect. Facilitating a collaborative alliance is more important than obtaining any specific information about the client or effecting any change. In this regard, one of the best ways to evaluate the success of an initial session is for therapists to ask themselves, "Do I feel like I made contact with the client and have a genuine feeling for who this person is?" Therapists are building an affective bridge to the client and creating the interpersonal context necessary for change, which will not begin until clients feel that the therapist is available, concerned, and sincerely trying to understand. If therapists provide this type of responsiveness throughout the initial session, they are already being helpful to the client. As noted earlier, anxiety and depressive symptoms may even abate as the therapist's attuned responsiveness provides a secure holding environment that contains the client's distress (Bowlby, 1988).

Finally, we must address the most important component of what it means to understand the client. Therapists must actively extend themselves to the client and communicate their interest in the client's well-being. That is, therapists must articulate their understanding of the client's experience in a way that also communicates their compassion.

This ability to articulate the client's experience in an accurate and caring way is illustrated elegantly in a classic article, "Ghosts in the Nursery," by Selma Fraiberg and her colleagues (Fraiberg, Adelson, & Shapiro, 1975). The therapist in this case study is working with a depressed young mother whom social services has judged to be at high risk for physically abusing her infant daughter. Early in treatment, the therapist is disconcerted as she observes the mother holding her crying baby in her arms for five minutes without trying to soothe it. The mother does not murmur comforting things in the baby's ear or rock it; she just looks away absently from the crying baby. The therapist asks herself the question, "Why can't this mother hear her baby's cries?"

As the young mother's own abusive history began to come out in treatment, the therapist realized that no one had ever heard or responded to the mother's own profound cries as a child. The therapist hypothesized that the mother "had closed the door on the weeping child within herself as surely as she had closed the door upon her own crying baby" (p. 392). This conceptual understanding led the therapist to a clinical hypothesis and basic treatment plan: When this mother's own cries are heard, she will hear her child's cries.

The therapist set about trying to hear and articulate compassionately the mother's own childhood experience. When the mother was 5 years old, her own mother had died; when she was 11, her custodial aunt "went away." Responding to these profound losses and the mother's resultant feeling that "nobody wanted me," the therapist listened and put into words the feelings of the mother as a child:

> How hard this must have been. . . . This must have hurt deeply. . . . Of course you needed your mother. There was no one to turn to. . . . Yes. Sometimes grown-ups don't understand what all this means to a child. You must have needed to cry. . . . There was no one to hear you. (p. 396)

At different well-timed points in treatment, the therapist accurately captured the mother's experience in a way that gave her permission to feel her feelings. As a result, the mother's grief and anguish for herself as a cast-off and abused child began to emerge. The mother sobbed; the therapist understood and comforted. In just a few more sessions, something remarkable happened. For the first time, when the baby cried, the mother gathered the baby in her arms, held him close, and crooned in his ear. The therapist's hypothesis had been correct: When the mother's own cries were heard, she could hear her baby's cries. The risk for abuse ended as this beginning attachment flourished.

This poignant case study illustrates how the therapist can use the therapeutic relationship to resolve the client's conflict. A corrective emotional experience occurred when the therapist responded to the mother's pain in a caring manner. This case study illustrates how powerful it is when the therapist articulates an understanding of the client's experience in a way that also communicates the therapist's compassion for the client.

Using role models such as this, beginning therapists need to explore their own ways of communicating that they are moved by the client's pain and are concerned about the client's well-being. Clients' inability to care about themselves is central to many of their conflicts, and most clients cannot care about themselves until they feel someone's caring for them (Gilligan, 1982). Therapists provide this care when they recognize what is important to the client, express their genuine concern about the client's distress, and communicate that the client is someone of worth and will be treated with dignity in this relationship. These are the therapist's goals in the initial stage of therapy.

Closing

The interpersonal process approach tries to resolve problems in a way that leaves clients with a greater sense of their own self-efficacy. This independence-fostering approach to psychotherapy can be achieved only through a collaborative therapeutic alliance. The client must be an active participant throughout each phase of treatment—not a good patient who waits to be cured or told what to do by the doctor. This process dimension of how the therapist and client work together is more important than the content of what they discuss or the theoretical orientation of the therapist. In this chapter we also have explored the profound therapeutic impact of listening with presence and understanding what is most important to the client. Offering the deep understanding described here is a gift to clients—one of the most important interventions that therapists of any theoretical orientation can offer. Perhaps because they are so simple, these basic human responses are too easily overlooked. They are the foundation of every helping relationship, however, and the basis for establishing a collaborative therapeutic alliance.

Suggestions for Further Reading

1. Attachment theory and its core concept of internal working models (templates, cognitive schemas) helps therapists identify the patterns and themes that link together clients' symptoms and problems, and helps therapists find a treatment focus. Readers are encouraged to read the excerpt "Representational Models" in Chapter 2 of the Student Workbook (pp. 31–40).

2. For a marvelous explication of the therapist's basic values and attitude toward the client, see Chapter 3, "The Therapist's Stance," in *Psychotherapy in a New Key*, by H. Strupp and J. Binder (New York: Basic Books, 1984).

3. *Presence* is an important but elusive therapeutic dimension. For a compelling illustration of a therapist's presence with an adolescent sexual abuse survivor, see F. McClure's "Case Study of Sheila: A 15-Year-Old African American Female," in Chapter 4 of *Child and Adolescent Therapy: A Multicultural-Relational Approach*, edited by F. McClure and E. Teyber (Fort Worth: Harcourt Brace, 1996).

4.   Readers are encouraged to read the chapter, "Therapist Variables," by E. Teyber and F. McClure in *Handbook of Psychological Change: Psychotherapy Processes and Practices for the 21st Century* edited by C. Snyder and R. Ingram (New York: Wiley, 2000). This chapter summarizes the research literature on the working alliance and characteristics of effective therapists.

5.   Illuminating ideas about the effects of multigenerational family relations on individual personality are found in two classic articles on triangular coalitions: J. Haley's "Toward a Theory of Pathological Systems," in *Family Therapy and Disturbed Families,* edited by G. H. Zuk and I. Boszormenyi-Nagy (Palo Alto, CA: Science & Behavior Books, 1967), pp. 11–27; and especially M. Bowen's, "The Use of Family Theory in Clinical Practice," *Comprehensive Psychiatry, 7* (1966): 345–376. Both of these pioneering articles will help therapists apply family systems concepts to the practice of individual psychotherapy.

6.   Practical guidelines for identifying themes, using self-involving statements, and creating immediacy are provided in Chapter 9 of G. Egan, (1994) *The Skilled Helper,* 5/e (Pacific Grove, Brooks/Cole, 1994).

CHAPTER THREE

 Honoring the Client's Resistance

Joan was nervous about seeing her first client, but the initial session seemed to have gone well. The client talked at length about his concerns and, to her relief, she found it was easy to talk with him. The client shared some difficult feelings and Joan felt that she understood what was going on for him. She was thinking they had begun a good working relationship; the client seemed so friendly and appreciative when he left. One week later, however, Joan received a brief telephone message from the client saying only that he was "unable to continue therapy at this time." Confused and dismayed, Joan sat alone in her office wondering what had gone wrong.

Conceptual Overview

Just when the therapist feels that something important is getting started, some clients put their feet on the brakes. The client cancels the second appointment, shows up 25 minutes late, or asks to reschedule for Sunday at 7:00 A.M. This *resistance* is puzzling and frustrating for the novice therapist: "Why didn't that client return? We had a great first session!" Although most clients will not be resistant in this particular way, other forms of resistance will occur regularly throughout therapy. The purpose of this chapter is to help beginning therapists recognize, understand, and respond to client resistance.

All clients have both positive and negative feelings about entering therapy, although the positive feelings are usually more apparent at first. Clients seek therapy in order to gain relief from their suffering. We must look further into the complexity of the client's feelings, however. Often, as clients seek help and genuinely try to change, they simultaneously resist or work against the very change they are trying to attain. How can we understand this paradox? For many clients, *shame* is associated with having an emotional problem they cannot solve on their own. In some cultures, revealing problems to people outside one's family meets with disapproval or is seen as *disloyal*. Other clients feel *guilt* about asking for help or meeting their own needs. Still other clients, because of their relational templates, have *anxiety* evoked by the expectation that others will respond in hurtful ways. Thus, if clients have a problem or need to ask for

58

help, then shame, guilt, anxiety, and other unwanted emotions may be evoked. Although most beginning therapists may be unaware of this paradox, we will also see that conflicted feelings will often be triggered when clients feel better and make progress in treatment. In order to formulate useful case conceptualizations (see Appendix B), therapists need to identify the specific issues that make it difficult for each particular client to approach her problems. Following the principle of client response specificity, specific concerns must be clarified for each individual client, but the following are common themes:

If I let myself depend on the therapist, he might leave me, or take advantage of me, or try to control me, as others have done when I needed them.

I cannot ask for help because I must be independent and in control all the time. Besides, I don't really deserve to be helped anyway.

I cannot need anything from others because I must be perfect or I will bring shame to my family.

Asking for help is admitting that there really is a problem, and that proves there really is something wrong with me.

I am afraid of what I will see or what a perceptive therapist will learn about me if I stop and look inside myself.

If I cannot handle this by myself, it means that I really am weak, just as they always said.

If I don't see a therapist, my spouse might leave me.

Many clients are struggling with feelings of *debasement* at the prospect of losing control over their emotions or of being controlled by another. Although there will be many different reasons, clients' resistance and defenses are driven most frequently by shame.

It is easy to respond to the approach side of clients' feelings—the pain or need that motivates them to seek help and enter therapy. If you gently scratch the surface, it is usually plain to see, and the client wants you to respond to it. In contrast, the other side of the client's feelings—the reactive emotions and the interpersonal consequences of having a problem and needing help—may not be so accessible. These concerns over what it means to enter therapy or ask for help, even though the client may not be aware of them, act as a countervailing force to the client's motivation to enter therapy. If unaddressed, these ambivalent feelings will draw some clients out of therapy prematurely, as in our opening vignette.

Chapter Organization

Resistance is not a welcome concept to most beginning therapists. In the first section, we examine why it is difficult for both therapists and clients to address this important dimension of therapy. In the second section, we look at a working definition of resistance, distinguish between psychological resistance and reality-based constraints on the client's ability to participate in treatment, and observe that both may be operating simultaneously for the client. We also see how therapists can formulate *working hypotheses* to better understand their client's resistance, which, in turn, enables therapists to respond more effectively. The third section presents intervention guidelines for responding to resistance and further information to better understand the client's resistance throughout treatment. Common expressions of client resistance are illustrated at three different stages: during the initial telephone contact; at the end of the first session; and during subsequent sessions. Sample therapist-client dialogues illustrate effective and ineffective responses at each stage.

Most of this chapter outlines practical guidelines to help manage resistance effectively in the early stages of therapy. These guidelines provide trainees with tools to use in keeping their clients from dropping out of treatment prematurely. Resistance and defense will occur at every stage of treatment, however, and the final section includes guidelines for conceptualizing and responding to resistance later in treatment. For now, readers should be aware that the clients most likely to drop out of treatment after the initial telephone contact or first session are people of color, men, and clients who have been vague or evasive about their problems.

Reluctance to Address Resistance

Imagine your client has missed, come very late for, or twice rescheduled the second appointment. Perhaps this has no significant psychological meaning at all. Cars do break down; traffic jams occur; children get sick; and employees get called for work at the last minute. However, if this behavior reflects the client's ambivalence about some aspect of being in therapy, and not just reality-based constraints, the therapist needs to find an effective way to address this or the client is far more likely to terminate prematurely. Initially, therapists usually do not know whether the client's behavior is reality-based, psychologically motivated, or both. As we will see, however, by finding nonthreatening ways to explore this behavior with the client, therapists can greatly increase the chances of the client's continued participation. In these explorations, therapists' curious, nonjudgmental manner is as important as what they say. Otherwise, the client may misunderstand their observations as blame or criticism, which never help and only serve to exacerbate client resistance.

More specifically, the model proposed here will suggest that therapists first enter supportively into clients' frame of reference, affirm or validate the reality-based constraints that they perceive, and do whatever possible to flexibly accommodate or help resolve the problem. Only after taking seriously clients' concerns, as they see them, can therapists begin to inquire about other conflicted feelings or meanings that entering treatment may hold. The therapist can begin this joint exploration by wondering aloud, in a tentative manner, whether entering treatment may be evoking other more psychological concerns as well. If therapists approach resistance without following this two-step sequence, most clients will feel misunderstood or blamed, impeding their ability to engage in treatment.

The Therapist's Reluctance

To a greater or lesser degree, every client will be ambivalent, defensive, or resistant. This push-pull occurs at the beginning of therapy and will continue to wax and wane throughout treatment. Resistance will often be a greater issue in managed care settings or Employee Assistance Programs (EAP) where clients are assigned therapists and cannot exercise choice and select their own therapists. Further, resistance will be enormous when clients are mandated by courts or others to attend treatment. Although some mandated clients may be helped, they evoke too much inefficacy in beginning therapists to be appropriate training cases. With every client seeking treatment, however, therapists need to address signs of potential client resistance. Let us examine three reasons why many beginning—and experienced—therapists may find this difficult.

First, many beginning therapists—especially if they have never been in therapy themselves—are unaware of the multiple meanings and conflicts associated with the decision to enter therapy. These therapists are surprised to find that, in the initial sessions, clients actually resist the help they are overtly seeking. Some therapists don't want to work with resistance because they believe it has to be addressed in a way that puts therapists in a superior position and denies the validity of the client's own experience—which they never want to do. This occurs, for example, when a therapist says to a working, single mother of three, "Hmm, I notice that you are 5 minutes late for our appointment today." Resistance does not have to be associated with this type of hierarchical therapist-client relationship. Therapists can work with resistance and still be sensitive to the social context or reality-based constraints of the client's life. For example, poor clients who use public transportation will not be able to arrive on time consistently. Clients from other cultural contexts may not think that being 5 minutes late is of any significance. For some, it may even be impolite to come on time; arriving late allows the host extra time to prepare for the visit. Therapists can educate clients about the bounds and procedures of therapy and, as we will see, respond to resistance in ways that strengthen the client's commitment to therapy, enhance the therapeutic alliance, and empower rather than invalidate the client.

A second reason why therapists are reluctant to address their clients' resistance is more personal. Most novice therapists have strong needs for their clients to like them, to find them helpful, and to keep coming to therapy. If the client does not show up or comes late, beginning therapists, in particular, may feel that they have failed. When this occurs, therapists may become increasingly concerned about their ability to help, place even more performance demands on themselves, and become overly invested in the next client's satisfaction with therapy.

Third, therapists often do not inquire about signs of potential resistance in order to ward off unwanted and/or feared criticism. Like their clients, most therapists are not eager to approach issues that arouse their anxiety. Thus, the beginning therapist may be hesitant to invite clients to express their negative reactions to the therapist's own behavior or their conflicted feelings about asking for help or needing to be in treatment. Although it is difficult to invite critical feedback and approach conflicts, it is necessary to do so. This is especially important if the client alludes to some conflict about being in treatment or suggests that there may be some dissatisfaction with the therapist. The paradox is that, if therapists allow their own anxiety to keep them from approaching clients' resistance, their clients will be far more likely to act on these unexpressed concerns and drop out of treatment prematurely.

An effective approach is to make a process comment that suggests the possibility that something about entering treatment may be difficult for the client, and proposes that the therapist and client explore this possibility together. Beginning therapists will soon find that it is easy to do this in a way that does not make the client feel accused or blamed, but actually enhances the working alliance.

Therapist:
You seem more discouraged than before, and you've had some trouble getting here. I'm wondering if something isn't going right between us. Any ideas?

Client:
What do you mean?

Therapist:
I'm wondering if something about our relationship—or being in treatment—isn't working well right now. We seem to be missing each other in some way. Any thoughts about what could be going wrong between us, or what I could do to make this work better for you?

Client:
Well, maybe you could talk a little more or give me some more feedback. You're pretty quiet. I don't really know what you're thinking most of the time.

Therapist:
I'm so glad you're telling me this. Sure, I'll be happy to share more of my thoughts with you. Let's check in on this together soon and make sure it's changed. Can we talk some more about my being quiet, too? I'm wondering, what's it like for you when you don't know what I am thinking?

The Client's Reluctance

Unfortunately, the client usually shares the therapist's reluctance to address resistance. Clients are often unaware of their resistance and externalize it to others or outside events. For example, the client may say, "Yeah, I'm late for our sessions, but the traffic is always so bad." While there may be truth to this statement, the consistency or frequency of the behavior suggests the need for further exploration. In order to acknowledge both the possible reality-based constraints and the probable psychologically based resistance, the therapist can ask if leaving earlier for the appointment is possible or if an alternative session time needs to be negotiated. The therapist also asks if they can discuss any thoughts or concerns the client might be having about treatment (for example, Therapist: "What were you feeling about our session last week as you walked out to your car?" or "Did you have any thoughts about me or our work together during the week?"). Even if the client is not ready to address whatever psychologically based resistance may be operating, the therapist has invited the client to consider the possibility of other psychological factors and laid the groundwork for further discussions. This is part of educating the client about the treatment process and making it safe to talk about the relationship (for example, Therapist: "I know other people don't usually talk this way, but one of the best ways I can help you is if you talk about the feelings you have about our relationship and any thoughts you have about treatment or the way we're working together").

In contrast, when clients see themselves resisting, they often feel confused or frustrated by their own contradictory behavior. For example, clients often exclaim with dismay:

> Why do I go to all the trouble and expense of coming here to see you, when I can't think of anything to say as soon as I walk in the door?
> Why would I forget our next appointment after we had such a great session last week? It doesn't make sense!
> Why do I keep asking you for advice, and then say, "Yes, but . . . ," to whatever you suggest? What's wrong with me?

When it is clear to clients that they are sabotaging their own efforts, they typically evaluate themselves and their own resistance harshly—as being bad in some way—and assume that the therapist shares this critical attitude. Thus, when the therapist begins to inquire about resistance, most clients want to avoid the topic because they are afraid the therapist is going to blame them— for not really trying, being unmotivated, and so on. Because it is never therapeutic for the client to feel blamed or criticized, the therapist must make this self-critical attitude overt and correct it. The best way to do this is to help clients reframe their critical attitude toward their resistance. First, both the therapist and client must honor the client's resistance, because it originally served a self-preservative and adaptive function. *The therapist can help the client appreciate the fact that this coping strategy was the best possible response to an unsolvable conflict that the client had available at an earlier time in his development.* Second, the therapist then helps the client to realize that he no longer needs this coping strategy, at least in the present relationship, but can develop more flexible response styles. This is a big concept, let's examine what it means.

Clients' resistance and even their symptoms once served as a survival mechanism that was necessary, adaptive, and often creative. For example, anxiety signals danger—perhaps not a current threat but a historical danger that really did exist. Depressive symptoms may reflect a client's attempt to cope with painfully unfulfilled wishes for protection or love from a care-giver and/or a way to cope with the exploitation or derision that she came to expect from others. In other words, past experiences with caretakers and current interactions with significant others have given clients good reasons for not wanting to ask for help, share a vulnerable feeling, or risk trusting someone. If the therapist and client explore patterns in how significant others are responding now, and have responded in the past, the client's resistance will make sense and become understandable. The feelings that underlie the client's resistance always make sense developmentally, although they may no longer be necessary or adaptive in many current relationships.

Suppose that a client is having trouble entering therapy. The therapist might ask a question such as the following:

How have others responded to you in the past when you have asked for help or needed someone?

If you and I begin to work together on your problems, what could go wrong between us?

What might I do to hurt you or make things worse if you seek help from me?

How could our ethnic differences be a problem for us in therapy?

Among other things, such questions will help the therapist and the client identify the aversive consequences that followed when the client asked for help in the past. Routinely, clients will have very specific answers to these questions and describe in detail the interpersonal scenarios that have typically transpired in the past. When they asked for help in their family, for example, they were ignored and felt powerless, were made fun of and felt ashamed, or were told that they were selfish and too demanding. They may have been told that they didn't really need or want what they asked for. They may have felt that anything they received brought unwanted obligations or seemed to burden their caregiver. In this way, therapists and clients can clarify together the very good reasons why clients originally learned to respond in the ways they do.

The first thing the therapist can do is help clients recognize their problematic relational patterns with others, understand why they originally needed to defend themselves in these ways, and clarify how they may be continuing to do so now with the therapist. Second, the therapist validates the protection this resistance once provided. The maladaptive resistance in the therapeutic relationship was once an adaptive response. The original defensive behavior (withdrawing, provoking, avoiding, pleasing, intellectualizing, and so on) minimized the hurt and allowed the client to cope with the unwanted responses he received. Third, therapists must differentiate their own responses in the current relationship from the aversive ways that significant others have responded in the past.

Therapist:
In the past, you learned that if you allowed your parents to help or take care of you, they wanted you to remain a child and continue to need them. I would like to respond to your need, but I'm different from them. I don't need you to remain dependent on me. I would like to help you achieve your own independence, so that you will not need me any longer, and will be able to go on with your own adult life.

By following this three-step sequence, therapists will provide the client with the experience of no longer needing to use the old, relational pattern:

1. Identify the original transactional pattern.
2. Validate the client's defensive adaptation to it.
3. Differentiate the therapist's current response from the old pattern.

This sequence will need to be repeated with each manifestation of resistance in the course of treatment.

We have seen, then, that clients are far more likely to become stalled in therapy, or to drop out prematurely, if their resistance is not addressed. One of the most important ways to keep clients in treatment, and to identify conflict areas in clients' lives, is to invite clients to discuss any problematic reactions they may have toward being in therapy or anything the therapist does. *The guiding principle is that if clients talk about their ambivalence or discontent, they will be less likely to act on it and drop out.* Before going on to illustrate specific ways of implementing this principle, we must learn how to identify clients' resistance and differentiate it from reality-based constraints that must be affirmed.

Conceptualizing Resistance

Early in their training, most therapists will have clients that drop out of therapy within the first few sessions. The therapist might conclude that the client wasn't really motivated or wasn't ready to look at her problems, or that the therapist unwittingly made some irrevocable mistake in the previous session. More likely, clients prematurely terminate because they are acting on, rather than talking about, their conflicted feelings about entering therapy. In order to help clients remain in treatment, the therapist must be prepared to identify when resistance is occurring; to approach this issue in a noncritical manner that allows clients to express their concerns more fully; and to help them resolve these concerns by responding flexibly whenever possible.

Identifying Resistance

How does the therapist know when a client is resisting treatment? Therapists may never know what any behavior means for a particular client, but resistance may be operating when clients consistently have difficulty participating in treatment. For example, after reality-based constraints have been considered, psychologically based resistance is probably occurring when any of the following occur repeatedly or in combination:

> The client misses the appointment or comes late.
> The client needs to reschedule appointments frequently.
> The client has very limited hours available for therapy.
> The client cannot make a firm commitment to attend the next session.

It is important to stress that the same behavior often means very different things to different clients. However, when clients have trouble attending sessions, some form of resistance is often at work. In such cases, the therapist should generate working hypotheses about the possible meaning this resistance holds for the client.

Formulating Working Hypotheses

Therapists must formulate case conceptualization to guide intervention strategies, and write treatment plans for third-party or HMO insurance payments. The first step in doing this is to formulate tentative working hypotheses about the client's maladaptive relational patterns, faulty expectations and pathogenic beliefs, and core conflicted feelings. One systematic way to do this is to begin formulating answers to three questions during the initial contact. (Further guidelines are provided in Appendixes A and B.)

1. What does the client elicit from others?   Even in the initial telephone contact, the client's interpersonal style—and what it tends to *elicit* from others—is an important source of information. For example, if the client sounds helpless and confused on the telephone, might he adopt a victim stance and invite others' rescuing behavior? If the client presents in an angry and demanding way, might others often withdraw or tend to become embroiled in power struggles? If so, perhaps this client is testing whether the therapist can tolerate the angry challenges and still see the underlying need or pain. As therapy progresses and the therapist learns more about each client, many of these initial hypotheses will prove inaccurate and will need to be discarded. As therapists learn to trust themselves and attend to the feelings and reactions evoked in them by the client, however, they will find that many hypotheses do indeed fit the client and can be further developed and refined. Thus, in order to begin conceptualizing the client's dynamics and developing initial treatment plans, therapists can follow these three steps:

a. Formulate working hypotheses about the potential meaning of the client's behavior.
b. Evaluate these tentative initial hypotheses as the therapist learns more about the client and they interact further.
c. Revise these hypotheses over time to more accurately reflect each client's personality and problems.

Although these steps may feel challenging right now, each of these important concepts will be discussed further in future chapters.

2. What is the threat?   On the basis of information the therapist has started to gather about the client's current and past relationships, the therapist can begin speculating about the different ways that entering treatment could be aversive for the client. Therapists can formulate working hypotheses such as the following about the issues or concerns that are likely to be difficult for each client.

Is it guilt-inducing for this highly responsible parent, who grew up taking care of her own depressed mother, to ask for something for herself?
Does this Christian client believe that her inability to resolve her problems through prayer means that she has failed in her faith?
Is it incongruent for this older Latino male to seek help from a younger female therapist?

Is it shameful evidence of his own inadequacy if this blue-collar worker has an emotional problem that he cannot solve by himself?

Is it disloyal for this Asian client to discuss personal problems with someone outside of the family?

Therapists may also keep in mind that resistance and defense are attempts to manage unwanted feelings of shame, guilt, and anxiety. Although clients may be unaware of it, they often struggle with the worry that, if they continue in treatment, the therapist will hurt them in a way similar to how others have in the past. (In contrast, some clients who were more traumatized may worry that they will hurt the therapist—and others—as they were once hurt.) Over time, the therapist can help clients clarify such concerns and encourage them to assess whether the therapist is responding in these familiar but unwanted ways, or is offering a new and safer relationship than they have come to expect.

3. How will the client express resistance?   Finally, therapists should formulate a third set of hypotheses that anticipate how clients will enact these concerns with the therapist. For example, suppose that a therapist has observed that her depressed client feels guilty and believes he is selfish whenever he does something for himself or someone does something for him. The therapist then hypothesizes that this client may withdraw from treatment as soon as he starts to feel better or emotionally disengage from the therapist when he realizes that she genuinely cares about their relationship.

To illustrate further how the therapist generates working hypotheses on the basis of these three questions, let's consider another example. Suppose that a young male client whose problem is substance abuse telephones the therapist to schedule an initial appointment. During this telephone call, the client tells the therapist that his alcohol and drug usage is more problematic than it has ever been and that his mother believes that he needs to be in therapy. On the basis of the helplessness and confusion expressed in his telephone call, and his disclosures that his mother wants him in therapy and how problematic his drug usage now is, the therapist begins formulating several tentative hypotheses. First, the therapist hypothesizes that an important part of the client's interpersonal style is to let others know that he is hurting, to elicit help from them, but then to avoid taking responsibility for his own behavior.

Second, the therapist hypothesizes that this client may become resistant to treatment when the therapist addresses the issue of how he meets his emotional needs, or when the client is given reality-based feedback about how he eschews responsibility for his own decisions (for example, to enter therapy or to continue drinking). Third, the therapist hypothesizes that this client is impulsive and may act out his resistance by bolting out of therapy as soon as these anxiety-arousing issues are addressed (just as, perhaps, he has used alcohol and drugs to avoid dealing with certain feelings, interpersonal problems, or responsibilities he feels inadequate to meet).

The therapist knows that these initial hypotheses may not be accurate and is ready to revise or discard them as more is learned about the client. *However, the therapist must try to anticipate the concerns that might cause each client to drop out of treatment and the way each client is likely to express these concerns.* Prepared with working hypotheses such as these, therapists will be able to respond more effectively, and they will diminish acting out by being able to help some clients anticipate their own resistance before they act on it.

As therapy proceeds, the therapist will learn much more about the client and many of these initial hypotheses will need to be discarded. Some will be accurate, however, and can be elaborated to better understand this particular client. In this way, the therapist identifies the enduring issues that arise for the client. The therapist will then be better prepared to center treatment on these repeated themes, which provide a focus for the ongoing course of treatment. As we will see in Chapter 7, this is part of the continuing process of formulating and refining case conceptualizations and treatment plans.

Responding to Resistance

In this section, we see how therapists can respond to the common types of client resistance that occur at the beginning of therapy. Sample therapist-client dialogues illustrate effective and ineffective responses at three critical points: during a telephone conversation in which the client has difficulty scheduling the initial appointment; at the end of the initial session; and during later sessions with the client who has difficulty keeping appointments. As before, student therapists should find their own words to express the principles embodied in these dialogues.

Resistance During the Initial Telephone Contact

In trying to schedule the initial appointment, the therapist wishes to obtain a firm commitment from the client. The client, however, may be ambivalent about entering therapy and express this during the initial telephone contact. The therapist should be prepared to respond directly to this issue.

An uncertain commitment   In the following example, the client is only somewhat uncertain about attending the first session.

Therapist:
It's been good talking with you, and I'll see you on Tuesday at 4 P.M.
Client:
OK, I guess I'll see you then.
Therapist:
You sound a bit uncertain; maybe we should talk a little more about how it feels to come in?
Client:
Well, maybe a bit uncertain, but I do want to come.

Therapist:

> Good, I'm looking forward to meeting with you. If you would like, we can talk together then about therapy and any questions you may have about me or treatment. See you next Tuesday at 4 P.M.

An effective response to the client's ambiguity ("I guess . . .") is modeled in this dialogue. The therapist hears the indecision in the client's commitment and addresses it directly. An ineffective response would be for the therapist to let it pass by on the assumption that it probably doesn't mean much. Perhaps it doesn't, and chances are that the client will still arrive at the appointed hour. However, the chances of the client not arriving are far greater if the therapist does not directly address the signs of uncertain commitment.

Further, the therapist should mentally note this indecision as a possible sign of resistance to treatment. This is an instance in which the therapist can begin to generate working hypotheses—derived from the first of our three orienting questions—about the possible meaning of relevant client behavior.

A more ambivalent client   Now imagine the same telephone conversation, but with a more difficult client.

Client:

> OK, I guess I'll see you then.

Therapist:

> You sound a bit uncertain; can we talk a little more about coming in?

Client:

> Well, in my family, we don't really talk to others about family problems.

Therapist:

> I respect your concern about being loyal to your family. Maybe one of the things we can talk about is how you can ask for help from others and get your own needs met without feeling that you are being disloyal.

Client:

> I'd like that. My family is really important to me and some people don't understand that.

Therapist:

> Yes, we can work together to try and find ways to sort out your own needs along with respecting your feelings about your family. Also, we can agree to meet only one time on Tuesday at 4 P.M. After talking together a little more, you can better decide for yourself whether you'd like to continue.

An ineffective response would be to accept the client's uncertain commitment ("OK, I guess I'll see you then") and avoid the possibility that an important issue is being played out. In this dialogue, instead, the therapist has tried to understand the client's concern first, and only then helped the client to make a firmer commitment by limiting the commitment only to attend the first session. By responding to the client's ambivalent feelings, the therapist has done much to help this client take the first step and enter treatment.

Why would the therapist take such a stance? Wouldn't it be more supportive to let clients leave the appointment a little bit tentative, if that is what they need to do? No. Little therapeutic change will occur until clients take responsibility for the decision to enter therapy and work on their problems. Therapists must be flexible, and clients can make a very limited commitment—for one session or just a few sessions—to get a better sense of whether this is for them. Without this commitment, however, the therapist has little to work with. In fact, it is better for the client to remain out of therapy than to enter without commitment and have an unsuccessful therapeutic experience. As Yalom (1981) emphasizes, the therapist must be concerned about preventing failed hope in certain clients. If clients have one or more unsuccessful therapy experiences, it may discourage them from trying to seek help with their problems again.

Clients who try to make the therapist responsible   Let's return to the telephone conversation between the therapist and a prospective client. Some clients will try to make the therapist take responsibility for their decisions.

Client:
I'm not sure if I should start therapy or not. What do you think I should do?

Although it is important to reach out to clients and welcome them into treatment, the therapist does not want to resolve clients' ambivalence by assuming responsibility for their decision or by trying to provide the necessary motivation for therapy. With certain clients, a helpful response might be as follows:

Therapist:
You're really hurting right now, and I would like to try and help. Would you like to come in for one appointment and see how it feels to talk with me? As we discuss your concerns, I think you'll get a better sense of whether you'd like to continue.

OR

Therapist:
I can't offer you any guarantees, of course, but yes, I do believe that therapy may be of help. I'd like to try and work with you. Let's agree to just one session. After we've talked together a little more, I think you'll be able to see for yourself if I can be helpful.

As we would expect on the basis of client response specificity, such direct invitations may be useful at times. When clients have felt unwanted while growing up, or believe that others are uninterested in them, it may be more important for therapists to express genuine interest in working with them. Even with clients such as these, however, therapists must still respond in a way that ensures clients take responsibility for their own decision to enter treatment; therapists do not want to cajole, coerce, or win clients into therapy. At the same time, therapists do not want to be aloof or indifferent. The therapist needs to meet clients where they are—working supportively with their ambivalence—but without taking responsibility for the clients. In many cases, an effective response to the client's question may be as follows:

Therapist:

It seems as if one part of you wants to be in therapy, but another part of you doesn't. Tell me about both sides of your feelings.

OR

Therapist:

It sounds like a part of you wants to try this and a part of you doesn't. Why don't we agree to meet for just one or two sessions. After we've worked together for a little while, you will be able to decide for yourself whether I can be of help.

Especially in the first response, the therapist is inviting the client to express all concerns about entering therapy. In both responses, the therapist is working with the client to reach a decision, but leaves the client responsible for her own decision. Here again, we are working with the process dimension. Although this process distinction is subtle, it will make all the difference in the eventual outcome of therapy, regardless of the theoretical orientation of the therapist.

The client in this dialogue is providing the therapist with potentially important information. As before, the therapist should begin generating working hypotheses about what this behavior may mean. For example, uncertain clients may not have been supported when they have tried to take responsibility for their own actions. These clients may now try to avoid the aversive consequences they have experienced in the past by getting others to take responsibility for them. If so, this could make it difficult for clients to initiate new activities (such as applying for college) or to follow through on their own interests (such as selecting the major or career they are most interested in). Being held back in this way may leave clients feeling resentful and frustrated.

Having generated such hypotheses, therapists can be alert for evidence of these themes in other areas of the client's life. If these prove to be enduring and pervasive issues, the therapist can use them as a focus to provide direction for treatment. The therapist must not become wed to these early hypotheses, however, and must be ready to discard them if further supporting data are not forthcoming. As therapists become more experienced, their initial hypotheses will become more fruitful. This developing skill will become an especially important aid to counselors as they begin working in short-term and crisis-intervention modalities that require quicker, more accurate assessment during the initial phase of treatment.

As emphasized in Chapter 1, the therapist is trying not to reenact the client's maladaptive relational patterns in the therapeutic relationship. However, this can easily occur in a metaphorical or encapsulated way in the initial negotiations between therapist and client. This is another reason why the therapist should not take over and tell the client seeking advice what to do. For example, the following response is not recommended:

Therapist:

On the basis of what you've told me, I think that therapy can help you, and I believe that you should give it a try.

Assuming such a directive stance often reenacts a problematic relational scenario for clients and confirms a pathogenic belief. Many clients feel that they cannot act on what they want, or that they have to act dependently and let others assume responsibility for their decisions. Although such clients feel they need to ask the therapist to tell them what to do, paradoxically, they resist the control they have just elicited by resisting the therapist because they feel their autonomy is being suffocated. As this scenario continues, clients come to feel badly about themselves—confused as to why they are rejecting the help they have just asked for; and guilty for rejecting the therapist who was trying to help. Therapeutic responses that give the client the responsibility of deciding whether to follow through prevent this reenactment from occurring and may provide clients with a new and healing response that begins to resolve an old conflict.

Resistance at the End of the First Session

At the end of the first session, no matter how well it seems to have gone, the therapist should ask clients how the session felt to them and whether they have any concerns about the treatment process or the therapist. Unless the therapist is willing to inquire, any potential ambivalence or concern is likely to remain unspoken. A client who is feeling some reticence will be far less likely to drop out prematurely if the therapist has given the client permission to express his concerns, has taken them seriously, and has tried to resolve them to the extent possible.

Addressing interpersonal conflict   Five or ten minutes before the end of the first session (to allow clients enough time to talk about their reactions to the therapist and the session), the therapist can check in with the client:

Therapist:
How was it to be here today?

Most clients will answer, "Fine." Thinking that the therapist may need some reassurance, clients may go on to say how helpful the session has been. If so, the therapist might respond as follows:

Therapist:
Good, I'm glad you've found our first session helpful. You've told me a lot about yourself today, and I feel we have gotten off to a good start, too. I would like for us to be able to talk about our relationship more directly than others usually do. Was there anything about coming today, or anything about our work together, that did not feel good to you? If so, it would help if we could talk about that.

Client:
Oh no. I was eager to begin, and you have been very nice and understanding.

Therapist:

Good. In the future, though, if you ever have any uncomfortable feelings about our relationship or being in therapy, I would appreciate it if you could talk with me about them. Would that be comfortable for you?

In this vignette, the therapist is trying to establish an important set of expectations for what is going to occur in therapy. The therapist is telling the client that, in contrast to many others, the therapist wants to hear about the client's dissatisfaction and address problems between them straightforwardly. Some clients may not be able to accept this offer initially because forthright communication was punished in their families of origin. With these invitations, however, the therapist is laying the groundwork for a different kind of relationship that is more direct and authentic. Therapists want to establish these important expectations with their actions, and not just their words, early in treatment.

A more assertive client   Now imagine that the therapist raises the same type of query, but this time the client is critical of the therapist.

Therapist:

How has it been for you to talk with me today?

Client:

Well, just fine for the most part, but I did feel you were trying to hurry me up a few times.

The therapist must learn to approach the conflict, rather than avoid it by moving on to another topic, truncating it with a punitive response, or expressing hurt and communicating that the client's criticism is unwanted. The best way to approach the conflict is to remain *nondefensive* and encourage the client to express her concern more fully.

Therapist:

I'm sorry that happened. Let's look at that together and try to understand what happened for each of us. Tell me more about your feeling of being pushed or hurried up.

Of course, it is often difficult for beginning therapists to do this. Such confrontations may arouse the therapist's anxiety, especially when the therapist has strong needs for the client's approval. If they are to approach such anxiety-arousing confrontations, therapists must first become aware of their own characteristic responses to interpersonal conflict. For example, some therapists will want to avoid the client's complaint and move on as if it wasn't said. Others will readily agree with the client and quickly apologize in order to abate the criticism. Still other therapists will automatically begin to justify themselves and offer lengthy explanations. Some therapists will counter with their own hostility. Therapists need to become aware of how they typically react to criticism or unwanted confrontations.

One of the best ways to do this is to recall how your own family of origin dealt with interpersonal conflict. Regardless of age, your initial reaction to conflict usually follows closely how your parents dealt with conflict in their marriage and how each parent dealt with conflict with you as a child. Rather than automatically following their initial response propensity, therapists can learn to approach conflicts with more openness.

After inviting clients to express fully their concerns, the therapist does not wish to rush to judgment and disagree with the client's perception. Perhaps the client is right and the therapist was rushing the client. The therapist must honestly examine her own behavior and see if the client's perception is accurate. If so, the therapist does not need to be afraid of acknowledging this to the client.

Therapist:
You know, I think you might be right. I was aware of the time going by, and I wanted to touch on a few more issues before we had to stop. I probably was hurrying you there, and I can see why that bothered you. I'm glad you made me aware of that; I'll try not to do it again.

Throughout their first years of training, many beginning therapists suffer greatly over their fears of making mistakes and failing. Sadly, many caring and responsible students do not enjoy their clinical training because of these worries. It is liberating for beginning therapists to learn that they do not need to be afraid of making mistakes; they need only to be able to acknowledge them. In this example, the therapist is responding to the client in a reality-based and egalitarian way. The therapist validates the client's perception, which tells the client that the therapist is willing to have a genuine dialogue. It also tells the client that the therapist will respond to the client's concerns with respect. Too often, therapists and clients enact a relationship in which the client is the sick and needy one and the therapist is the healthy one who does not make mistakes or have problems. The response suggested here discourages that illusion. The therapist's willingness to risk having an authentic relationship often has a powerful effect on clients and accelerates the exploration they can undertake.

In contrast, therapists can handle the confrontation in a way that keeps the relationship superficial or puts them in a superior position.

Therapist:
Oh, you thought I was hurrying you up, how was that for you?

OR

Therapist:
That's interesting. Do you ever feel this way in other relationships, too?

These responses are a profound misuse of the client-centered reflection or psychoanalytic neutrality. With this type of *deflection,* the message to the client is, "It's always your problem. I will not look at my own contribution here. This will not be an honest relationship." Such a response limits the relationship to a superficial encounter and sets up a power battle between client and therapist. Little therapeutic gain will be realized as long as the interpersonal process continues in this problematic mode.

But what if the therapist does not agree with the client's complaint? The therapist can still accept the validity of the client's perception, but without agreeing to the comment.

Therapist:

I'm sorry you saw me as being impatient with you; that wouldn't feel very good to me either. I wasn't aware of being in a hurry or trying to push you, but I'll be alerted to that in the future. If you ever feel that happens again, stop me right then and we'll look at it together.

Therapists do not want to disavow the client's perceptions, and it's not helpful for therapists to agree to something they don't believe is true. In this example, the therapist is telling the client:

I will take your concerns seriously even if I do not see it the same way.
We can have differences between us and still work together.
Your feelings about our relationship are important to me.

This type of response also tells the client that problems between them can be resolved, which lends hope that the therapist and client may be able to resolve the client's problems with others as well. By addressing conflicts between the therapist and the client directly, the therapist provides the client with the opportunity to have a new and more adaptive means of conflict resolution modeled in vivo. Especially for clients who have learned to avoid interpersonal conflict, such experiential relearning is empowering indeed.

Finally, it is also possible that the client is systematically misperceiving the therapist's behavior. For example, this client may readily experience the therapist and most other authority figures as demanding more than the client would like to produce and then being dissatisfied with whatever the client does produce. If so, this client may have grown up with a parent who repeatedly demanded that the client do everything on the parent's timetable and/or ignored the child's own wishes and accomplishments. Even if the therapist has gained further evidence to support such a hypothesis, this type of historical or transference interpretation should not be presented before the therapist and client have resolved the dispute in their current relationship. Therapists stand to lose a lot if clients see them as sidestepping a reality-based confrontation. However, the therapist can use this information to generate working hypotheses about their relational templates that can be utilized later.

Clients who test the therapist's adequacy   Because it is usually difficult for beginning therapists to respond to client confrontations, we will examine another example. In the following situation, the client challenges the therapist's competence. Although this taps into an anxiety-arousing area for the novice therapist, the therapist is able to approach the issue directly and invite the client to express the concern more fully.

Therapist:

I'm wondering if there is something about our relationship, or being in therapy, that just isn't feeling quite right to you.

Client:

> Well, you did say you were a student here. Isn't that what you said when we first spoke on the phone?

Therapist:

> Yes, I'm a second-year graduate student, working on my master's degree.

Client:

> Well, I don't really feel like a guinea pig or anything, but you really are just practicing on me, aren't you? I don't want to sound unkind or anything, but do you think that you've had enough experience to help me?

Therapist:

> Are you worried that if you go to all of the trouble of coming here and talking about difficult problems with me that I just won't know enough to be able to help you?

Client:

> Well, sort of. After all, you really are just a beginner, and I must be 15 years older than you. What do you think? Can you help me?

Therapist:

> I certainly cannot offer you any guarantees, but I will do my best. And as we continue to meet together a few times, I think you will be able to decide for yourself if I can help you or not. But for now, let's see if there are any questions I can answer for you about my training status, and how our age difference might get in the way.

In this vignette, the therapist was able to approach the client's concern directly. She responded effectively because she remained nondefensive, even though this was anxiety-arousing and difficult for her to do. The therapist did not act on her initial impulse, which was to try to reassure the client of her ability to help. Instead of responding to her own personal need to seem competent to the client, the therapist was able to respond to the client's need—by inviting the client to express reservations directly and fully. The therapist tolerated her own personal discomfort well enough to be able to discuss the client's concerns and, in so doing, behaviorally demonstrated her competence. This is always more effective than offering verbal reassurances, which would only sound hollow to the client and still leave the burden of proof on the therapist.

Clients' concerns about entering therapy   In the previous two examples, the therapist inquired about resistance and the client expressed concerns that personally challenged the therapist. In the following example, the therapist asks about the client's potential resistance, and the client expresses concerns about entering therapy. Such concerns are far more common than challenges to the therapist and easier for the therapist to address.

Client:

> It's hard for me to ask for help. I guess I'm not used to talking about myself and my problems.

OR

Client:
> I feel a little awkward talking to a stranger. In our family, we always kept problems to ourselves.

It is crucial to let clients know that you have heard their concerns, take them seriously, and are willing to be flexible and do something about their concern if possible. Again, the best way to begin demonstrating this responsiveness is to invite clients to discuss their concerns more fully.

Therapist:
> That really does sound like an important concern for us. Let's talk some more about it, so that we can both understand it better. What does it mean for you to talk to others about your problems or to ask for help?

OR

Therapist:
> I'm so glad you can tell me about this awkward feeling. I think you're right: It is hard to talk to someone you don't really know about personal things. Let's be sure to go at your pace in here. You decide when and how much you wish to share with me. I'd be very comfortable with that way of working. In fact, I'd even prefer it.

Listening and responding to the client's concerns is even more important when ethnic or cultural differences exist between the therapist and client (Pinderhughes, 1989; McClure & Teyber, 1996). Cultural differences may be a sensitive issue; neither client nor therapist may feel comfortable addressing them. It sometimes seems as if there is an unwritten societal rule against recognizing differences. As before, however, the best way to work with these important differences is to acknowledge them in an open-ended way that gives clients permission to express any concerns they may have. For example, suppose that the client is a person of color, and the therapist is not a person of color.

Therapist:
> Was there anything about coming today or about our work together that concerns you?

Client:
> Well, this is a little awkward to talk about, but I don't know if you can help me because you're white.

Therapist:
> I'm glad we're talking about this, because our cultural backgrounds really are very different. I think there probably will be important things about you that I won't understand as well as I want to without your help. Tell me some of the things you are thinking about now that could keep us from working together well.

The therapist responds by accepting the client's concern, validating his point of view, and being willing to explore it further. *Different* is an important word as therapists work with people of color, gay men and lesbians, and other clients who feel on the outside. For example, many people of color will feel different in a predominantly white school or workplace—especially if few attempts are made to reach out and include them. In some settings, people who are different may be regarded pejoratively as deficient. The therapist's willingness to acknowledge differences and invite clients to express any concerns or misgivings they may have will go a long way toward establishing a collaborative relationship and keeping clients who may feel different from dropping out prematurely.

In all three of the examples here, the therapist is able to approach the client's conflict. As the therapist and client discuss the conflict, some of the client's concerns will abate with a more complete understanding of the process of therapy. Other concerns will be resolved as the client's inaccurate assumptions about the therapist are corrected. However, some of the concerns that clients express cannot be assuaged so easily. The therapist can offer to be sensitive to these concerns and express a willingness to work with the client on them over the course of therapy. One of the best ways to do this is by enlisting clients' help in better understanding their background and experience. In other words, the therapist can invite clients to tell the therapist when they are not feeling understood. The therapist can then ask clients to share their experience more fully and begin a mutual dialogue that will bring the therapist closer to clients' actual experience.

As we have seen, when obvious ethnic, religious, or other cultural differences exist, this approach of enlisting the client's help is even more apt. Ethnic, class, gender, and religious differences are complex, and it is counterproductive for therapists to labor under the misconception that it is their responsibility to understand everything about a client from a different culture. However, therapists do have a responsibility to educate themselves about clients from different social contexts, for example, by reading and consulting with informed colleagues. The therapist can also invite the client to educate the therapist whenever the client feels that the therapist is misinformed or does not understand. By welcoming this clarification instead of feeling threatened by it, the therapist signals acceptance of their very real differences, openness to a true dialogue, and offers an invitation for a real relationship. In so doing, the therapist will gain the client's appreciation.

The therapist also needs to explore more fully with clients how they believe their cultural or social context influences their current problems and shapes their subjective experience.

Therapist:
Would you help me understand this problem in terms of the expectations from your culture?

Even when obvious cultural differences (of race or religion, for example) are not noted, therapists will be more effective and understand more fully the clients' worldview if they recognize that "all counseling is cross-cultural because all humans differ in terms of cultural background, values, or lifestyle" (Speight, Myers, Cox, & Highlen, 1991, p. 29).

To sum up, the therapist must encourage clients to express any concerns about treatment or the therapist that arise, and then sincerely try to accommodate those concerns. This will go a long way toward diminishing them and allow many more clients to enter treatment and receive help.

Resistance During Subsequent Sessions

Addressing resistance directly   Later in treatment, if clients begin to cancel frequently, do not show up, or come late to appointments, the therapist needs to consider the possible meanings that this behavior might hold for the client. As different treatment issues are discussed, clients will not be fully aware of all of the varying and often conflicted feelings that have been activated by a certain topic. When their feelings have been consistently invalidated and their perceptions disconfirmed, people eventually lose awareness of many aspects of their experience. Resistant clients are not lying or deceiving the therapist; they are simply unaware of the multiple and often contradictory feelings they have about seeking help and exploring difficult topics in treatment. For example:

> Getting help is relieving, but may also arouse detested feelings of humiliation.
> Feeling cared about is comforting, but can also evoke sadness about the many times this need went unfulfilled.
> Being listened to and understood is reassuring, but may trigger guilt over not having a right to one's own feelings.

Unless addressed and clarified, these conflicted emotional reactions will cause some clients to leave therapy prematurely. More commonly, when these concerns are activated later in treatment but go unaddressed, therapy will become stalled—repetitious or intellectualized—and clients will be unable to move forward and make progress with their problems. At the beginning of their clinical training, however, many therapists do not appreciate the ambivalent, push-pull nature of emotional conflicts. Let's examine this further by recalling Marsha from Chapter 2.

It was profoundly reassuring for Marsha when her second therapist approached her feelings directly and validated her experience. At the same time, however, this very positive experience also aroused other contradictory feelings that she "hated." Marsha now felt heard and understood, but she also felt:

- Sad, as years of unacknowledged loneliness welled up
- Angry at not being heard so many times
- Guilty and disloyal to her parents for feeling angry at them
- Anxious for breaking familial rules and talking to someone about what was wrong in her family

Marsha did not want to experience any of these feelings. She did not understand and felt threatened by them, but she could not make them go away. Fortunately, her therapist was comfortable with the push-pull nature of conflict and consistently helped her understand the different, contradictory aspects of her experience that emerged. His accepting response helped her to integrate these conflicted feelings and gradually resolve them. In these ways, the second therapist was providing Marsha with *containment* or a holding environment that provided her with the psychological safety she needed to address—explore, experience, share—these sensitive issues for the first time.

At different points in treatment, clients will often be unaware of, and threatened by, the conflicted feelings evoked in the change process. As a result, they may feel misunderstood or even blamed when the therapist inquires about potential signs of resistance. Clients may feel that the therapist's questions imply that they are doing something wrong or are not trying hard enough. Concerned that the therapist is frustrated or disappointed with them, clients may try to justify their good intentions (for example, Client: "Oh no, you don't understand. I really do want to see you and get here on time. It's just that . . . "). As we have already noted, therapists never want clients to feel blamed. Rather, the therapist is inviting the client to join in a collaborative exploration to understand the threat or danger that some aspect of treatment has aroused.

Thus, therapists can best approach potential resistance in a spirit of mutual exploration. Although beginning therapists are often concerned about being too blunt or being misperceived as critical, the problem actually goes the other way. *Because so many beginning therapists are too worried about hurting their clients or evoking their disapproval, they do not respond as clearly and forthrightly as they could and consequently mute their impact.* If therapists are worried that their explorations may be seem intrusive or unwanted to the client, they can shift gears and find another way to respond. Better yet, they can check out their concern with the client by posing the question directly.

Therapist:

Is it okay for us to be talking about this? I like talking about these issues that come up between us, I think it helps our work together. But I want to make sure that you feel comfortable with this. How is it for you to talk with me about these things so directly?

The therapeutic goal is to inquire about potential signs of client resistance in a collaborative manner that is nonthreatening and respectful. Therapists can just wonder aloud with clients about the possible meanings certain behavior may hold, there is no insistence on exploring a client's resistance or any other topic. Therapists can approach resistance without arousing defensiveness through a series of gradual steps that are progressively more direct: a three-step sequence is often effective.

In the first step, the therapist offers a *permission-giving and educative response* to encourage clients to talk about the positive and negative reactions that come up for them about therapy or toward the therapist. In particular, the therapist explains that problems or misunderstandings with the therapist are inevitable, and the client can help by bringing them up. When this invitation is insufficient, the therapist can take a second step and encourage the client to explore the potential threat these feelings hold; this focuses on the *defense.* (For example, Therapist: "What could the threat or danger be for you if . . . ?") If the client continues to show signs of resistance but cannot talk about it, the therapist can draw on previous working hypotheses and try to interpret the *content.* (For example, Therapist: "I wonder if it has been harder for you to get here since we began talking about . . . ?") Let's consider examples of these three steps.

Step One: Permission-Giving and Educative Response

Therapist:
> Can we talk about how you feel about coming to see me? I noticed that you were 20 minutes late today and had to reschedule your appointment last week. Maybe it doesn't mean anything, but I was wondering if something doesn't feel right for you. Any thoughts come to mind?

Client:
> Oh no, I really do want to be here. And you've been very helpful already. It's just that my boss called.

Therapist:
> OK, but if in the future you ever find yourself having any concerns about the way we are working together, I would like you to tell me about them— especially any problems or misunderstandings that might come up between us. How would it be for you to tell me about something I did that you didn't like, or something about being in counseling that was uncomfortable for you?

Client:
> Well, that might be kind of hard for me to do.

Therapist:
> What would be the hardest thing about that for you?

At this point the therapist and client are off and running together. The therapist can help clients work through the issues that would prevent them from expressing dissatisfaction or disagreement (for example, clients' concern that the therapist will not like them or others will leave them unless they are always nice). Not only will this help keep clients engaged in treatment; it opens up important new conflict areas that need to be addressed.

In this example, the therapist offered a permission-giving and educative response in the hope that the client would then feel free to talk about any ambivalent or conflicted feelings that emerge. If the client continues to show signs of resistance without being able to talk about it (missing, canceling, rescheduling, coming late, discussing only superficial issues), the therapist moves to the second step by addressing the defense.

Step Two: Explore the Danger/Identify the Threat

Therapist:
> It's important that you can talk to me about any difficulties you have about our work together because they're bound to occur. Would you be able to tell me if something about our relationship or work together was troubling you?

Client:
> I think so.

Therapist:
> Well, maybe so, but I guess I have to say that I'm not convinced of that yet. I'm wondering if it might be pretty difficult for you to tell me if I did something that you didn't like, or if there was something about being in

therapy that was uncomfortable for you. Sometimes people believe good clients or nice people never have frustrated or disappointed feelings, or at least they don't ever express them.

Client:

Well, yes, that's probably true.

Therapist:

Let's try to figure this out together. It seems that you are having some trouble being in therapy, but don't seem to feel comfortable talking with me about that yet. I'm wondering what the threat or danger might be for you if you were less worried about being nice and took the risk of telling me what doesn't feel right? What's gone wrong with others in the past or what might happen between us?

In this example the therapist is trying to identify the threat that keeps the client from expressing conflicted feelings. *The therapist is not trying to find out why the client is not showing up for therapy, but why the client is having trouble talking about it.* In most cases, this approach will prove effective. If focusing on the defense does not work, however, it's time to move to the third step. As a last resort, the therapist can try to interpret the content of the client's resistance. For example, if the therapist thinks the client is about to pull out of treatment to avoid a difficult topic and the client cannot talk about the difficulty, the therapist can draw on working hypotheses and interpret the resistance.

Step Three: Tentatively Interpret the Reason for the Defense

Therapist:

It's been hard for you to get here; you've missed the last two sessions. Maybe you just forgot, but I'm wondering if there is something about being in therapy, or something that we have been talking about, that just doesn't feel safe to you. On the basis of some other things we have talked about, I'm wondering if . . . ? What do you think is going on here?

Why does the therapist take this more direct interpretive stance only as a last resort? The interpretation, accurate or not, is the therapist's issue and puts the ball in the therapist's court. Whenever possible, it is better to try to follow the client's lead or focus on the therapeutic process rather than to pull the client along in the therapist's direction. In other words, it is usually more effective to explore why the client is resisting, or make a process comment and address the current interaction, than to make interpretations or tell the client what to do. The goal is to maintain a working partnership and keep the relationship as collaborative as possible.

Understanding clients' inability to change   Resistance, defense, and ambivalence will occur throughout treatment. As we would expect from client response specificity, the reasons for clients' resistance to entering treatment, and then to change later in treatment, are complex and varied. Later in treatment, when therapy has stalled or clients have stopped making progress on a problem, therapists should consider three possible sources of resistance: the therapist and client are reenacting maladaptive relational patterns, aversive feelings of shame have been evoked, or guilt has been activated. Although there are many exceptions, we will see that these three factors sometimes emerge for clients in a temporal sequence.

First, therapists should consider the possibility that in some actual or metaphorical way, the same problem that the client is having with others has been activated or is being played out between the therapist and the client. That is, the client's resistance is occurring because the interpersonal process between the therapist and client is reenacting rather than resolving some aspect of the client's problem (for example, the client thinks the therapist has been critical of her or become disappointed in her—just as her spouse has been). Using process comments or finding other ways to change the maladaptive relational pattern in the therapeutic interaction will allow the therapist and client to restore their working alliance and resume productive work. Early in their training, most beginning therapists will not possess the objectivity necessary to identify these predictable but subtle reenactments and will need assistance from their supervisor. All therapists can also benefit from personal exploration and their own therapy, so that they can become more aware of their own response styles and can better anticipate how these styles may interact with their clients' interpersonal patterns.

When such relational reenactments are occurring between therapist and client, clients do not have the *interpersonal safety* they need to explore their conflicts and treatment becomes stalled on a more superficial or intellectualized level. Once the ship has been righted and the therapeutic relationship is enacting a solution to clients' relational patterns, rather than a recapitulation of them, clients regain the interpersonal safety they need to reenter their conflicts more fully.

The second most common source of resistance is shame, which often begins to appear as clients enter more fully into their conflicts. Only in recent years have therapists begun to fully appreciate the pervasive role of shame in clients' symptoms and problems. Shame is more apt to generate resistance and defense than any other affect and, in an unfortunate interaction, it is also the feeling that therapists are most likely to avoid. As counselors read and learn more about this central issue, they will gradually develop the ability to hear the shame motifs in their clients' narratives and begin to recognize the many different faces of shame. Soon, they will be able to detect it routinely in clients' concerns about entering therapy, as well as in the issues and concerns clients explore during treatment.

Shame can be differentiated from guilt. With guilt, clients feel only that they have done something wrong or bad to hurt another. Shame, on the other hand, is more primary and pervasive. Clients often feel worthless, inadequate, or defective in the core and totality of their being. Clients often will be without words to communicate their experience when they are suffering a profound shame reaction. A full-blown shame reaction is agonizing—making some clients wish they could die during these annihilating moments.

It is universal to feel embarrassed in awkward situations or even more deeply shamed in certain moments. In contrast to these situation-specific feelings, however, clients who have been repeatedly shamed throughout their lives develop a shame-based sense of self and are *shame-prone.* Humiliating feelings of being profoundly diminished, exposed, or inadequate are readily triggered in many situations that would not evoke such strong reactions in others. In response to the intense feelings of shame that are continuously evoked, shame-prone clients employ a wide variety of coping strategies to keep their shame concealed from others and from themselves. Routinely, such clients defend against their own feelings of shame by:

Acting arrogantly and self-righteously toward others
Intimidating, controlling, or inducing shame in others
Adopting perfectionistic standards around their own work, cleanliness, or religious practices
Developing eating disorders or other addictions

Therapists will observe other maladaptive relational patterns that clients have developed to cope with these dreaded feelings. Perhaps most disturbing to witness is the *shame-rage cycle* that characterizes most battering and physical-abuse episodes. The individual explodes in rage immediately after feeling diminished or demeaned by someone, in a desperate but futile attempt to disprove their shame-worthiness and artificially restore a sense of personal power. Clients who have grown up expecting contempt from others and who have developed a shame-based sense of self may not remember original developmental experiences (for example, the client's drunken father made him lick the kitchen floor while siblings looked on helplessly) or recognize the current shame-evoking experience.

For other shame-prone clients, intense anxiety may be evoked in a host of different situations that hold the ever-present threat of revealing their basic unworthiness, core unlovability, or inherently flawed inadequacy. As the dreaded feeling of shame is evoked yet again in this new situation, (for example, asking someone for a date, applying for a job, needing to ask for help with a flat tire, getting lost, or making a simple mistake) therapists will commonly witness *shame-anxiety.* Strong anxiety is evoked that may look like an anxiety attack or generalized anxiety disorder, signaling that the client's defenses against her shame are being threatened and raising the spectre that her shame might be revealed to herself or others. To manage this anxiety and protect against the potential exposure of her shame, the client may cope by withdrawing or avoiding, which becomes a characteristic or habitual response pattern.

As we will explore later, therapists can help clients recognize and stop these self-defeating defenses against their shame (for example, hitting and yelling, withdrawing and avoiding, controlling and inhibiting, complying and disassociating) and begin to allow themselves to actually experience the self-hatred or contempt they feel toward themselves. However, therapists often have as much difficulty witnessing and responding to the raw experience of shame as clients have in bearing it. Guidelines for responding to shame and helping these clients contain their painful self-hatred will be provided in Chapter 5. For the time being, therapists need to begin listening for the emergence of shame and to allow themselves to consider the possibility that resistance often serves to protect clients from having seemingly unacceptable parts of themselves revealed.

After clients have been able to resolve or integrate their shame and other conflicted feelings that are central to their problems, they begin to make healthy new changes. At this point, some clients will exhibit the third most common source of resistance: binding feelings of guilt or disloyalty over success in treatment. Therapists will observe that some clients retreat from progress in treatment or undo successful changes they have just made. Often, clients cannot sustain positive changes they have just achieved because the healthy new behavior is inconsistent with internal working models and threatens their attachment ties to internalized caregivers. Becoming stronger or improving in therapy makes some clients feel cut off from parental approval and affection, disloyal to caregivers, or guilty about hurting, leaving, or surpassing the parent in some way. These clients may report feeling either: (1) alone, empty, or disconnected following meaningful change; or (2) guilty, selfish, or bad. When these reactions occur, clients need to be reassured of the therapist's continued presence and support for this stronger self, or else they will have to restore insecure attachment ties by sabotaging their own success. These three commonly occurring sources of resistance will be elaborated throughout the chapters that follow.

Closing

Listening to the client with concern and respect is the most important intervention the therapist can make in the beginning of treatment. Such empathic listening is the basic tool the therapist uses to begin a working alliance and provides the context for further interventions. The next step is to address ambivalence or resistance to therapy in order to maintain the relationship that has just begun. Resistance will be more of an issue with some clients than with others, but it will occur to some extent for every client. As we have seen, if the client can talk about these conflicted feelings, and the therapist responds affirmingly, the feelings will be far less likely to pull the client out of therapy. Resistance will wax and wane throughout therapy, often reemerging as clients enter more deeply into difficult issues. To sustain progress in treatment, therapists need to continue responding effectively to the many types of resistance that occur. If therapists formulate working hypotheses about potential client resistance, it will help them to anticipate when it is likely to occur and to respond more effectively. In sum, the therapist's goals are: to encourage the client's lead, to foster a working alliance, to remain alert for potential signs of resistance, and to begin formulating treatment plans by generating working hypotheses.

Resistance, ambivalence, and defense provide a window to certain inner aspects of the self. Each provides an opportunity to observe the fascinating workings of internal conflict. People do not possess a unified self; we are so complex and multifaceted that it often feels as if one part of ourselves is working against another part. By resolving an internal conflict, a person becomes a little more integrated or whole. Integrating disparate parts of the self is usually an important part of resolving problems and the avenue to personal growth and change. This early success in dealing with resistance can show clients that they possess the internal resources, and have the help they need, to resolve the bigger issues that lie ahead.

Suggestions for Further Reading

1. Shame plays a bigger role in resistance and defense than any other issue, but it is the most sensitive affect for therapists to approach. A case study illustrating how therapists can respond to a client's shame is provided in Chapter 3 of the Student Workbook: "Shame Dynamics: Case Study of Hank."

2. Beginning therapists are encouraged to read further about resistance. One classic reference in this area is Ralph Greenson's *The Technique and Practice of Psychoanalysis, Volume 1* (New York: International Universities Press, 1967). Chapter 2 provides a lucid introduction to the topic of resistance and offers informative case studies to illustrate effective therapeutic responses. The reader does not have to be psychoanalytically oriented to be informed by this discussion.

3. Beginning therapists will find it useful to examine other approaches to working with resistance. The family therapy literature, in particular, has highlighted therapeutic responses to resistance. Students interested only in individual psychotherapy still may find it useful to examine such basic texts in this area as *Mastering Resistance: A Practical Guide to Family Therapy,* by C. M. Anderson and S. Steward (New York: Guilford Press, 1983). A highly informative article about shame intervention in couples therapy is "The Systemic Treatment of Shame in Couples" by D. Balcom, R. Lee, and J. Tager, *Journal of Marital and Family Therapy,* 21, no. 1 (1995): 55–65.

4. A process-oriented approach to multicultural issues in counseling can be found in Chapter 1 of *Child and Adolescent Therapy: A Multicultural-Relational Approach,* edited by F. McClure and E. Teyber (Fort Worth: Harcourt Brace, 1996). An excellent discussion of ethnographic and cross-cultural issues in counseling can be found in H. Pinderhughes's *Understanding Race, Ethnicity, and Power* (New York: Free Press, 1989).

CHAPTER FOUR

An Internal Focus for Change

Conceptual Overview

The first stage of therapy is complete when therapist and client have established a collaborative relationship and begun to work together on the client's issues. This collaborative alliance is a necessary prerequisite to the second stage of therapy: the client's journey inward. In order to change, clients must become less preoccupied with the problematic behavior of others and begin to explore their own internal and interpersonal responses. Clients need to stop focusing exclusively on historical events, past relationships, and the problematic behavior of others in their lives. Instead, clients must begin to clarify their own thoughts, feelings, and reactions to the troubling events they are experiencing. Why? Clients will usually fail in their attempts to change others in their lives, whereas they can often resolve problems by understanding and changing themselves and their own role in problems. In this process, clients can examine their habitual response patterns in problematic situations, evaluate their usefulness in their current lives, and begin redefining how they want to construct future relationships. This internal focus on identifying and changing their own responses, rather than trying to change the other person, gives clients greater self-control and the freedom to act more purposefully in ways they choose.

The therapist's task is thus to help clients make the transition from seeing the source and resolution of problems in others to adopting an internal focus for change. This is a twofold process. First, the therapist helps clients to begin looking within. That is—the therapist—without invalidating clients or using language that makes them feel "blamed" focuses clients away from complaining about or trying to change others, and toward identifying, understanding, and changing their own reactions. Second, therapists must help clients become active agents in their own therapeutic work. This begins to occur as clients become aware of their old templates and usual styles of responding. With greater awareness of their own internal and interpersonal responses, clients begin to recognize that they do not have to respond in their old maladaptive ways and always to receive the same hurtful or disappointing responses from others. This new awareness, coupled with the therapist's support for the clients' own

87

self-direction and initiative within the therapeutic relationship, will empower them to find new and more effective responses to old problems. Self-efficacy develops out of this collaborative process between therapist and client. Enduring change results when clients feel ownership of the change process. However, it may be difficult for some clients to shift their focus from others and begin looking at their own role. In order to take this rewarding journey inward, clients need a secure and supportive relationship with the therapist. In this chapter, we will see how therapists can provide that safety.

Chapter Organization

This chapter has four sections. In the first section, we examine the need for clients to shift from an external to an internal focus for change. We describe how clients often externalize their problems onto others, which perpetuates their feelings of frustration, helplessness, or depression. Thus, we discuss how therapists can help clients achieve this inward focus and gain a greater sense of legitimate or internal control; and we examine how some clients can misunderstand the therapist's shift to an internal focus at first. Some clients initially misperceive the internal focus as implying that they have to accept unwanted blame for problems with others, or as denying the reality of others' problematic behavior. We will see how therapists can resolve this misunderstanding if it occurs.

In the second section, we examine ways in which clients can become active agents in resolving their own problems. We explore how therapists can utilize the therapeutic relationship to foster the client's own initiative and self-efficacy; and we explore specific ways of intervening that facilitate the client's ownership of the change process. Sample therapist-client dialogues illustrate effective and ineffective ways to achieve these goals.

In the third section, we see how therapists can actively enlist clients in understanding their own problems by extending the collaborative alliance into the "working through" or problem-solving phase of treatment as well. A case study illustrates how therapists can help clients share responsibility and remain contributing agents throughout each phase of treatment.

In the final section, we will examine how therapists can help the client adopt an internal focus by tracking the client's anxiety. Anxiety is a signal that can guide therapists closer to the heart of the client's concerns and indicate when it is most productive to focus the client inward.

Shifting to an Internal Focus

A Prerequisite for Change

In the initial stage of therapy, many clients see the source of their problems in others. Clients often want to spend more time describing others' problematic behavior than discussing their own experience of and response to the problems.

For example, many clients begin the first few therapy sessions by announcing that the problem is really with another person:

> My husband won't pay any attention to me.
> My wife is always on my back.
> My children are impossible; they won't do anything I say.
> My boss is a demanding tyrant. If he doesn't back off, I'm going to have a heart attack.
> My mother won't stop criticizing me; I can't do anything right according to her.
> I'm 27 years old, and my father treats me like a child.
> My boyfriend has a drinking problem.

It is essential for the therapist to affirm these complaints as genuinely troubling concerns, and communicate directly that the therapist wants to help the client improve this situation. Even if these complaints are overreactions or include strong transference distortions, they also contain much truth from the client's point of view. As emphasized in Chapter 2, the therapist validates that the behavior of others is indeed troublesome. However, when that is not genuine, the therapist can at least affirm the subjective reality of the client's perceptions of what the problematic behavior of others is doing. If the therapist does not provide this affirmation first, many clients—especially those who have suffered much invalidation in their lives—will feel that the therapist doesn't understand or care. This is counterproductive and will often reenact the client's developmental conflict. Thus, to keep this potential invalidation from occurring, the therapist's first responsibility is to hear and validate the client's concerns. Only after making this genuine empathic connection, which often requires the therapist to de-center and see issues from the client's subjective viewpoint, can the therapist take the next step.

The second step is to begin focusing clients back on their own thoughts, feelings, and reactions to the problematic behavior of others, rather than joining the client in focusing exclusively on the external problem. Often, it will be easy for therapists to pair the first two steps together.

Therapist:
I'm sorry your father does that to you. I think it would be upsetting, too. What do you say and do when he ... ?

OR

Therapist:
It sounds so frustrating to have your boss respond that way over and over again, no matter what you do. How do you feel and what are you thinking inside when she ... ?

The same principle applies when working with people of color, women, gay men and lesbians, and others who accurately explain the social inequities underlying their problems. Once again, the therapist must join such clients in the reality of their situation, as they experience it, before focusing inward on their reactions to, and ways of coping with, these inequities. Why is therapy more productive when therapists focus clients inward? In many cases, clients who feel anxious, distrustful, angry, or helpless because of the behavior of another person will try to enlist the therapist in blaming, criticizing, or trying to change the other person. However, therapy will not progress very far if the therapist merely joins clients in focusing on the other person's behavior, no matter how problematic this behavior may be. Why? As the existential therapists inform us, clients' attempts to change the other person will usually fail; clients are much more likely to resolve the problem by changing their own way of responding (Wheelis, 1974). Thus, the therapist's task is to expand clients' focus beyond the other person and include themselves as well.

Note that the therapist should be *flexible* in adopting an internal focus. Some clients will benefit greatly from the therapist's help in understanding the behavior or potential motivations of problematic others, other clients will not. For example, if a client has grown up befuddled by a parent with a borderline or narcissistic personality disorder, it may be deeply validating for the client to read the descriptions of these disorders in the *Diagnostic and Statistical Manual* (DSM-IV) and to realize that all of this was not his fault and did not happen solely to him.

How can therapists increase clients' awareness of their own thoughts, feelings, and reactions in problematic situations? Recall the externalizing comments cited at the beginning of this section. Each of the following questions can be used to focus those clients back on themselves.

What do you find yourself thinking when your husband is ignoring you?
How do you feel when your wife is nagging you?
What do you do when your children disobey you?
What is it like for you when your boss is so demanding?
What thoughts were you having as your mother was criticizing you?
How do you react to your father when he diminishes you like that?
What is most troublesome for you about your boyfriend's drinking?

Simple inquiries of this type serve two important functions. First, they tell clients that the therapist is listening to their concerns and is taking them seriously. The therapist has not changed the topic or brought up something that was discussed 10 minutes ago, but is responding directly to their concerns. Second, while inviting clients to say more about their concerns, the therapist is also shifting clients' focus away from others and encouraging them to look more closely at themselves. Inviting clients to become more aware of their own reactions is a critical step toward understanding and resolving their problems. As we will see, the simple response of focusing clients on their own experience is a powerful intervention that will elicit strong feelings and important connections.

In many cases, clients will welcome the therapist's offer to talk more about themselves. When clients begin to focus inward and learn more about themselves, they begin to change. How? Clients' active exploration of their own expectations and responses in problematic situations will lead to increased awareness of the narrow range of responses they tend to repeatedly employ and, with the therapist's help, of the wider range of options that are actually available to them. This increased self-awareness of their old response patterns, coupled with more flexible or varied response options, leads to more problem-resolution skills and increased feelings of self-efficacy. However, not all clients will respond so positively to the therapist's initial invitation to look more closely at themselves. Clients may avoid or actively reject an internal focus and continue to talk about the problem out there in others. In that case, the therapist continues to inquire about the personal meaning that this particular situation holds for the client.

Therapist:
Where would you like to begin this morning?

Client:
My wife is impossible to live with. She complains constantly; nothing ever pleases her.

Therapist:
It sounds like you've had a difficult week with her. What's been the hardest thing for you?

Client:
Do you know how hard it is to live with an angry, demanding wife who keeps trying to tell you what to do all the time?

Therapist:
You're very angry at her. And I can see why; it would be hard to have someone after you like that. Tell me, how do you respond to her when she is doing this?

Client:
I don't know; I just hate it. I guess I yell back sometimes, or just try to get away from them. It's not just her, you know; her whole family is like that.

Therapist:
They really are very critical of you, and I can imagine how hard that would be to live with. But it seems like criticism, in particular, really gets under your skin. What's it like for you to be criticized so much?

Client:
I hate it. I just hate it. They make me feel like I can't do anything right— that I'm doing it wrong and failing all the time.

Therapist:
Doing it wrong and failing—ouch! Sounds like her family's constant criticism taps into your own painful feelings about yourself of not measuring up or not being enough. And having those shameful feelings of inadequacy aroused all the time would be infuriating.

Client:
Yeah, I hate them for making me feel this way. If I could just make them stop, everything would be okay.

Therapist:
This is very important for you, and we need to work together on it. I would like to understand better what you do when they criticize, so that I can help you learn some more assertive, limit-setting responses. But your own feeling of not measuring up is also a part of this problem that we need to work on, too, so that you can stop charging at the red flag they are waving.

If we can change the internal part of you that overreacts as if you believe what they are saying is true, it would be much easier for you to handle this than it has been in the past.

Client:

What do you mean, "overreacts"?

Therapist:

Sounds like their criticism makes you feel you can never do anything right. Is that accurate?

Client:

Yes, but that's not true. They just expect too much and I can't do it all. I can never do it good enough for them.

Therapist:

No matter what you do they will never be pleased. But their expectations still upset you. It's the way you end up feeling inside, about yourself, that becomes the problem?

Client:

Yes.

Therapist:

So being able to work on your feelings about yourself would be helpful?

Client:

I see what you mean, but I'm not sure what to do.

Therapist:

Tell me more about your feeling of not measuring up.

Client:

Well, I guess I've always sort of felt like I'm not really good enough.

In this dialogue, the therapist validated the client's experience while at the same time encouraging the client to look at his own reactions and role in the conflict. Although the client kept trying to focus on the problematic behavior of his wife and her family, the therapist's repeated attempts to have the client look at himself as well eventually served to mitigate his defensive, externalizing stance. As a result, the client moved closer to his own problematic feelings and beliefs about himself that contributed to the marital conflict. By gaining a better understanding of his own feelings of inadequacy, the client will become less shame-prone and reactive to others' criticisms. This, in turn, will be an important part of learning to respond more assertively and set limits more effectively, rather than merely yelling back or withdrawing.

Many clients have not been encouraged to focus on themselves before. As in this example, some clients may resist this internal focus initially, perhaps because it asks them to come face to face with difficult feelings that they have not been able to understand or resolve alone. The paradox is that, as long as clients avoid the internal or personal aspects of their relational conflicts by externalizing their problems onto others, they will feel powerless, dependent, frustrated, and out of control.

Many clients who are struggling with conflicts are inappropriately invested in changing others as a means of managing their own problems or insecurities. In attempting to shape or direct others, the client is seeking an external solution to a problem that can best be changed internally. The reality is that, beyond clearly expressing our preferences and personal limits ("I would like . . ."; "I will not . . ."), we cannot readily influence how others think, feel, and act, but we can change our responses. For example, clients routinely fail in their attempts to make their spouse stop drinking, smoking, or overeating. Similarly, some clients try for decades to win the approval or recognition from others that they never received from a parent. Others try unsuccessfully for years to have their grown offspring choose a different mate, religion, or career. As a result of these failed attempts to change others, many clients enter therapy with feelings of helplessness and depression.

To regain a sense of mastery over their own lives, and become less depressed, a therapist can offer clients the more productive alternative of decreasing their attempts to change others and gaining more understanding and control over their own reactions instead. This internal focus is a necessary prerequisite before clients can adopt new, more effective responses to old problems. Only when clients begin to focus on understanding and changing themselves will they begin to feel in charge of their lives and capable of change. Clients experience these feelings as empowerment. Thus, as we have seen, a consistent therapeutic response is to patiently focus clients back onto themselves. The following types of questions will help clients explore their own responses:

What is the main feeling you are left with when . . . ?
What were the thoughts you were having when . . . ?
What was the most difficult thing for you when . . . ?
How would you like to be able to respond when . . . ?

Focusing Clients Inward

Focusing clients on their own behavior often reveals how they are contributing to, or participating in, their own conflicts. Often, clients will not have realized their own role in these conflicts. As we will see, clients who can focus on themselves and see their own participation in a conflict are usually motivated to change their own part in it. This, in turn, may allow the other person to respond differently as well. To illustrate, let's return to our example of the husband who complained about his critical wife and in-laws.

Although the client did begin to explore his own feelings of inadequacy and the ineffective ways he responded to his wife, he kept complaining about his "obnoxious" wife and "superior" in-laws. As before, the therapist was empathic, validated his experience, and affirmed his anger. At the same time, however, the therapist did not join the client in focusing exclusively on his wife or blaming her as the sole source of his problems. Instead, the therapist continued to focus the client away from his preoccupation with his wife's behavior and toward his own reactions to her.

What would have happened if the therapist had not done this? If the therapist had responded to the client's eliciting pull to blame his wife, the client would have remained an angry but helpless victim. In contrast, if the therapist had emphasized just the client's contribution to his marital conflict, without first validating his experience, it would have repeated a maladaptive relational pattern. Exploration of the client's feelings of inadequacy revealed that these shameful feelings stemmed in part from rarely being supported by his parents and consistently being blamed for whatever went wrong. Thus, by validating the client's experience first, and not prematurely focusing him inward, the therapist helped the client find a new and more satisfying response to the old conflict of constant critical intrusions. By coupling this validation with a consistent invitation to adopt an internal focus, the therapist allowed the client to become more aware of how his own current reactions to his wife contributed to the ongoing conflict in his marriage.

For example, the client learned that he was quiet and unresponsive to his wife when he arrived home from work. The therapist helped him see that his silence escalated his wife's attempts to gain his attention and, in response, he became even less communicative. Their problematic interaction worsened as the wife became more insistent and demanding of his attention and he withdrew even further—by taking a nap, for example. By focusing on his own emotional reactions at that moment, the client realized that his wife's increasing requests for attention also made him feel overwhelmed by the prospect of trying to meet all of her needs. He had held the unrealistic belief that if he did not always respond to her, he was failing as a husband and again being inadequate.

By encouraging the client to become more aware of his own internal and interpersonal responses to his wife, the therapist gave the client an opportunity to change his part in the marital conflict. Specifically, the husband began to express more directly to his wife when he would like to be close to her and when he would prefer to be alone. He explained to his wife that he wanted 30 minutes alone to unwind when he returned home from work, after which they could sit down and share their day together.

More importantly, as the husband gained a better understanding of his tendency to feel overwhelmed by others' demands, he became less reactive to his wife's requests. He learned that he felt shamefully ruled by her because he could neither say "no" to her requests nor express his own wishes. If his wife wanted something from him, he believed that he was failing if he didn't respond immediately. And because he couldn't set limits or ask for what he wanted without feeling selfish, he resented the "unfairness" in their relationship. The client also learned that he expressed this resentment to his wife indirectly—in his own critical and withholding ways. Over the next few months, his new awareness allowed him to respond to his wife's requests in the manner that felt most appropriate to him; he no longer felt he must comply with them, and his resentment abated. As the client set these limits, he began to feel more control over himself in their relationship and was able to make some tentative steps toward greater closeness with his wife.

Thus, as clients begin to focus on their own behavior, they often start to feel less "stuck" or powerless. As they start to see more alternatives available to them, clients begin to feel more hopeful and in charge of their lives. Thus, the therapist's task is to discourage clients from passively complaining about others or from trying to manipulate and control others' behavior. However, therapists must do this gradually, after affirming clients' legitimate concerns about the problematic behavior of others.

Reluctance to Adopt an Internal Focus

Therapists' reluctance   Most beginning therapists feel comfortable trying to assuage their clients' fears or concerns. However, focusing the client inward often means that the therapist is drawing out—rather than reassuring—the client's fears and concerns. For many different reasons, therapists may have difficulty in approaching and clarifying client conflicts rather than trying to diminish or avoid them—as occurs in most social interactions. For some, such directness goes against unspoken familial rules or cultural norms. Other therapists may equate being forthright with being intrusive or exposing because they have only seen forthrightness utilized in these hurtful ways. Some therapists are reluctant to address clients' internal conflicts directly because they believe that good people don't make others hurt or feel bad. These therapists try to make clients feel better by reassuring them about their insecurities and by emphasizing only their strengths and successes. However, moving away from client conflicts in these ways is usually an ineffective response. Although well-intended, it keeps clients from fully experiencing and exploring their problems. In this way, therapists rob clients of the opportunity to find solutions to problems and develop the self-efficacy that comes from being able to resolve problems. Still other therapists may avoid an internal focus in order to please clients and maintain their approval. These therapists fear that clients will become angry as the therapist invites them to explore the uncomfortable feelings or difficult conflicts they have been avoiding. Therapists are also concerned at times that they may not be able to remain present and responsive to the client when strong emotions are expressed. This issue will be addressed more fully in subsequent chapters.

For these and other reasons, therapists often join with clients in looking away from the internal aspect of the conflict or the client's participation in the problem. Therapists collude with the client in externalizing problems when they repeatedly:

- Give advice and tell the client what to do or how to respond to others
- Interpret or explain what the other person's motives or behavior may mean
- Reassure clients that their problems will go away or are not something to be concerned about
- Disclose what the therapist has done to cope with a similar person or problem

As we would expect from client response specificity, each of these responses will be effective at times. However, if they come to characterize the ongoing course of treatment, it will prevent most clients from exploring their own experience and, ultimately, from being able to resolve their own conflicts. Thus, throughout the course of treatment, therapists must repeatedly help clients expand their awareness of their internal experiences and interpersonal responses.

As we will see, however, this can be challenging. A consistent focus on self and internal experiences may arouse clients' anxiety as unwanted feelings, pathogenic beliefs, or expectations of hurtful responses from the therapist and others may all be evoked.

Clients' reluctance   Why do clients tend to avoid an internal focus? As we have seen, a shift away from their focus on others can make some clients feel that the therapist does not really understand them or is not sympathetic to their concerns. Other clients fear that, if they give up their attempts to change others, they will either have to accept the blame for the problem or remain forever resigned to "defeat." Still other clients may not yet feel psychologically held by the therapist's compassion and understanding; they do not have the secure base they need to approach the difficult feelings or choices that looking inside entails. Clients will also be unable to enter certain issues if the therapist and client are reenacting some aspect of the client's conflict in their interpersonal process. When this is occurring, however, both are usually unaware of it.

Cultural factors may also impede an internal focus. For example, Native American, Asian, and other clients from communally centered cultures may initially react negatively to looking within because it sounds self-centered. If clients perceive a difference in values and goals, they may question the therapist's credibility. Therapists can help with this by educating clients about the treatment process and, more importantly, by responding in a manner that is congruent with the clients' worldview.

Therapist:
How does your response to this affect everyone else?

OR

Therapist:
Let's look at their role in this—and your role, too—and see if we can make more sense of what's going on for you.

When clients are not ready to look at their own issues, this presents another opportunity to work with resistance. Here, instead of insisting the client focus on his own behavior, the therapist can work with him to explore his reasons for not wanting to look within. How can the therapist best respond when clients resist an internal focus? One thing the therapist can do is to follow clients along for a while, while offering clients repeated invitations to reveal a little more about themselves. In most cases, clients will gradually accept the therapist's invitation and soon come to welcome the opportunity. If the client is unable to respond, another option is to give the client permission to disclose or proceed at her own pace (for example, Therapist: "It doesn't seem comfortable for you to share very much of yourself yet. I respect that, and believe it is important for you to wait until you feel safe and you choose how much you share."). If the client's externalizing stance does not begin to change, however, the therapist can also make a process comment that simply describes their interaction.

Therapist:

> I've noticed that you talk very easily about your husband and your daughter, but you don't say very much about yourself. Are you aware that that happens?

OR

Therapist:

> I think we've been missing each other the last two sessions. I keep asking what you were thinking about or trying to do in a particular situation, and you keep responding by telling me more about the other person. What do you see going on between us?

These process comments will help clients to become aware of their externalizing style with the therapist and to learn that they probably respond in the same way to other people. For example, clients who cannot share themselves in a personal way with others will often be perceived as boring or aloof. In that case, part of the clients' presenting problem—for example, clients' sense of loneliness or lack of meaningful contact with others—may also be enacted in their relationship with the therapist. Although it is often discouraging for clients to realize that they are recapitulating their conflict in the therapy session, this reenactment provides the opportunity to resolve the problem by changing the distancing pattern within the therapeutic relationship. The therapist helps clients become aware of how they interact with others and how others experience them, and then offers a new and more effective way of relating in the therapeutic relationship. For example, the therapist can describe the current interaction and offer to develop a more meaningful relationship:

Therapist:

> I feel that I am being held away from you when you talk about others rather than yourself. I feel like I am missing you, and I don't want that. I would like to learn more about you, or your concern in talking about yourself. Can we work together on this?

When the therapist uses questions or process comments to focus clients inward, clients will often reveal a variety of new and important concerns that they have not talked about before.

> Well, I guess there really isn't very much about me that people would want to know.
> I'm not used to telling people what I'm thinking or feeling.
> Every time I try to get close to someone, I get hurt in the end.
> You wouldn't like me very much if you knew what I was really like.

In this way, the therapist's process comment has brought out important new concerns that can now be addressed in therapy. The therapeutic relationship has deepened and access to significant information about the client has increased. *This process of helping clients to focus inward and to explore their resistance to looking within will provide some of the most important material to be addressed in therapy.* These important new concerns are not likely to be brought up for treatment unless the therapist uncovers them by using process comments and focusing clients inward. Thus, one of the most important reasons for adopting an internal focus is to evoke the central conflicts that are operating in the client's life and to do so in a way that makes them accessible for treatment.

Placing the Locus of Change with Clients

The first component of the internal focus is to help clients look within and become more aware of their own responses. A second component is to help the client adopt an internal locus for change and to encourage the client to begin to act from within. In this section, we examine how clients can gain a greater sense of *effectance* in their lives by becoming active agents in their own change process (Bandura, 1982; 1997).

First, we will see how the therapist can use the therapeutic relationship to foster the client's own initiative. Second, we will examine specific types of therapeutic interventions that encourage the client's participation in the change process. Throughout, the therapist's goal is to respond to the client's interests and initiative, and to thereby support the client's increasing sense of self-efficacy.

Using the Therapeutic Relationship to Foster Clients' Initiative

In the course of treatment, a therapist has the opportunity to convey to clients the broader message that they are able to take charge and effectively manage their own lives. Effective therapists of every theoretical orientation nurture clients' own sense of personal dignity and self-efficacy. Beyond verbal encouragement of clients' initiative, therapists can help clients gain an experience of *acting* more effectively during the therapy session. Once clients experience greater control in their relationship with the therapist, it is often relatively easy to help clients generalize this greater sense of personal power to other relationships beyond the therapeutic setting (Bandura, 1997). Let's see how therapists can help clients assume an active, initiating role in therapy.

The first way to help clients feel responsible for and capable of change is to repeatedly encourage clients to respond to their own agenda in therapy and to talk about whatever they feel is most important right then. Once clients begin pursuing their own concerns in this way, the therapist can be an active participant who helps clients to explore their own concerns more fully, to understand their problems better, and to generate potential solutions and behavioral alternatives.

Why is it essential to encourage the client's own initiative, even in the time-limited modalities that increasingly dominate the therapeutic milieu? In one way or another, most clients are unable to act on their own interests and goals. In past relationships, these clients have not had significant others support their own interests or respond affirmingly to their own genuine feelings—especially of sadness and anger. As a result, these clients will feel encouraged, but anxious, too, if the therapist supports their own feelings and self-direction. Regardless of treatment length or the therapist's theoretical orientation, therapy becomes a more intense and productive experience when the therapist can successfully engage clients in pursuing their own interests and exploring the issues that matter most to them.

When the therapist can help clients explore and understand the material they choose to focus on, rather than merely direct them to the therapist's own agenda, change will usually begin to occur. When clients provide the momentum and direction for therapy, they often experience a corrective relationship where they can have their own opinions, act more effectively, and be more in charge of their own lives. Once a client has begun to initiate, the therapist can actively participate by providing the types of responses that this particular client seems to utilize most productively. Commonly, this will include interventions from a wide variety of theoretical orientations—contributing information, explanations, or interpretations; constructing behavioral alternatives; suggesting possibilities about what problematic others may be doing or intending; providing interpersonal feedback about how others may be perceiving or reacting to the client; and so forth. Whatever interventions or techniques the therapist uses, however, they will be far more effective when given in response to the clients' own interests or concerns, rather than the agenda of the therapist, employer, or spouse. This process dimension is one of the most important characteristics of the therapeutic relationship.

Avoiding a hierarchical relationship   We have already noted that clients may resist the therapist's attempts to make them active participants in treatment. Why? Frequently, clients will be conflicted about expressing their own wishes, acting on their own initiative, and achieving or enjoying success. Because so many clients have not been supported in their own individuation and feel guilty or anxious if they exhibit self-direction or success, they continually elicit advice and direction from the therapist. Therapy will not be productive if this helper-helpee mode comes to characterize the therapeutic process. This mode shifts responsibility away from the client and onto the therapist and creates the hierarchical relationship discussed in Chapter 2. Clients will not be able to feel greater efficacy and adopt a stronger stance in their lives as long as this dependency is fostered—their *compliance* is being reenacted in the therapeutic relationship rather than questioned.

Too often, therapists unwittingly comply with the clients' subtle or overt request to tell them what to talk about in therapy and what to do in their lives. It is flattering to think that we know what is best and can tell others what to do; it appeals to the narcissism in every therapist. As we would expect on the basis of client response specificity, there will be circumstances in which directives, advice, and so forth are facilitative for a client. However, when such responses characterize the ongoing process of therapy, clients will usually remain dependent on helpers to manage their lives. Because this problematic process occurs subtly (but frequently), we need to examine further this dependency-fostering approach to therapy.

Supporting clients' own autonomy and initiative   Effective therapy of any treatment length should foster the client's self-efficacy. Therapists cannot just talk with clients about choice, responsibility, and personal power, however; they must co-create a relationship in which clients behave in healthier ways. To do this, the therapist must first give clients permission to follow their own interests and actively encourage them to introduce the material that is most relevant to them. As detailed in Chapter 2, the therapist then tries to identify the subjective meaning in this material and encourages clients to elaborate it further. In response to clients' increasingly specific exploration, the therapist actively helps clients clarify and understand their experience. Together, they begin to generate more effective ways to respond. When this mutual interchange continues throughout several sessions, most clients become deeply committed to treatment and report that meaningful changes have begun to occur. *Facilitating this growth process may be the primary challenge and satisfaction of being a therapist.* Let's see how these ideas might sound with a client:

Client:
    I need more direction from you. What should I do here? You're the expert.
Therapist:
    I have a couple of ideas to share with you, but I think it would work best if I heard yours first.
Client:
    If I knew what to do, I'd just do it, and wouldn't ask you or be coming here to see you!
Therapist:
    All right, let's swap ideas. You tell me what you think is going on; and then I'll tell you what I see occurring; and let's see what we can put together.
Client:
    Like I just said, I don't know—I don't really have any ideas.
Therapist:
    Let's wait for just a minute and see if anything comes to you. If not, I'll be happy to go first and start us off.
Client: (*20-second pause*)
    Well, maybe I'm afraid of being alone or something like that.
Therapist:
    That's very interesting, and fits with what I'm thinking about. Tell me more about what it means to be "alone"; it sounds important.
Client:
    I think I've always been worried about that.
Therapist:
    Well, it makes me think that maybe your parents were physically present for you, but that they cut off from you emotionally whenever you disappointed them. I wonder if this created the feeling of being all alone, even though others were present—which would have been very confusing to a child.

Client:

Yeah, I think that's right, and the worst thing about it was that it seemed like it was all my fault. They were going away because I let them down. What do you think?

Therapist:

I think you have very good ideas, but something often seems to hold you back from expressing them in the strong way you have just been doing with me. I have noticed that happening in here with me sometimes, and I hear it in your relationships with others as well. I wonder if there is some connection between holding yourself back in this way and being afraid of being left. What do you think?

This collaborative interaction, in which the therapist and client each build on what the other has just produced, is an *independence-fostering* approach to psychotherapy. Clients have the experience of sharing responsibility for the course of treatment, as the therapist actively encourages them to define and address their own concerns. As therapists support clients in achieving this more active stance, however, they are simultaneously contributing their own information and guidance in a way that creates a working partnership.

A common misconception is that longer-term, dynamic, or relationship-based therapies are dependency-fostering, whereas short-term, problem-solving, or strictly behavioral approaches are not. Actually, whether the therapy fosters dependence or independence is determined by the therapeutic process and not by the length of treatment or the theoretical orientation of the clinician. In short- or longer-term therapy, the client's dependency is inappropriately fostered when the therapist repeatedly directs the course of therapy, gives advice, and prescribes solutions for the client. Let's examine an effective middle ground between the nonproductive polarities of directive and nondirective control.

Shared control in the therapist-client relationship   As we have begun to see, therapy will be most successful when the therapist and client share control over the direction of therapy. It is overly controlling for the therapist to play the predominant role in structuring and directing the course of treatment, and it is ineffective to nondirectively abandon clients to their own confusion and conflicts. In a more productive relationship, therapists will encourage the client to take the lead but will actively contribute their own thoughts and suggestions about the material that the client has produced. It will be therapeutic for many clients to experience a relationship in which control is shared by both participants, rather than a relationship in which control is held by one party or the two parties compete for control. This therapeutic collaboration is the key to empowerment and provides the client with the support to act in new and stronger ways that modify old templates. In Chapter 2, for example, we stressed that in the beginning of treatment the therapist should encourage the client to take the initiative and bring up whatever she would like to discuss.

Therapist:

I would like to begin each session by having you bring up what you want to talk about. I think it's important that we work on what is most important to you.

Some therapists will be frustrated by this approach and direct clients to solutions and action, especially if they are feeling pressed to fix the problem quickly because of the limited number of sessions available. Although a highly prescriptive stance will be helpful at times, it imposes several important limitations. First, as we have seen, a directive approach prevents the therapist and client from uncovering conflicts that were not evident in the client's presenting problem. Second, the goal of therapy is not merely to fix the client's presenting problem, but to do so in a way that leaves clients with a greater sense of their own personal resources and ability to effectively negotiate their own adult lives. Working collaboratively, the client gains an increasing sense of self-efficacy by participating in treatment progress, which directive approaches do not provide.

Responding to these arguments, the action-oriented therapist might ask, "What if you follow the client's lead and it takes you nowhere?" Certainly, a nondirective approach requires an unrealistic amount of time from the therapist and often results in a disorganized therapy lacking in focus. Moreover, a purely nondirective approach is likely to reenact problematic relational patterns for clients who grew up in a permissive home, just as a purely directive approach may reenact such patterns for clients who grew up in an authoritarian home. As we have seen, however, there is an effective middle ground of shared therapist-client control, although it has not been delineated well in the counseling literature.

Therapists need treatment plans and intervention strategies with short- and longer-term goals. However, these must develop out of a collaborative interaction with the client, who will then feel ownership of the treatment process. In order to implement this interpersonal process, the therapist must be able to tolerate ambiguity and to refrain from subtly controlling the interaction. Instead, the therapist must try to create opportunities for clients to voice their own concerns and act on their own initiative. The therapist must then be able to decenter, enter into the client's subjective worldview, and focus on the central meaning or feeling that this issue holds for the client.

This approach requires an active therapist who is neither directive nor nondirective. In this model, therapists respond actively by:

Communicating their understanding and affirming the client's experience

Joining the client in expanding and exploring further what the client has produced

Using self-involving comments or their own experience of the client to provide interpersonal feedback about how the client effects the therapist and others

Enthusiastically affirming client behaviors that show strength or effective change

Providing a focus for treatment by highlighting the repetitive relational patterns, core conflicted affects, and pathogenic beliefs that recur in the client's narratives

Furthermore, the therapist is using process comments to address treatment impediments—such as resistance and an external focus—and to prevent relational reenactments by describing or asking about what is transpiring between them in their current interaction. That is, *the therapist responds actively by taking the client's conflicts out of the abstract and creating immediacy by looking for opportunities to link the client's problems with others to what is occurring between the therapist and the client.*

As we will see in Chapter 6, an authoritative parenting style that encompasses both parental warmth and firm limits reflects a more effective middle ground between the better known and more widely adopted authoritarian and permissive parenting styles. In parallel, therapists will find that shared control offers most clients a more productive alternative than either directive or nondirective approaches. Let's examine some therapeutic interventions that foster a strong working alliance and illustrate this middle ground of shared control.

Therapeutic Interventions That Place Clients at the Fulcrum of Change

As we have seen, the therapist's goal in the interpersonal process approach is to encourage the client's lead while still participating actively in shaping the course of therapy. The critical technique is for the therapist to intervene without taking the impetus away from the client. The following examples illustrate effective and ineffective ways to do this.

Ineffective interventions   Suppose that the client is filling the therapy session with seemingly irrelevant storytelling. The therapist cannot find a common theme to any of the client's narratives or understand the personal meaning that these vignettes hold for the client. It seems as if nothing significant is occurring. At this point, it is easy for the therapist to stop the client and direct her toward a specific topic that the therapist thinks would be more fruitful. Although this may be effective at times, therapists will succeed more often if they revitalize the therapeutic interaction without shifting the impetus away from the client and onto the therapist. In the following dialogue, the onus for therapy comes to rest with the therapist, and the client loses an internal focus for change.

Therapist:
I'm wondering where this is taking us. Have I lost the focus here?
Client:
I'm not sure where I'm going with this either. What do you think I should talk about?
Therapist:
You've had trouble asserting yourself in the past, and I think we need to look more closely at that. Last week, you said you wanted to ask your boss for three weeks of vacation instead of two. How are you going to handle that confrontation?

Client:

I'm not sure. What do you think I should say?

Therapist:

To begin with, you need to arrange a face-to-face meeting with him. It is important that only the two of you are present, so that you can have his full attention and there is less threat for either of you to lose face. Then use the "I" statements we have practiced to directly state what you want.

Client:

Sounds good, but what would you say to him first?

This exchange will not be productive, even though the therapist has moved the client to a more salient topic and provided useful information about effective confrontations. The fulcrum of therapeutic movement has tipped from the client to the therapist, and a hierarchical teacher-student process has been established. Most clients will not be able to utilize the therapist's useful information until they pick up the momentum and begin to actively participate again. Furthermore, because the client remains in a passive role as the therapist continues to inform, the client does not gain the increasing sense of mastery that comes from participating in successful therapeutic movement. A more productive intervention might be to ask the client if there is another topic he feels might be more relevant, and to wait until the client becomes actively involved again before offering additional information or further shaping the direction of treatment. However, if the client does not become actively engaged again, therapists do not want to wait nondirectively. Instead, they can intervene by simply observing this with a process comment (for example, Therapist: "I'm having difficulty following you right now. Wherever you go, it just seems to trail off and get lost for me. What comes to mind as I say that?").

Effective interventions   Let's look at how the therapist can reorient the client toward more productive material in a way that keeps the momentum for therapy with the client. Process comments that make the current interaction between therapist and client a topic for discussion are often effective in these circumstances.

Therapist:

I'm wondering if you're using our time the way you want to right now? I don't have the feeling that what you're talking about is really very important to you.

Client:

Yeah, I'm not sure where I'm going with this either. What would you like me to talk about?

Therapist:

I think that we should try to identify what would be most important to you, and talk about that. What would that be right now?

Client:

I'm not really sure.

Therapist:

Let's just sit quietly for a moment, then, and see what comes to you.

Beginning therapists often feel uncomfortable with silences and may find themselves filling them. If therapists can refrain from doing so, however, these are opportune moments for clients to establish their own agenda or introduce new material. This type of process comment also invites the client to approach more substantial material but without taking the impetus away from the client. The therapist has directly intervened by sharing her observation and asking about the client's perceptions of their current interaction, but has still left the client an active participant in revitalizing the discussion. Routinely, clients produce far more significant material when they are invited to lead in this way rather than to follow.

Revealing new conflicts   Important new issues can be identified for treatment when the therapist intervenes without taking the locus for change away from the client. Imagine the following situation: A 19-year-old client in a college counseling center had been telling his therapist about his attempts in early adolescence to observe his mother undressing. The client had been detailing his voyeuristic efforts at great lengths, but the therapist did not feel that this was genuinely of much concern to the client. Unless the therapist could find the relevant meaning for the client, he wanted to move on to other, more salient issues in the client's current life. In the following dialogue, the therapist uses a process comment to refocus the client in a way that keeps the locus for change with the client and opens up an important new issue.

Therapist:
Does this feel like an important topic that you want to discuss with me? As I listen to you, I don't get the feeling that you are really very interested in what you are telling me. Is that the case, or am I not understanding the meaning this holds for you?

Client:
I thought therapists were interested in this oedipal stuff. I figured you would want to hear about it.

Therapist:
I'm struck by the fact that you are telling me what you think I want to hear, rather than working on what is most important to you. Maybe that's something we could look at in therapy. I wonder if you find yourself doing this in other relationships as well—trying to sense other people's needs or responding to unspoken demands at the expense of expressing your own interests or concerns?

Once again, the therapist has used a process comment to revitalize treatment without taking the impetus away from the client. In this vignette, the therapist has also used their current interaction to uncover a potential new problem area for the client. Because this new issue arose from their joint interaction and not from the therapist's own agenda, most likely the client will be interested in exploring it. Thus, the therapist has effectively focused the client internally, identified a new conflict area to be explored, and left the client an active participant in directing the course of therapy.

As the client begins to work with the new issues that have developed out of the therapeutic relationship, important inroads will be made on the client's presenting problems as well. In most cases, significant progress cannot be made on presenting problems until movement is made on other related conflicts that are not evident in the initial sessions. As we have seen, adopting an internal focus will identify salient issues that clients did not realize were relevant to their presenting problems. We will return to this issue of re-framing clients' conceptualizations of their problems in Chapter 7.

Enlisting Clients in Resolving Their Own Conflicts

In this section, we bring the internal focus one step further into the therapeutic relationship. Once clients have begun to look within and share their inner world, the therapist can further clients' sense of efficacy by engaging them in understanding and resolving their own conflicts. *A common misconception about psychotherapy is that the therapist assumes responsibility for figuring out what is wrong and what clients should do.* Beginning therapists can have unrealistic expectations that they must be very wise and possess insightful solutions to clients' problems; they then suffer under these performance demands. Here again, this misconception places the impetus for change in the therapist's lap and gives clients the role of passive recipients, waiting to be enlightened. A more effective approach is to extend the collaborative alliance into the problem-solving phase of therapy as well.

Therapists will be more effective if they elicit clients' active involvement in resolving their own conflicts instead of merely setting the stage for the therapist to announce what it all means, give advice and explanations, and prescribe solutions. Clients' participation in a working alliance frees the therapist to respond more fully to what clients produce and gives clients the opportunity to become more capable of managing their own lives. Rather than trying only to provide answers, therapists can develop the skills required to help clients explore their own problems and exercise their own strengths and personal resources. Real change has occurred when therapy has not only resolved clients' presenting problems but has fostered their sense of self-efficacy as well.

To illustrate this, let's look at two different ways of responding to a client's dream. Although the interpretation of the dream remains the same, the therapeutic approaches differ and enact very different interpersonal processes with the client.

Recapitulating Clients' Conflicts

Anna, a 22-year-old client, lives at home with her embittered and chronically embattled parents. For several years, Anna has been struggling with the developmental transition of emancipation from her family of origin. She cannot establish her own adult life. Anna has few friendships, dates little, and has no serious educational or career involvements. Anna's "martyr-ish" mother and alcoholic father bicker constantly, and Anna feels it is her responsibility to stay home and help her mother cope with her failed marriage. Anna entered therapy complaining of depression.

After several sessions, Anna recounted to her therapist a dream that was of great importance to her. As the dream began, Anna was riding a beautiful horse across an open savannah. She felt as one with this graceful animal as they glided effortlessly across the broad grasslands. They sped toward a distant mountain, past sunlit rivers, birds in flight, and herds of grazing elk. Anna felt strong and free as she urged the tireless animal onward.

The distant mountains held the promise of new life amid green meadows and tall trees. Anna felt their promise quicken inside her as she urged the horse onward. But as the mountains drew near, Anna and the horse began to slow. The horse's legs became her own and grew heavier with each step. Anna desperately tried to will them on, but their footing became unsure and they began to stumble. At that moment, she was surrounded on all sides by menacing riders, ready to overtake and capture her. As she awoke, Anna choked back a scream.

"What do you think my dream means?" Anna asked.

Her bright, concerned young therapist offered a lengthy and insightful explanation. The interpretation focused on Anna's guilt over leaving home. The therapist suggested that, if Anna went on with her life and pursued her own interests, she would feel powerful and alive—just as she had in the dream. But before Anna could reach her goal and experience the satisfactions of having her own adult life, she would have to free herself from the binding ties of responsibility that she felt for her mother and her parents' marriage. Her loyalty to her mother threatened to entrap her in guilt and prevent her from having her own independent life.

Anna: (*enthusiastically*)

Yes, I do feel as if I am bad and going to get in trouble whenever I leave my mother and do what I want. Is that why I dreamed that?

Therapist:

We've been talking about whether you are going to move into an apartment next month. I think the dream reflects your conflict over taking this big step on your own.

Anna:

I want to move out, but I can't leave my mother with my father. She says she will divorce him if I move out, and it'll be my fault. What should I do?

Therapist:

I can't make that decision for you. You have to be responsible for your own decisions.

Anna:

But I don't know what's best to do, and you always do. You're so much smarter than me. I thought about that dream a lot, and I didn't know that's what it meant.

Throughout the rest of the session, Anna continued to plead for advice and expressed her discouragement about being able to resolve her own problems.

The therapist kept refusing to tell her what to do and gave a lengthy explanation about autonomy, independence, and the need for Anna to find her own solutions. At the end of the hour, Anna felt agitated and depressed, and the therapist was still explaining the need for Anna to make her own decisions.

Although the therapist was astute in linking Anna's dream to her separation guilt and the current manifestation of her conflict over leaving home, this was an unproductive session. How did their therapeutic process go awry?

In this session, Anna experienced a relational pattern with her therapist similar to the pattern she was struggling with at home. In her family, Anna was trained to be dependent and to believe that she didn't have the wherewithal to succeed on her own. Further, she was made to feel guilty whenever she did act independently or on her own behalf. In the first 15 minutes of their session, the therapist acted as the knowing parent who gave all the necessary answers to the needing child. This interaction behaviorally communicated the message that Anna will be dependent on the therapist's superior understanding. On another level, however, this message was contradicted by the therapist's verbal message about independence. This mixed message about independence from the well-intentioned therapist immobilized the client. The therapist's direct interpretation could have worked well for many clients, but (recalling the concept of client response specificity) such an approach was problematic for Anna.

Anna reacted so intensely because she needed permission from the therapist to become more independent yet simultaneously expected that the therapist needed her to remain dependent—just as her mother did. When the relational pattern of having to remain dependent on the authority was reenacted with her therapist, Anna's pathogenic belief that she was incapable of making her own decisions and having her own life was confirmed. Anna remained depressed until the therapist successfully reestablished a collaborative alliance. With the help of his supervisor, the therapist did this in the next session by soliciting Anna's ideas about a problem they were discussing. The therapist expressed his genuine pleasure in watching her be so insightful and, following this, shared some related ideas of his own. This time, Anna readily picked up the ideas and used them to further her own thinking.

Providing a Corrective Emotional Experience

We now examine a different type of response to Anna's question, "What do you think my dream means?" Responses that engage clients in exploring issues with the therapist will usually be more effective than any explanations provided by the therapist, no matter how accurate. Therapists can choose to respond in many different ways and can engage clients in exploring their own dreams (and any other material they present) as follows:

> It sounds like a very important dream to me, too. Let's work on it together. Where should we begin?

What was the primary feeling you were left with from the dream? Can you
connect that feeling to anything going on in your life right now?

What was the most important image in the dream for you? What does that
image suggest to you?

Let's exchange ideas. I'll give you a possibility I'm thinking about, and you
share one with me. It'll be interesting to see what we can come up with
together. Who should go first?

Each of these varying responses encourages the client to participate
actively with the therapist. Ultimately, the therapist may want to give the
same interpretation of the dream. Even though the therapist provides an
explanation, however, the process will be significantly different and far more
enlivening for clients if it evolves out of their joint efforts and is integrated
into their continuing collaboration. This gives clients a relationship in which
they are not one down or dependent on the therapist but are encouraged to
exercise their own abilities. When this occurs, clients feel more ownership of
the therapeutic process, their motivation to work in therapy increases, and
their sense of personal power is enhanced. Facilitating such self-efficacy provides a corrective emotional experience for clients like Anna, who get the
permission they need to grow out of their dependence and act capably in
their relationship with the therapist—something they may not have experienced in the past.

Tracking Clients' Anxiety

Tracking clients' anxiety is a Sullivanian concept that will help therapists determine when and where to focus clients inward. Although the subjective experience of anxiety is uncomfortable, and may be painful, anxiety is still the therapists' ally. It serves as a signal or signpost to help therapists identify clients'
central or generic conflicts. Therapists do not fully understand clients' problems—or what needs to occur in therapy to resolve them—until they understand what makes each client anxious. Why do certain situations or interactions
make a particular client anxious, and how has the client learned to cope with
recurrent anxieties? Therapists need to formulate working hypotheses to these
questions in order to conceptualize client's dynamics and formulate treatment
plans. As we will see, focusing clients inward on their anxieties will lead to the
heart of clients' concerns.

Therapists can use a four-step sequence to track clients' anxiety and better
use the guidelines for conceptualizing clients provided in Appendix B. In this
sequence, the therapist (1) identifies manifest and covert signs of client anxiety,
(2) approaches signs of client anxiety, (3) notes the topic presently under discussion and considers the interpersonal process currently transpiring that may
have precipitated the client's anxiety, and (4) focuses the client inward to explore the meaning of the client's anxiety.

Identifying Signs of Clients' Anxiety

Clients will become anxious in the session when the issues they are discussing touch on their generic (central) conflicts and activate their relational templates, pathogenic beliefs, or core conflicted affects. Clients develop certain interpersonal coping strategies that they employ over and over again, in different situations, to ward off the anxiety evoked by developmental conflicts and maladaptive relational patterns. If the therapist can identify the interpersonal scenarios, subjective needs, or situational demands that make the client anxious, the therapist is better prepared to recognize the unifying themes or repetitive patterns that have triggered the client's anxiety. Thus, throughout treatment, therapists are continually asking themselves: "What makes this client anxious? When does the client feel threatened or insecure? Where does safety lie, and where does danger lie?"

In order to conceptualize the generic conflicts or identify the repetitive patterns that are troubling clients, the therapist must actively attend to the anxiety (or anxiety equivalents) that they express. For example, the therapist must be alert for signs indicating that something clients are experiencing is making them anxious. There are a thousand different ways for clients to express their anxiety—nervous laughter, nail biting, hand gesticulation, agitated movement, stuttering, hair pulling, speech blockage, and so on. These anxiety equivalents will be expressed in endlessly varied ways across clients, and therapists will have to learn how each particular client tends to express anxiety. (This will be especially important when working with clients from cultures in which certain behaviors—such as gaze aversion—may be normative rather than a signal of anxiety.) When they become anxious, most clients utilize the same interpersonal coping strategies in a repetitive or characteristic way. These strategies may include pleasing others, becoming critical or controlling, seeking help or acting confused, withdrawing from others, and so forth. We will examine these interpersonal coping strategies closely in Chapter 7. For present purposes, it is enough to note that the first step in tracking the client's anxiety is to observe and mentally note when the client becomes anxious.

Approach Clients' Anxiety Directly

In Chapter 3, we suggested that therapists draw out the feelings and concerns that emerge for clients as they enter treatment. Similarly, therapists are encouraged to approach clients' anxiety about an issue being discussed or about what may be occurring with the therapist at that moment. Therapists can focus clients on their immediate experience in an open-ended way (for example, by asking, "What are you feeling right now?") or by labeling the anxiety more directly (for example, by saying, "Something seems to be making you uncomfortable right now. Any idea about what that may be?").

By focusing clients inward on their anxiety as they are experiencing it, and inviting them to explore it further, the therapist is leading clients closer to the source of their problems. As a result, some clients may be ambivalent about the therapist's request to look within and explore their anxiety further. On the one hand, clients may welcome the opportunity to share their anxiety with the therapist—and the confusion and pain that often follows close behind. On the other hand, clients may also recognize that approaching their anxiety will bring them closer to pathogenic beliefs about themselves and unwanted expectations of others that often evoke painful feelings.

Thus, while at times a client may experience the therapist's focus on his anxiety as validating and reassuring, at other times it may intensify anxiety and even arouse resistance. As we discussed in Chapter 3, therapists need to respond to the client's resistance when it emerges, by exploring why it is threatening or does not feel safe to approach this anxiety-arousing topic. This approach will be far more effective than ignoring the client's reluctance or trying merely to push through it by persuading the client to talk about the conflict. Before the therapist and client can address the conflict itself, they must explore why it is threatening for the client to share it with the therapist. For example, perhaps the client fears that the therapist will not like or respect the client anymore. As we have already noted, the therapist must honor the client's resistance. The therapist does this by (1) trying to understand the original, aversive experiences that led the client to behave in this particular way; (2) helping the client appreciate how this "resistant" or "defensive" response was once a necessary and adaptive coping strategy; and (3) providing the client with a different and more satisfying response than the client has come to expect, as in the following example.

Therapist:
No, I don't find you "needy" or "demanding" for asking me if I would call you before you make this big presentation on Tuesday. It would be easy for me to take a few minutes to do that, and I'm honored that you are willing to risk sharing your need with me.

Observe What Precipitates Clients' Anxiety

As we have seen, the therapist first observes when the client is anxious, and then helps the client focus inward more directly on the experience. At the same time, the therapist has a third task: to try to recognize (1) what issue was just being discussed (content) or (2) what type of interaction between the therapist and the client may have precipitated the client's anxiety (process). If the therapist can identify what it was that made the client anxious, the therapist will be better able to identify the pathogenic belief, maladaptive relational pattern, or core conflicted affect that may have generated the anxiety. Therapists must consider this precipitant carefully in order to formulate effective case conceptualizations and treatment plans.

The client will be able to discuss many different issues comfortably with the therapist. When the client becomes anxious, however, the therapist must try to identify the issue or theme that has just precipitated the anxiety. What was the client just talking about? Was it sex, death, intimacy, money, vulnerability, success, divorce, inadequacy? *The answer will highlight the issues that are linked to the client's core conflicts and need to be explored further.* As we know, therapists can help the client explore this more effectively if they have already formulated working hypotheses about the possible issues that may be generating the anxiety. Keeping process notes, as suggested in Appendix A, will help therapists formulate their working hypotheses more effectively.

More importantly, the client's anxiety is often a signal that the interpersonal process between the therapist and client is currently reenacting a developmental problem or maladaptive relational pattern for the client. For example, the therapist might observe client anxiety in the following circumstances:

1. As the therapist was expressing confusion about what was going on in the session at that point, the client became anxious, perhaps because her alcoholic father would start demeaning her whenever he felt uncertain or inadequate.
2. The client had just made an important insight and the therapist acknowledged his achievement. The client then became anxious, perhaps because he believed that he would always have to be so insightful and perform so well, as a parent had always expected of him.
3. The therapist had just been supportive, and perhaps the client felt anxious because strings had always been attached to what she had been given in the past. Following her relational templates, the client may have become concerned that the therapist would make her pay for his support in some unwanted way, just as a caregiver always used to do.

The therapist should generate working hypotheses such as these about how the current therapist-client interaction may have activated a developmental problem or unwittingly reenacted a maladaptive relational pattern for the client. These hypotheses can then be used to help clients explore the issues that evoked their anxiety and better understand the more primary concerns that generated it.

Focus Clients Inward to Explore Their Anxiety

As we have seen, the therapist is alert for signs of client anxiety and approaches it whenever it occurs. Simultaneously, the therapist tries to identify what precipitated the client's anxiety—especially what has just transpired between them—and begins to generate working hypotheses about the relational patterns, pathogenic beliefs, or accompanying feelings that are the source of it. Finally, the therapist must focus the client inward, to explore more specifically what the threat or danger is for the client. If the therapist can help the client focus inward and clarify the thoughts and feelings that are associated with the anxiety, the client's vague discomfort will usually become more specific. Consequently, the client's basic concern will be revealed more clearly, making it easier for the therapist to provide a better response to it than the client has come to expect. The following dialogue illustrates this process.

Therapist:

It seems like something's going on for you right now. I wonder what's happening?

Client:

You know, this doesn't sound very nice to say, but I don't think my mother really wants me to change very much. I don't think she's a mean person or anything, but I do think I'm getting better in therapy, and she's not completely happy about that.

Therapist:

How so?

Client:

Well, I haven't been depressed for a while now, and I've actually been feeling pretty good the last month or so. Maybe it's just coincidence, but it seems that, as I've gotten better, my mother has withdrawn and been harder to talk to. And I think this has gone on between us before.

Therapist:

You know, I think you're right. You have been doing very good work in here and getting better, and that may be kind of hard for your mother sometimes.

Client:

(*fidgets, begins picking at her nail, and then starts talking about another topic*)

Therapist:

Maybe something just made you feel a bit uncomfortable?

Client:

I don't know. How much time is left?

Therapist:

I'm wondering if something we're talking about, or something that might be going on between us, could be making you uncomfortable?

Client:

(*pause*) Well, this is silly, but maybe I'm afraid that you're going to go away or something?

Therapist:

That sure makes sense to me. We were just talking about how your mother seems to act hurt and withdraws when you are feeling stronger, and I just told you that I thought you were doing very good work in here.

Client:

Well, I guess so, but that doesn't really make me feel very good either, because you're not going to stay with me either if I get better.

Therapist:

What do you mean?

Client:

If I get better, then we stop. Isn't that the point of counseling anyway? Then I won't have my mother, or you, or anybody.

Therapist:

I'm glad you can risk telling me this so clearly. I didn't understand at first, but you're right, the issue we are struggling with right now is like what's happened before. To feel better and act stronger has meant that you have had to be alone. In the past, in order to be close to others, you have had to be sad and depressed and needy. I think you've been struggling with this dilemma most of your life and, right now, it feels like it has to be the same old story again for you and me.

Client:

Yeah, it kind of does.

Therapist:

But I wonder if we can do it differently for once. Can we work together and try to find a way to make it come out another way this time?

Client:

It's not working very well so far.

Therapist:

Well, one difference I'm thinking about is that we're talking about it—and that hasn't happened before.

Client:

What difference does that make?

Therapist:

Maybe it's different because I can see what's happening for you and feel for you in this predicament. It's been very discouraging for you to be undermined in this way so many times.

Client:

I suppose that's true, but these don't really seem like huge differences to me.

Therapist:

OK. I see these differences as more significant than you do right now, but let's keep talking about this.

In this way, tracking the client's anxiety can help the therapist and the client focus inward and clarify the client's most basic concerns. In particular, *it will often reveal how the client's maladaptive relational patterns have been activated in the therapeutic relationship.* By clarifying how aspects of the client's conflicts with others seem to be recurring between the therapist and client, the therapist can begin to change or differentiate the therapeutic relationship from the problematic patterns that have occurred with others. We will return to other aspects of these basic orienting constructs in the chapters to follow.

Closing

The overarching theme in the interpersonal process approach is that clients will not be able to resolve their problems and change unless they enact a new and more satisfying solution to their conflict in their real-life relationship with the therapist. Helping clients to adopt an internal focus for change is one important way in which therapists provide clients with a different type of relationship than they have experienced and come to expect with others. By codetermining what issues will be covered and by participating in finding solutions to their problems, clients are far more likely to uncover the feelings and beliefs that constitute the source of their problems, and gain an increasing sense of self-efficacy in the process. Perhaps Bowlby (1988) has captured this interpersonal process most succinctly by saying that the basic therapeutic stance toward the client is not, "I know; I'll tell you" but, "You know; you tell me."

As clients begin to reflect on their own experience and become less preoccupied with others, not just their anxiety but their entire affective world opens up and becomes more accessible. This *affective unfolding* is a pivotal step in the process of change. However, the strong feelings aroused by pursuing an internal focus can be distressing to clients and intimidating to therapists as well. The next chapter examines how therapists can help clients come to terms with the deep-seated emotions aroused by looking within.

Suggestions for Further Reading

1. As we are seeing, beginning therapists need to develop certain interpersonal skills. To help develop these basic skills, a self-assessment measure is provided in Chapter 4 of the Student Workbook. Completing this assessment will help readers systematically evaluate their current level of skill in working along several important therapeutic dimensions.

2. Albert Bandura argues convincingly that perceived self-efficacy is the underlying mechanism of change across all treatment approaches; see his classic article "Self Efficacy: Toward a Unifying Theory of Behavioral Change," *Psychological Review*, 84, no. 2 (1977): 191–215 or his book *Self-Efficacy: The Exercise of Control* (New York: Freeman, 1997). He emphasizes that little enduring and generalized change results from verbal suggestion or persuasion. Instead, clients change when they have authentic experiences of enactive mastery—that is, when they have the experience of performing adaptive new coping responses with the therapist.

3. Alan Wheelis discusses the internal process of change in his book *How People Change* (New York: Harper & Row, 1974). In particular, see the chapter "Freedom and Necessity," which uses the language of responsibility to clarify how clients must focus on themselves rather than others in order to change. Beyond the issue of responsibility, beginning therapists can learn much from other existential psychotherapists; see, for example, I. Yalom's *Existential Psychotherapy* (New York: Basic Books, 1981).

4. Therapists working with clients from cultures other than their own must be especially aware of normative verbal and nonverbal behavior. For a discussion of how to assess anxiety in Asian clients by focusing on nonverbal behavior, see S. Tsuneyoshi's case study "Rape Trauma Syndrome" in *Child and Adolescent Therapy: A Multicultural-Relational Approach*, edited by F. McClure and E. Teyber (Fort Worth: Harcourt Brace, 1996). Additionally, readers may wish to examine Chapters 1 and 2 of *Counseling the Culturally Different* (3rd ed.) by D. Sue and S. Sue (New York: Wiley, 1998), which address language differences among cultural groups.

CHAPTER FIVE

Responding to Conflicted Emotions

Conceptual Overview

Painful feelings lie at the heart of enduring problems. Change in therapy is a process of affective relearning, and this change occurs when the therapist responds effectively to clients' conflicted (difficult, unwanted) emotions. When the therapist focuses clients on themselves and they look within, clients feel more intensely, and express more directly, the conflicted emotions that accompany their problems.

This affective unfolding is a pivotal point in therapy because it reveals the emotional basis of clients' problems. On the one hand, clients will welcome the promise of having therapists make contact with their problems on a deeper, more personal level. On the other hand, clients also want to avoid feelings that are too painful, shameful, or just seem hopelessly unresolvable. Thus, the therapist's response to clients' emerging feelings will be an important influence on treatment outcome.

Therapists facilitate change by providing a more helpful response to the client's feelings than he has come to expect from others. Further, when therapists avoid or do not respond to the client's emerging feelings, the therapeutic relationship loses its vitality and treatment is reduced to an intellectual pursuit. Thus, the purpose of this chapter is to help therapists respond effectively to the feelings aroused when clients begin to look within.

Chapter Organization

This chapter is divided into five sections. The first section provides basic intervention guidelines for responding to clients' feelings. Therapists need to approach clients' feelings and help expand the affective component of their problems. The second section continues this theme of drawing out the client's conflicted affect and focuses on the predominant affect that clients tend to reexperience in many different situations.

116

In the third section, therapists will be introduced to the interrelated sequence of emotions that often comprise clients' conflicts. For many clients, three related feelings recur in an ordered sequence; therapists can identify and respond to each feeling in the sequence. Therapists can help clients resolve conflicted emotions by addressing each of the distinct feelings that make up this affective constellation. Once therapists are successful in evoking the client's conflicted emotions, they need guidelines to help them respond effectively. Thus, in the fourth section, we introduce how the therapist can provide an interpersonal holding environment that helps contain conflicted emotions and allows clients to resolve them.

In the fifth section, we examine personal characteristics of the therapist that influence the therapists' ability to respond to clients' emotions. This is one of the most important issues in clinical training because therapists' own personalities, needs, and current life stressors can keep them from responding to the wide ranging feelings and sometimes challenging emotional reactions that clients present. All therapists need to seek consultation at times in order to manage their own reactions to the evocative material that clients often present. Otherwise, they will be less capable of responding effectively.

Finally, we must begin this chapter with a note of encouragement and an eye toward realistic performance demands. The material presented in this chapter is more personally evocative and challenging for many beginning therapists than the material presented in other chapters. The issues discussed here often evoke significant personal concerns for therapists-in-training, and you may not be able to apply these principles as readily with clients. Understanding these processes intellectually is a realistic goal for first-year therapists; being able to utilize them effectively in ambiguous therapeutic interactions is not. Be patient: In a year or two many of these responses will be your own—it comes with experience and practice.

Responding to Clients' Conflicted Emotions

Many of the problems that clients present can be resolved by trying out new coping strategies or behavioral alternatives, challenging faulty beliefs, or recasting problems in a new framework. These brief interventions are sufficient to help with many of the problems that clients present, and they constitute an important aspect of change for most problems. For the more enduring and pervasive problems that clients often present, however, the conflicted emotions that accompany problems need to be addressed as well.

At times, clients may be reluctant to explore the conflicted emotions that emerge in treatment. On the one hand, clients hope that the therapist will be able to help with their emotional reactions; on the other, they may want to avoid the difficult feelings that they have not been able to resolve alone. There are many different reasons why clients may be reluctant to share certain feelings in the beginning. It may seem pointless to reexperience the same painful feelings again when they have scant hope of receiving help, to suffer shame by revealing feelings of inadequacy or need, to risk unwanted but expected judgment or scorn from the therapist for having feelings that seem unacceptable, or to struggle with their own fear of losing control. As a result of these and other concerns, clients sometimes resist the therapist's attempts to touch the emotional core of their experience, although at the same time they often long for this contact.

Once the therapist recognizes the client's relational templates and understands how significant others have characteristically responded to the client's vulnerability in the past, the fear or shame that motivates her resistance will make sense. As outlined in Appendix B, therapists can generate working hypotheses and try to conceptualize the clients' potential resistance to conflicted feelings and attempt to work with those feelings rather than avoiding them. If not, therapy will be reduced to an intellectual exercise and will not effect enduring change. The therapist can appreciate the client's reluctance to approaching certain feelings, but also keep in mind the client's simultaneous wish to be responded to in this deeply personal way.

Although difficult emotions are central to most clients' problems, the therapist's affirming response to these emotions also provides the avenue to resolution and change. The corrective emotional experience that the therapist provides to clients—the experience of sharing profound feelings with a caring other who remains connected and validating—loosens the hold of pathogenic beliefs and faulty expectations. Rigid templates expand and become more flexible, and clients are empowered to make the changes they desire. At this critical juncture when the client experiences a reparative response to core conflicted feelings, the treatment process is propelled forward. As we will examine later, clients become more receptive to a wide range of interventions from varying theoretical orientations at this point.

In the next section, we examine intervention guidelines to help therapists approach and draw out their clients' conflicted emotions and begin to set this process in motion.

Approaching Clients' Affect

In their moment-to-moment interaction with clients, therapists repeatedly choose how to respond to what the client has just said or done. For example, the therapist might respond by seeking more information, clarifying what the client has just said, relating this material to other issues the client has discussed previously, or linking it to what is occurring between the therapist and client at that moment. As a guiding principle, however, *the most productive response is one that directly addresses the feeling that the client is currently experiencing.* In other words, the therapist's first priority is to acknowledge and approach the affective component of the client's response.

Imagine that the therapist and client have just sat down to begin their first session:

Therapist:
>   Tell me, Mike, what is the difficulty that brings you to therapy?

Client:
>   I'm having a lot of problems with my 15-year-old son. We disagree about everything and can't seem to talk to each other anymore. He doesn't do what I ask him to, and I don't like his values. I guess I'm pretty angry at him. His mother and I are divorced, and I'm thinking it may be time for him to go live with her. Do you think it's OK for a teenaged boy to live with his mother?

Therapist:
>   I don't think either of us understands what's going on between you two well enough to make any decisions yet. You said you were pretty angry at him. Tell me more about your anger.

The client has presented many different issues that the therapist could have chosen to pursue. Following the general guideline, however, the therapist approached the primary affect that the client presented. The therapist could have responded differently and inquired further about the issues that the father and son disagreed about, worked on values clarification with the father, obtained more background information about the father-son relationship or the divorce, provided research findings about the effects of mother-custody versus father-custody on boys, and so on. Although these and many other responses will often work well, *responding to the client's feelings will usually produce the most information and intensify the therapist-client exchange.*

In this dialogue, the therapist responded immediately when the client spoke about his anger. Additionally, therapists need to respond to the *covert* or unverbalized feelings that clients experience. For instance, the therapist in our example might have chosen to respond instead to the more covert affect by acknowledging, "It sounds like you're feeling really discouraged about being a dad right now—maybe so discouraged that you just feel like giving up sometimes."

Important feelings are often aroused when the client begins discussing her problems. Such nonverbal affective signs as tearing, grimacing, or blushing are signals to the therapist that the client is addressing a significant issue. Responding to these nonverbal cues, the therapist draws out the affect that this situation holds for the client, as in the following example. Imagine that the client has been discussing her marital problems:

Client:
>   I don't know if I should stay married or not. I haven't been happy with him for a long time, but I can see how hard he's trying to make our relationship work. And our 4-year-old son would be devastated if we broke up. I don't know what to do, and it's so important that I make the right decision.

Therapist:
>   As you speak about this, your face tightens. What are you feeling right now?

When the therapist responds to the client's feelings in this open-ended way, it often serves to clarify the client's central concern. For example, clients may respond by expressing more specific concerns:

I'm so sad about hurting the people I love.

I'm afraid that everyone will think I'm selfish for leaving and they'll blame me for the divorce. They'll think I haven't been a good mother.

I don't want to be alone. I don't want to be married to him anymore, but I'm afraid I can't leave and make it on my own.

I'm furious that I'm the one who has to make this decision. I'm responsible for every decision we make, and I always pay for it. It's not fair.

Throughout each session, clients have emotional reactions to the issues that concern them most. Sometimes the client's affect will be presented forthrightly; at other times, it will be subtle and elusive. Cultural, class, and gender factors will further shape how each client displays affect. For example, some clients may overtly state their reactions, whereas those from other cultural or familial backgrounds may simply divert their eyes or become noticeably quieter when certain feelings are evoked. However, if the therapist can find collaborative ways to explore the client's affect, and invite it more fully as in the preceding examples, it will evoke further disclosure from the client and clarify the client's concerns. This, in turn, provides a clearer treatment focus.

Expanding and Elaborating Clients' Affect

We must look further at how the therapist can best respond to clients' feelings. At times, it can be helpful to anticipate or label a client's emerging feeling (for example, by saying, "You must have been furious" or "You're frightened right now"). As a general guideline, however, a more effective response is to give clients an open-ended invitation to explore the feeling further. For example, the therapist may encourage clients to explore what they feel is central in their affective experience. A simple response—"What are you feeling right now?" or "Tell me more about that feeling"—is remarkably effective in eliciting the client's affect. An open-ended invitation is more effective than trying to label the client's feeling in an either-or fashion—for example, by saying, "Were you feeling _____ or _____ in that situation?" Such a choice may restrict the client rather than invite a free range of responses. The client often experiences something that the therapist has not anticipated; moreover, the client may have several feelings about the situation. The open-ended query is more effective in drawing out exactly what the client is feeling, and it facilitates the collaborative alliance. We will discuss how to deal with those clients who lack a vocabulary for feelings.

Also, this type of open-ended response is usually more effective than asking clients why they are experiencing a particular feeling. Although *why* questions may be effective at times, as a rule of thumb, therapists are advised to minimize them. Clients often don't really understand why they feel, think, or do something. Asking them why may only make them feel inadequate or resentful. The therapist will usually elicit more information and collaboration from clients by offering an open-ended invitation—"Help me understand your anger better" or "Tell me about your anger"—than by asking, "Why were you angry about that?" By inviting the client to express more fully whatever feeling he is having, therapists are giving the message that they are interested in the client's own subjective experience and they are comfortable sharing whatever the client is experiencing. *Most clients have not had permission to share their emotions so genuinely in the past,* which is why this is one of the most important responses therapists can offer. As they discover that they can be authentic with the therapist and still remain connected, clients become more willing to take risks, try out new behaviors, and invest further in the therapeutic relationship.

Clarifying clients' subjective feelings   Giving clients an open-ended invitation to explore their feelings will also help them to clarify what each particular feeling means to them. Too often, therapists assume that a particular affective word, such as *angry* or *sad,* means the same thing to the client as it does to them. Therapists should not assume that they understand a particular affective word without clarifying what it means to this particular client. Although this may be especially apt when there are cultural differences between therapist and client, it applies with every client. In a process of mutual exploration, therapists actively encourage clients to clarify the personal meaning that a feeling holds for them. This is especially important if clients repeatedly use the same affective word to describe themselves, such as boring, overwhelming, and so forth. Having the client elaborate his experience in this way will also safeguard therapists against becoming overidentified with the client and misperceiving the client's situation as being the same as their own, as often occurs.

Because they are often uncomfortable with many of their emotional reactions, clients themselves may be unfamiliar with their own feelings. As a result, *many clients need to clarify the broad, undifferentiated feeling states that they often experience.* Therapists can help clients learn more about the emotional reactions that are linked to their problems by asking questions such as the following:

> Can you bring that feeling to life for me and help me understand what it is like for you when you are feeling that?
> Do you have an image that captures that feeling or goes along with it?
> Is there a particular place in your body where you experience that feeling?
> Is this a familiar or old feeling? When is the first time you can remember having it? Where were you? Who were you with? How did the other person respond to you?
> How old do you feel when you experience that emotion? Can you attach an age to it, such as 7 years old or 13 years old?

Some of these exploratory questions will work well with a particular client and not at all with another. Following client response specificity, therapists will need to find what works best for each client. In general, however, these responses are all ways of approaching clients' emotions and clarifying their subjective meaning. These types of responses also tell clients that the therapist is concerned about their feelings—who they are and what is most important to them. Further, they behaviorally inform clients that the therapist is different from many other people they have known because the therapist is comfortable sharing whatever emotions they have and wants to know their experience.

When the therapist works with clients to clarify their emotions, it creates an opportunity for the therapist to see and respond to the clients' most personal experience. An important aspect of most clients' problems is that significant others have repeatedly invalidated, ignored, or disdained certain feelings that the client had (for example, being sad, mad, or afraid). As old expectations of judgment, invalidation, and so forth are disconfirmed by the counselor's affirmation, the client receives an important experience of change.

Experiencing feelings versus talking about feelings   Little change will occur until clients are able to stop talking about their emotions in an intellectualized manner and actually feel their feelings in the presence of the therapist. The exploratory responses that we have been discussing will intensify clients' experience and they will feel more fully the subjective meaning of their own words.

Therapists may also draw out the client's affect by addressing any incongruence that they perceive between the client's narrative and the accompanying affect. In telling a story that the therapist finds poignant or disturbing, the client may display no feeling at all or an incongruent affect, such as laughter. Therapists are subtly but powerfully pulled to match the client's incongruent affect, and often feel they are violating unspoken social rules if they do not do so. However, clients are usually relieved when the therapist risks violating the social rules, makes a process comment, and acknowledges this discrepancy—and, in so doing, implicitly invites the client's true feelings or congruent experience.

Therapist:
I'm a little confused. What you are saying seems very sad, yet you are smiling as you tell this to me.

It is important to help clients register or connect with the feelings that accompany their actual experience. Many problems are based on the client's inability to feel what they feel: to be sad about what was lost; afraid of what was frightening; or angry about what was hurtful. *Simply allowing clients to be able to have the feelings that are commensurate with what happened to them is a lifelong gift to many.* Recalling the concept of invalidation, it allows them to claim themselves and their own authentic voice.

The greatest opportunity to help clients change occurs when they are experiencing the full emotional impact of their problems. If therapists can communicate their compassion to clients as they are experiencing difficult emotions, it will help them resolve the affective component of their problems and move forward toward behavioral change. As the following case example illustrates, however, clients cannot make such progress when they are psychologically alone with their feelings or can't let their true feelings emerge.

During one session, Jean had been telling her therapist about her hopelessness. She described her life as a merry-go-round of endless ups and downs that always returned to the same discouraging spot. This past week she had dropped out of school, as she had done "so many times before," and had given in to pressure from her selfish and undermining boyfriend to let him move back into her apartment again. Jean described her hopelessness about ever being able to do or have anything for herself. At times, she acknowledged, the only way out of her misery seemed to be suicide.

The therapist could feel Jean's despair and was deeply moved by it. The therapist tried to work with Jean's feelings by acknowledging her pain and validating her experience.

Therapist:
> I can see your hopelessness right now, and how painful it is to feel so defeated. You haven't been able to follow through and do what you wanted to do for yourself again, and that's so discouraging. As you've done so many times before, you felt that you had to meet someone else's needs rather than say "no" and do what you wanted for yourself.

As in all of their previous sessions, however, Jean was not able to let the therapist be emotionally connected to her while she was experiencing the sadness. Because this critical connection to the therapist was missing, Jean still felt helplessly alone with her feelings and hopelessly alone in her life.

The setbacks of the previous week had intensified Jean's distress. The therapist recognized that this crisis provided an opportunity to make contact with Jean's feelings for the first time:

Therapist:
> It just seems hopeless.

Jean:
> *(long pause, nods)*

Therapist:
> You don't want to try anymore. You don't want to hurt anymore. You just want it to stop hurting, and sometimes suicide seems like the only way out.

Jean:
> *(nods)*

Therapist:
> You are so sad. Maybe you've been sad inside most of your life?

Jean:
> *(looks at therapist and nods)*

Therapist:
> I can see how painful this is for you right now.

Jean:
> *(looks away, puts hands over eyes)*

Therapist:
> I'm feeling with you in your sadness right now. I can't take it away, but you don't have to be alone with it any more.

Jean:
> *(looks at therapist sorrowfully, and slowly begins to cry)*

Therapist:
> Yes, I see your sadness there and how much this has hurt you. I'm so honored that you are willing to share all of this with me.

Jean:
> *(begins sobbing)*

For the first time, Jean was able to let the therapist be available to her while she experienced her despondency. The therapist acknowledged the risk that Jean had taken, and they discussed how it felt for her to share her sadness so openly. Jean said that she felt the therapist cared about her and that she felt "lighter." At their next session, Jean commented that her depression had been better that week.

Identifying and Punctuating the Predominant Affect

Clients often feel confused by their emotions; their emotions seem to occur for no reason and often are stronger than, or different from, what the situation calls for. Clients' inability to understand their own emotional reactions undermines their self-confidence and engenders self-doubt. Helping clients make sense of their emotional reactions is an important pathway to their greater self-efficacy. One way that therapists can do this is by working with the client's *predominant affect*. As we will see, clients' experience often revolves around a central affect, such as being shame-prone or guilt-prone, and therapists need to identify and punctuate this predominant feeling.

Clients usually enter therapy in response to one of two situations. First, a current life crisis echoes an emotional conflict that originally occurred years before; the current stressor taps into an *old wound*. Second, clients may enter therapy in response to feeling overwhelmed by *too many stressors* in a short period. These stressors have overwhelmed the clients' usual coping strategies and symptoms develop.

An Old Wound

Many clients enter therapy because of an old wound. For example, a client who has suffered a painful loss through divorce or death will find it more difficult to cope with the stress of that loss if it also evokes painful feelings from previous losses in the client's life. The current loss may be associated with a predominant feeling—perhaps regret, loneliness, or sadness. *That feeling becomes understandable, and far more manageable, when it can be linked to similar feelings that have been aroused by other losses in the client's life.* Thus, the therapist needs to identify the central affect that the current crisis has aroused and to link it to the original wound. Linking this primary affect to the way clients felt in the past enables them to make sense of their seemingly irrational feelings, to feel more accepting toward themselves, and to better appreciate the bigger problem they are actually dealing with, and thereby to gain greater self-efficany. Although this is essential in all treatment modalities, it is often one of the most significant interventions that therapists can make in crisis intervention or short-term therapy.

Multiple Stressors

Most individuals have adaptive coping mechanisms that allow them to manage a single stressful event, unless it taps into the type of preexisting vulnerability discussed above. If a second or third stressful event closely follows the first, however, it becomes much harder for individuals to cope. The client's usual coping strategies fail, and the cumulative stressors precipitate symptoms and maladaptive defensive maneuvers.

When subjected to multiple stressors, clients may describe themselves as "overwhelmed," "defeated," or even "broken." They tend to repeat a *compacted phrase* that encapsulates their emotional reaction to the stressful events:

I want to go away and just be alone.
It's too much for me; I can't stand it.
I don't care anymore; there's no point in trying.

As noted in Chapter 2, therapists must look for *recurrent affective themes* and try to unravel the emotional reactions that are encapsulated in these compacted sentences. Underlying each repetitive, compacted sentence are one or two central feelings that capture the basic impact these stressors have had on the client. Generally, the various stressors are arrayed around one or two integrating feelings, like spokes around the hub of a wheel. For example, in response to losing too many people or things, the client may have an overriding feeling of being alone or powerless, or the client may be afraid of any further losses, changes, or new commitments. When therapists are able to articulate and capture the underlying feeling that links the impact of several different crisis events, clients feel deeply reassured (for example, Therapist: "As I listen to what happened over the weekend, it seems that in each case—with the soccer coach, the telephone call from your boss, and the argument with your husband—you were not heard and became so upset because you were never being taken seriously."). When the therapist has been accurately empathic and grasped the integrating theme, clients feel seen and understood. In turn, they feel more in control of themselves and invest further in the working alliance.

To succeed in short-term therapy, in particular, therapists need to develop this skill in order to make an empathic connection and establish a strong working alliance early in treatment.

A Characterological Affect

As we have seen, it is important for therapists to respond to the client's moment-to-moment affect. Therapists can also listen for a recurrent feeling that pervades the client's life—a *characterological affect*.

As therapy proceeds, the therapist can often identify a predominant feeling that captures the client's current conflict and also characterizes her life. That is, if the therapist is accurately empathic and can see things from the client's point of view, the therapist can often find a central feeling that reflects the fabric of the client's life. Clients often have one or two core conflicted feelings that are continuous throughout their lives. For example, when the therapist can identify these core feelings that capture recurrent life experiences, clients often respond emphatically:

Yes, that's how it has always been for me.
That's what it's like to be me.
That's who I really am.

Clients often seem to experience these core feelings as their fate, because it feels as if they have always been there and seems like they always will be. One of the most significant interventions therapists can make is to accurately identify and reflect these characterological feelings to clients.

For example, after listening to the client for awhile, and grasping the integrating affective theme, the therapist may respond:

It seems as if you've always felt *burdened* by all the demands you feel you have to meet.

I'm getting a sense that you've always been *afraid* that people are going to find you out and point up your true inadequacies.

It seems as if you've always felt *resentful* that, no matter how much you do, it's never enough.

I'm getting a sense that you've always felt *wary* that others are trying to put you down or take advantage of you.

I'm wondering if you're *worried* that if people really knew the true you they couldn't possibly love you?

Therapists are prepared to respond to all of the varying emotions that clients present in the course of therapy. However, therapists will be most effective when they can identify the repetitive feelings that have recurred throughout a client's life and that the client considers to be central to her sense of self.

Clients' Affective Constellations

It is deeply rewarding but sometimes challenging to work with clients' conflicted emotions. Some clients want to avoid their emotions; others may talk at length about the same feelings without any change resulting. When clients are not progressing in treatment, one reason may be that therapists have only responded to the client's single presenting affect.

If the therapist acknowledges the client's current affect and invites the client to explore it further, *a sequence of interrelated feelings will often occur together in a predictable way.* This constellation of emotional reactions is central to the client's problems and each successive affect in the sequence needs to be addressed. Although different sequences occur for each client, the therapist can often identify a triad of interrelated feelings that cycle repeatedly when the client's conflicts are activated. Making a similar point, Lazarus (1973) speaks of tracking a client's "firing order." The process of change involves coming to terms with each of the feelings in this affective triad. In this section, we examine two affective constellations that commonly occur. We will see that therapists help clients change by responding to the entire sequence of feelings, rather than just responding to the first feeling in the sequence—the client's primary presenting affect.

Anger-Sadness-Shame

The first affective constellation we will examine consists of anger, sadness, and shame. The predominant feeling state for some clients is anger. These clients often experience and readily express angry feelings, including irritation, impatience, criticism, and cynicism. However, the client's anger is usually *reactive*; it is a secondary feeling that occurs in response to an original experience of sadness, hurt, or pain.

The therapist must acknowledge the client's anger, locate who and what the client is really angry at, and help the client find appropriate ways to express it (for example, telling someone what you don't like rather than yelling, demeaning, or losing your temper). If the therapist stops at this point, however, the client will still remain stuck in an angry feeling state and substantial change will not result. Instead, *the therapist can help the client identify and stay with the primary emotional response of sadness or hurt that led to the reactive feeling of anger.* This primary feeling is more threatening or unacceptable for the client than the anger that it prompts.

To reach the original affect, the therapist might wait for clients to spontaneously ventilate their anger and immediately afterward invite them to examine what they are feeling. Following the expression of the reactive feeling, there is often an open window to the primary feeling of hurt or sadness. For example, the therapist might say, "You've just expressed how angry this made you. Can you stop here for a moment and check in on what are you feeling right now?" Often, the original feeling (for example, vulnerability, embarrassment, helplessness) will often surface next. By joining the client in this feeling and working together to accurately name it, the therapist can often dislodge the client from the familiar (but safer and nonproductive) reactive feeling of anger.

If this approach does not work, the therapist can inquire directly about the original experience that led to the client's anger.

Therapist:
Something must have hurt you very much for you to remain so angry. Tell me how you felt when that first happened.

If the client again begins to express anger or indignation about what oc-curred, the therapist focuses the client back on the original experience:

Therapist:
   Yes, I can see how unfair that was and why it made you so mad, but how did it feel when . . . ?

If, in response to this query, the client's original feeling of sadness or hurt is expressed, and the therapist responds affirmingly, the client comes in contact with the internal aspect of the conflict. Recalling our internal focus for change, it is this internal aspect of the conflict that has kept the client from being able to change.

As we have seen, when this type of client experiences anger, sadness often follows. Although the anger is easily experienced, the client has learned that it is not safe to experience the original hurt or vulnerability. To avoid the original painful feeling, the client defensively returns to the reactive feeling of anger in a repetitive manner. This defensive coping pattern often leads others to regard this client as an angry person.

The therapist's treatment goal is to help such clients experience the sadness, hurt, or vulnerability that precipitates their repetitive anger. However, as soon as the original hurt is activated, the client will defend against this feeling and re-flexively return to the safer expression of anger. Thus, the therapist must ex-plore the client's resistance to the original painful affect. Often, clients avoid the original painful affect because it arouses a third, aversive feeling of anxiety, guilt, or—most frequently—shame.

For example, if the therapist inquires about clients' resistance to the origi-nal hurt, they will often say something like this:

   If I let the pain be there, it's admitting that she really did hurt me.
   If I acknowledge that it hurts, it means he won.
   If I feel sad, it shows that I really am weak.

Thus, if the therapist draws out the original pain that underlies the anger, a third feeling of shame will also be aroused. To defend against the original sad-ness, and the shame associated with being sad, the client will reflexively return to the stronger anger.

Although shame may be the most common reaction, other clients will ex-perience anxiety or guilt in response to feeling the original hurt. For example, some clients will say that, if they let themselves feel sad, hurt, or vulnerable, then "no one will be there," "others will go away," or "I will be alone." These clients feel painful separation anxieties upon experiencing their hurt or pain and avoid this anxiety by reflexively returning to their anger. Still other clients will say that, if they let their sadness be real, they would be admitting that they really do have a need, which would make them feel selfish or demanding. For these clients, guilt is the third element of their affective constellation.

This triad of feelings exists for many clients: Frequent anger defends against unexpressed sadness, which, in turn, is associated with shame, guilt, or anxiety. Enduring change results when clients can come to terms with each feeling in the triad. More specifically, clients resolve such conflicted emotions when they can (1) allow themselves to experience or feel each feeling; (2) share the feeling with the therapist so that they are no longer alone with it or hiding it; and (3) with the therapist's support, be able to stay with the difficult feeling for a moment or two (for example, tolerate, hold, or contain it) rather than having to move away from it, as they have in the past.

Clients integrate their conflicted emotions when they can progress through these three steps for each feeling in their affective constellation; they are internally resolving their conflict. With this resolution, clients no longer need the faulty coping strategies, symptoms, or defenses that they have employed in the past. Further, clients are then prepared to adopt new, more adaptive responses with others in their lives that they were previously unable to incorporate. Clearly, this is a pivotal experience in the process of change.

Before going on to examine another common affective constellation, we must pause to highlight the special role of shame. Although much has been written about anxiety, depression, and guilt, shame has long been ignored. Correspondingly, clients could be anxious, guilty, or depressed with their therapists, but not ashamed. Thankfully, shame is no longer the taboo topic it has been; indeed, it is now regarded by many theorists as the "master emotion" ("Shame," 1995). The cardinal role of shame in most symptoms and defenses has been illuminated by the groundbreaking work of Helen Block Lewis (1971) and others.

As we began to explore in Chapter 3, new realms of understanding will open up to counselors as they allow themselves to begin hearing the family of shame-based emotions that routinely pervade their clients' narratives—self-consciousness, embarrassment, low self-esteem, self-hatred, shame-anxiety, shame-rage cycles, and so on. When counselors are able to break cultural taboos and begin approaching shame reactions directly but compassionately, profound client movement often follows. Because most counselors have been socialized to avoid these shame-based affects, however, they will need assistance from their supervisors and additional reading (see Suggestions for Further Reading at the end of this chapter) to begin responding effectively.

Sadness-Anger-Guilt

Whereas some clients lead with their anger and defend against their hurt, others characteristically lead with their sadness and avoid their anger. This type of client may present with an undifferentiated feeling state of sadness, helplessness, vulnerability, or depression. These clients do not experience or express anger, they avoid interpersonal conflict, and tend to respond to others' needs at the expense of their own. To illustrate this affective constellation, we return to the case of Jean.

Earlier in this chapter we examined a dialogue in which, for the first time, Jean was able to let the therapist respond to her sadness and be emotionally connected to her while she was experiencing it. Subsequently, the therapist explored why it had always been so difficult for Jean to share her vulnerable feelings with others. Jean responded that her painful feelings were all sickeningly more real if someone else saw them. Jean had always experienced these feelings, but they had always been ignored or dismissed in her family of origin. Over time, Jean had internalized her family's taboo against unhappy and angry feelings, and she now disdained her own feelings in just the same way as her caretakers did. As part of this, she denied to herself the reality of the emotional deprivation and disparagement that she had suffered throughout her childhood. Following her relational templates, Jean continued in adulthood to reenact the affective themes and relational patterns of her childhood in the problematic love relationships that she kept reconstructing.

When Jean returned to therapy the following week, she had made a significant change. For several months, she had been involved with yet another critical but irresponsible man who took her for granted and took advantage of her responsiveness. That week, when the boyfriend borrowed her car and arrived late to pick her up from work, she expressed her anger for the first time. Jean told him that she deserved to be treated with more respect and that she had the right to have her own needs considered. Jean had remained in this relationship because she was afraid to be alone. At that point, though, she realized that this relationship was not good for her and soon ended it.

Once the therapist was able to touch Jean's sadness and hurt, her anger was activated and subsequently expressed in an appropriate manner. Jean had remained locked in her sad and helpless victim role, in part, because the direct expression of anger and the assertive expression of her own needs were unacceptable. As often happens, Jean began to claim her own personal power when her feelings were expressed and she received affirming responses to them. However, we must look more carefully at the sequence of feelings that unfolded for Jean.

Although Jean began the next therapy session by sharing her good news about ending the problematic relationship, she had difficulty staying with this feeling of strength. In the course of the session, Jean gradually retreated from her indignation at how badly this man had treated her, began to feel concerned about how she may have "hurt" him, and wondered if she had been selfish to end the relationship because he needed her so much. However, the therapist's working hypotheses had prepared her for the possibility that Jean would feel guilty upon experiencing her anger, setting limits, or responding to her own needs. With this awareness, the therapist was able to quickly recognize Jean's guilt and engage her in questioning its validity.

As with the anger-sadness-shame constellation, Jean's affective constellation included a third, aversive feeling—guilt. Jean defensively avoided her anger because it aroused guilt, which served to reflexively return her to her characteristic affect of sadness/helplessness. As the therapist helped Jean identify and consider all three of these interrelated feelings, she was able to relinquish her victim stance for increasing periods.

With the therapist's support, Jean could acknowledge the reality of the painful deprivation and disempowering invalidation that she had originally experienced in her family. As Jean stopped avoiding the reality of these painful memories, she stopped reenacting them in her current relationships; she no longer recreated frustrating relational patterns that evoked the same disheartening emotions. With the therapist, Jean began to protest the hurtful ways she had been treated—both as a child and in her current adult relationships. The therapist affirmed how unfair this had been, and Jean began to assert her own limits and preferences in the presence of others' competing needs or expectations, without becoming immobilized by guilt. Coming to terms with each component of this affective constellation eventually left Jean feeling stronger and more in control of her own life than she had ever been. Once again, significant life changes followed from Jean's internal resolution of her conflicted emotions. By responding to each feeling in clients' affective constellations, therapists help clients resolve their problems.

Let's step back from this discussion for a moment and consider how the previous topic, the predominant characterological affect, relates to the affective constellation. The predominant characterological affect will typically contain the leading edge of the affective constellation or triad. For example, clients who present affectively a predominant sense of helplessness can easily access and express sadness, which is first in their triad. Often, the first emotion in the triad is the one that was allowed in the client's family of origin; all other affects became subsumed under, or covered over by, this emotion. Oftentimes, the second affect in the triad was considered too threatening to a caregiver—and therefore was not allowed to be experienced or expressed. Over time, the client also came to see these legitimate feelings as unacceptable, and the client's own authentic voice was muted. The third affect in the sequence, shame or guilt in the examples given above, is often avoided because it is too personally disruptive for the client.

The therapist's intervention goal is to validate the legitimacy of each feeling in the affective triad and to remain empathically connected to clients while they are experiencing them. This gives clients the experience of expressing themselves authentically without expected but unwanted responses resulting (on the one hand, for example, the parent/therapist withdrawing, rejecting, or criticizing the child/client, or on the other hand, the parent/therapist feeling burdened, immobilized, or overwhelmed by the child/client).

Profound changes are set in motion for many clients as they find that their previously unacceptable feelings can be shared without the familiar but unwanted consequences always occurring. In particular, many begin to feel that they can center in themselves and "live inside their own skin," rather than being so anxiously preoccupied with how others see them. Thus, outgrowing these limiting familial rules about affective experience brings clients into a fuller personal identity. Feeling safer and stronger, they are able to risk revealing other aspects of themselves, are better able to utilize a variety of therapeutic interventions, and begin to make changes with others beyond the treatment setting.

Holding Clients' Pain

Having drawn out the full range of the client's feelings, the therapist's next concern is how to respond most effectively. Therapists can provide a safe holding environment to contain the client's feelings. In essence, such emotional containment requires the therapist to maintain a steady presence in the face of the client's distress. Let's clarify this important concept drawn from the attachment literature.

As we saw in Chapter 2, therapists are trying to empathize accurately with the client and to express that empathic understanding to the client. Therapists also wish to communicate their genuine compassion for the client's distress, verbally and by their presence. Accompanying this responsiveness, however, the therapist is also striving to be a participant/observer who simultaneously can maintain an *observing self*. This self provides a holding environment by conveying three essential messages to the client: (1) these emotions will not overwhelm the therapist; (2) the intensity of the pain will decrease; and (3) the upsetting event is not catastrophic but can be endured.

In this section, we explore how the therapist can provide such a holding environment, which gives clients a new and more satisfying response to their conflicted emotions than they may have received in the past. When an emotionally responsive and affirming therapist provides this reparative experience and contains clients' core conflicted feelings, clients are empowered to begin acting in new and more adaptive ways with others.

Clients Resist Feelings to Avoid Interpersonal Consequences

As we saw in Chapter 3, clients can be resistant to treatment; likewise, they also defend against difficult feelings at times. As before, the therapist does not simply want to push through these defenses in order to reach clients' sadness or other significant feelings. As before, that would be winning the battle but losing the war. Resistance and defense are more about interpersonal consequences than intrapsychic dynamics. They concern *expectations of receiving familiar but unwanted responses from others* that evoke fear, shame, or guilt. Therapists create a more affirming interpersonal process, with greater safety for clients, if they enlist clients' collaboration in trying to understand the very good reasons why it has been threatening to experience or share certain feelings. For example, rather than pressing the client to disclose a difficult feeling, the therapist may use responses such as these to explore the relational threat:

> OK, let's not talk about how you feel about this. Let's talk instead about what might go wrong if we did.
>
> If you did cry or let me see your vulnerability, what might go on between us? Is there something that has happened with others that you don't want to happen again here?
>
> What is the danger for you if you let me see your sadness? Help me understand how you have been hurt in the past.
>
> How did your parents respond to you when you were feeling _____ ? What did they say and do? Can you recall the look on their faces or what they were feeling toward you?

Thus, the therapist does not push for the content—the specific feeling—but works with the client to clarify the defense—the reality-based reasons for not wanting to experience or share the feeling. As in Chapter 3, the therapist is honoring the client's resistance by collaborating with the client in trying to understand why this defense was once a necessary and adaptive coping mechanism. The therapist's basic orienting assumption is that the client's fear or defense makes sense in terms of what has transpired in past relationships, although it is no longer necessary or adaptive in many current relationships. *If the therapist and client can clarify how the client's defense was necessary and adaptive at another point in time, and the therapist can make it clear that she will not respond in the unwanted ways that others have responded, it will soon be safe enough for the client's conflicted feelings to emerge.*

Let's look further at the relational aspect of clients' defenses. Though clients may tell therapists that they don't want to explore their underlying feelings because they are "too painful" or it "doesn't do any good anyway," this is not the whole story. Beginning therapists should always generate working hypotheses about the *interpersonal consequences* that clients may be expecting if they express certain feelings. Clients usually avoid conflicted feelings because of their expectations that certain unwanted interpersonal consequences will result if they share with the therapist feelings that have been unacceptable to significant others. For example, on the basis of their experiences with others, clients commonly have expectations of being ridiculed; being invalidated and having their feelings discounted; being ignored and unheard; seeing that others are hurt or burdened by their feelings; and so forth. Such interpersonal threats are usually far more important than their own subjective discomfort. Although it is hard for most beginning therapists to grasp this, *clients also hold these same expectations toward the therapist*—even when the therapist has never responded in any of these problematic ways! In sum, the discerning therapist can usually find very good reasons why clients learned that certain feelings were unsafe to experience or share. Therapists can also help clients differentiate then from now, and make overt how they are responding differently than others have in the past and present as well.

Of special interest, one purpose of clients' defenses against their feelings may be to protect their parents. Many clients have learned to avoid or deny painful feelings in order to protect their caregivers from seeing the hurtful impact they are having. This loyalty to the caregiver is carried out at the clients' own expense, however. As adults, these clients are still complying with familial rules and colluding with their caregivers in denying the impact of hurtful parental actions. By so doing, these clients are preserving insecure attachment ties at the cost of not acknowledging their own painful feelings. This protects the (internalized) parents from seeing what they are doing (or once did) that is hurting the client. If clients participate in this denial and avoid the conflict, homeostatic family rules are maintained and the clients remain emotionally connected to internalized parents. However, if painful feelings are permitted expression in therapy and the therapist affirms clients' experience without blaming or rejecting the parent, chronic symptoms such as lifelong dysthymia often lift. In the process, though, familial rules are broken and emotional ties to attachment figures are threatened—evoking separation anxieties and separation guilt which the therapist needs to attend to.

To sum up, interpersonal themes motivate clients' defenses against their conflicted feelings. By clarifying that these unwanted interpersonal consequences have indeed occurred in past relationships, but are not going to occur in the therapeutic relationship, the therapist provides the safety that clients need in order to resolve their conflicted emotions.

Providing a Holding Environment

As we have just seen, the best way for therapists to help clients resolve conflicted feelings is to provide a safe holding environment to contain their distress. When clients' core conflicted feelings are evoked, their relational templates or internal working models will be activated. In these affect-laden moments, *clients anticipate that they are going to receive the same problematic response from the therapist that they have received in the past.* It is often difficult for beginning therapists to believe that clients fear this, because they have never responded to the clients in this problematic way. As we have emphasized, a corrective emotional experience occurs when the therapist responds in a new and safer way that resolves, rather than metaphorically reenacts, clients' original conflict. However, unless the therapist clarifies or makes overt what the therapist is thinking and feeling as clients experience these conflicted feelings, *clients usually misperceive the therapist's response to fit their templates.* For example, therapists may clarify this by asking: "When you are feeling sad like this, what do you think I'm thinking about you?" Without such clarification, clients often believe that the therapist is privately disappointed, frustrated, burdened, and so forth, just as significant others were in the past. With these considerations in mind, let's examine further how therapists can provide a holding environment that contains the client.

While growing up, the client had certain developmental experiences that were painful, confusing, disempowering, and so forth. Enduring conflicts arose because the adult caregivers were not able to provide the affirmation and understanding the child needed to contain these emotional reactions. caregivers provide such containment when the child is psychologically held in a close interpersonal envelope of empathic caring, and the child's concerns are heard, taken seriously, and understood by a caregiver who remains emotionally present and responsive. In other words, the child's distress registers with the caregiver and is taken seriously, without being diminished or exaggerated, and the child feels secure that the caregiver will remain available to help manage the problem, if only by expressing interest and concern. Before children can develop the capacity to manage feelings on their own, someone must first provide this holding context for them. Too often, however, this essential developmental experience is not provided. For example, suppose the child is sad or hurt. In some families, the child's sadness

- Arouses the parent's own sadness, and the parent withdraws, which leaves the child emotionally alone
- Makes the parent feel guilty, and the parent denies the child's sad feelings and tries to cheer the child up, which leaves the child confused and alienated from her own authentic experience

- Makes the parent feel inadequate in the parenting role, and the parent responds derisively toward the child's sadness, which leaves the child ashamed of his vulnerability

In such scenarios, the child's sadness cannot be heard and responded to; as a result, these feelings cannot run their natural course and come to their own resolution. Young children cannot contain a strong, painful affect by themselves without support from an emotionally available caregiver. As a result, the child will have to find ways to deny or avoid this painful feeling. Years later, when this child is an adult client and risks sharing the unacceptable or threatening feeling with the therapist, the therapist's goal is to provide the holding environment that was missed developmentally. We have already seen two examples of this in Chapter 2: Marsha's second therapist heard and responded to her experience of emptiness, and the therapist of the depressed young mother sensitively articulated the mother's own unanswered cries. Both therapists provided a holding environment. Turning to specifics, we'll look first at what therapists should not do; then we'll examine further what therapists can do to help clients contain and resolve their conflicted feelings.

Too often, therapists avoid or move away from clients' painful emotions; in so doing, they reenact the clients' developmental conflict and leave clients disconnected from others while they are experiencing their emotions. The most common reason why therapists do this is because they assume too much responsibility both for causing the client's pain and for alleviating it. Beginning therapists may think to themselves, "I got the client into this, so now it's up to me to get the client out of it." Both parts of this belief are false. By responding in the ways suggested in this book, therapists reveal feelings that are already present in the client; they do not cause the pain. Further, it is patronizing for therapists to think they can fix clients or put them back together again. In actuality, the therapist never has the power to manipulate another's feelings in this way, and attempts to do so often recapitulate the parents' controlling or dependency-fostering stance with the client.

When therapists take on responsibility for the client's feelings, and place on themselves the expectation that they will fix or solve the problem, the usual outcome is feelings of inadequacy. In addition, therapists may be threatened by their own conflicted feelings that have been evoked by the client. In such situations, therapists are likely to respond to the client's affect in ineffective ways:

- To interpret what the feelings mean and intellectually distance themselves
- To become directive and tell the client what to do
- To reassure the client and explain that everything will be all right
- To become anxious and change the topic
- To fall silent and emotionally withdraw
- To self-disclose or move into their own feelings
- To diminish the client by trying to rescue him
- To become overidentified with the client and press the client to make some decision or take some action to truncate the feeling

When these ineffective responses occur, there is no holding environment. Therapists' avoidance, as reflected in these responses, often dynamically recapitulates the client's developmental conflict. For example, by trying to talk her out of her feelings, Marsha's first therapist reenacted her experiences in her family. Marsha's first therapist went even further to avoid Marsha's feelings by, in effect, sending her away: When Marsha got near her sad, empty feelings, the therapist suggested that she join a group and talked to her about seeing a psychiatrist for antidepressant medications. Although such referrals are certainly necessary at times, in this case they were used to distance the therapist from the client's feelings. This is, sadly, a common occurrence.

If these distancing and controlling responses are ineffective, what should the therapist do instead? The therapist needs to welcome the client's feelings and approach them directly. For example, the following would be a more effective response to Marsha's strong feelings.

Therapist:

I'm glad you're taking the risk of telling me about these sad and empty feelings. Let's keep talking, I want to understand them better.

Therapists also want to find genuine ways to express their care and concern. Therapists can do this in words, of course, but it is equally important to communicate nonverbally in tone and manner. In particular, therapists must be emotionally connected or present with clients while they are experiencing their feelings. There are many ways to do this, of course, and much of this communication is nonverbal or involves only a few words. We'll now consider some of the elements found in effective responses that create a holding environment. Having assimilated these guidelines, however, therapists will want to tailor their responses to fit each particular client and to make them congruent with their own personal styles.

Initially, the therapist's primary goal is to stay emotionally present and connected to clients while they experience painful feelings. As discussed in Chapter 2, the therapist may also want to reflect what clients are feeling and to affirm the reality of their experience. This can be done simply by acknowledging, "You're very sad right now." With some clients, it may be helpful to further validate their experience by saying, "Of course you are feeling sad right now. It hurt you very much when he did that." As an important part of developing their own professional identities, therapists also need to explore and find their own ways of expressing their compassion for clients and connectedness to them.

Therapist:

I can see how much this has hurt you and how sad you are feeling right now. I feel close to you, and honored that you choose to share such a special part of you with me.

In addition, therapists must demonstrate behaviorally that they can tolerate the full intensity of clients' feelings. In other words, they must communicate by their manner—and overtly in words with some clients—that they are in no way hurt, burdened, or undone by clients' feelings.

Therapist:

> I'm privileged that you choose to share this with me. I'm with you right now, and we're going to stay together in this until we have worked our way through it.

In order to undergo a new or reparative experience, clients must see that the therapist is not responding in the problematic ways that others have and that they expect from the therapist. It must be clear that the therapist is not threatened by their feelings or judgmental of them, does not need to move away from them in any way, and is still fully committed to the relationship. Beginning therapists often fail to appreciate how clients' relational templates are activated in these affect-laden moments and take precedence over the current reality.

As the client gathers his equilibrium from a strong emotion, the therapist may also want to describe what caregivers and others have done when the client felt this way. The therapist may want to clarify how their current relationship is different from the client's relationships with others (for example, the therapist remains emotionally present and affirming rather than becoming impatient or withdrawing), and emphasize that the therapist is accepting of this aspect of the client that has been revealed. It is important to *debrief* clients by asking them how it has been to share these feelings or personal disclosures with the therapist. Therapists often miss this important opportunity. Genuinely touched by the client, they fail to appreciate that it is at these intense, special moments that the client is most likely to misperceive the therapist's caring response in terms of earlier relational templates. To correct these expectable misunderstandings or transferential distortions, it is necessary to ask clients what they think the therapist was thinking, feeling, or going through while they were feeling sad, angry, and so forth. Long-standing transference distortions can be identified and resolved at these moments. For example, the therapist might clarify the client's distortion by means of a self-involving comment:

Therapist:

> No, I wasn't thinking that you "looked silly" and "sounded like a baby" at all. I was touched by your sadness. I think your feelings make sense and fit exactly with what happened to you.

When offered sincerely, simple human responses of validation and care mean much to clients and help to restore their dignity. Unfortunately, beginning therapists often believe that they have to do much more than this to respond adequately to the client's pain. These concerns may reflect the therapist's own unrealistic expectations that they have to respond in some precisely correct way. In reality, the client does not need the therapist to provide eloquent, crafted responses in a calmly self-assured manner. Such perfectionism is as unnecessary as it is unrealistic, and may simply immobilize the therapist. As Winnicott (1965) says, the child needs only "good enough mothering." Therapists provide a satisfactory holding environment when they find their own genuine ways to enact these guidelines. Clients will be appreciative of sincere efforts and find them comforting, even if they are expressed in a fumbling or halting way.

Change from the Inside Out

Clients improve markedly when the therapist provides a holding environment that allows them to contain feelings that have been too painful to integrate or come to terms with. How does this come about? A corrective emotional experience occurs when the therapist can be affirming and remain emotionally connected to clients' pain without explaining, reassuring, or otherwise moving away from it.

Some clients hold a primary pathogenic belief about their core conflicted feelings. In its most primitive expression, the fear about their feelings is either "I will be destroyed" or "You will be destroyed." Some clients learned in their families of origin that certain feelings are "too much" or "overwhelming" or that they "won't be able to stop crying," "will go crazy," or "won't be able to breathe." In contrast, some clients believe that the other—the therapist or attachment figure—will be burdened, impaled, or will "go away" because of their unacceptable feelings. The most conflicted clients are often double-binded and struggle with both pathogenic beliefs at the same time.

Therapists disconfirm these "grim, unconscious pathogenic beliefs" (Weiss & Sampson, 1986) and provide a corrective emotional experience when they can (1) compassionately accept clients' painful feelings; (2) stay emotionally present and connected to them; and (3) demonstrate that they have in no way been injured, burdened, or overwhelmed by these feelings. If the therapist does this, many clients will be profoundly relieved to disconfirm their pathogenic belief and find that neither they nor others need be hurt by their feelings. How does this translate into real change? When the therapist compassionately accepts clients' most painful or threatening emotions without reenacting the unwanted consequences they have come to expect, clients forge an authentic emotional connection to themselves and to the therapist. This reparative experience is empowering and allows clients to begin making meaningful changes in their coping styles, expectations of others, and beliefs about themselves. Clients now have greater freedom to pursue their own interests and goals and more actively choose how they want to shape their lives.

Moreover, in a supportive context, clients' feelings can run their course and come to a natural close, often in only a matter of minutes. Clients then recover by themselves and readily reestablish their own internal equilibrium. Through this corrective emotional experience, clients disconfirm pathogenic beliefs about themselves and learn that their feelings—and they themselves—are not disgusting, bad, and so forth. As they have the experience that their previously unacceptable feelings do not rupture this relationship, their feelings, although still a part of them, no longer compel them to be isolated or ashamed. As a result, clients no longer need to maintain long-held defenses, symptoms, or elaborate coping strategies—the feelings that had to be sequestered away can now be experienced, understood, and integrated. When the therapist can hold clients' pain and allow them to tolerate experiencing and sharing it, they, in turn, will become more able to contain their feelings on their own. *This is change from the inside out.* This interpersonal process often propels a far-reaching trajectory of enduring behavioral change, and long-standing symptoms often abate at this point. Ironically, when clients risk exposing their pain, vulnerability, or shame, and the therapist responds with compassion, they feel more powerful. One of the most satisfying aspects of being a therapist is to facilitate such self-efficacy. To illustrate the principles discussed in this section, let's look at a brief case study.

Sherry, a first-year practicum student, was struggling to assimilate all of the new ways of thinking about therapy that she had been learning. It was an exciting but difficult year as her entire worldview was shifting. Part of the difficulty was that her practicum instructor and her individual supervisor were suggesting very different ways to approach her clients. Having grown up with combative, divorced parents, she had conflicts about trying to please both sides, which were being aroused in this situation—making the already complex challenge of clinical training even more difficult. At times, Sherry was immobilized as she tried to sift through these competing ideas, find what made sense to her, and come up with something that she thought would be useful to say to her clients. Two-thirds of the way through her first practicum, she felt more self-conscious and inhibited about responding to people than she had been when she started the program!

Despite her insecurity, Sherry was learning a lot and had been responding quite helpfully to her client, Maria. Near the end of one of their weekly sessions, Maria began to disclose a painful memory. With tears in her eyes, Maria recounted the sickening experience of a date rape 15 years earlier, when she first went to college. Listening to Maria's story, Sherry felt sad and angry. However, as she began to think of all the things she should and should not say, and what both of her supervisors would want her to say, she felt frozen and could not find the words to say anything at all. Although it was less than a minute, the silence grew painfully awkward as Maria finished speaking and waited for Sherry to respond. Unable to speak, Sherry just sat there. Shaken by Maria's painful story and immobilized by thinking of all the different ways she "should" respond, she couldn't find any words.

After an excruciating pause, Maria tried to take care of Sherry and started talking again, but this did not last long. After a few minutes, Maria chided, "Don't you understand! He hurt me a lot when he did that!" Sherry had made a "mistake," but she was able to recover by validating Maria's anger and affirming her experience.

Sherry:

Yes. He did hurt you very much. It was frightening and painful for you. I'm very sorry that happened. And I understand why you are angry at me for not responding. I was touched by what you said, but I just couldn't find the right words for a minute.

Maria:

You couldn't?

Sherry:

No, I wanted to, because I felt for you. But I just couldn't find the right words, and the longer I searched for them, the worse it got. I'm really sorry this happened. I feel embarrassed, and I know it must have been hurtful to you.

Maria:

It's OK. I don't know what to say either sometimes. I don't know why, but this makes me think of my daughter when she was little. If I would take the time to stop and pick her up and really listen to her when she needed me, we would get through her problem pretty quick. My husband called it "collecting her." But if I was in a hurry and tried to ignore her or speed her up, she would escalate, and it would go on for a longer time.

Sherry:

You're a very good mom. I think you're remembering that right now because that's what you needed from me—to be attended to, cared about, collected. You needed me to give you what you gave your daughter.

Maria:

Yeah, I think that's just exactly what I needed. *(tearing)* You know, I never dreamed of telling my mother what happened to me—you know, about the date rape thing. She never seemed to listen to me when . . .

Sherry recovered from her "mistake" and began providing the holding environment that the client herself had articulated so well with her own young daughter. As in this example, validation of the client's dissatisfaction with the therapist often leads the client to become stronger; to clarify related issues in other relationships; to achieve meaningful insights that lead to behavior change; to bring up significant new material; and so forth. Moreover, by recovering and responding affirmingly to Maria's feelings at the second attempt, Sherry gave Maria a corrective emotional experience that helped her to progress in treatment. We're too afraid of mistakes—they can be undone.

To sum up, when the difficult feelings associated with the client's generic conflict emerges, therapists have the best opportunity to help the client resolve problems and change. The most important corrective emotional experiences usually occur around the therapist's new, more satisfying response to the client's emotions. However, the client's maladaptive relational patterns are also most likely to be reenacted with the therapist in these affect-laden moments. Why? Vulnerability activates our relational templates or internal working models. Consequently, clients expect the same unwanted response to their feelings from the therapist that they have received in the past, and readily distort or misperceive the therapist's benevolent response to fit their expectations. Compounding this, the therapist's own personal issues or countertransference reactions are also most likely to be aroused in these intimate moments. Thus, we need to address countertransference issues, and turn now to factors in the therapist's own life that make it difficult to respond to clients' emotions.

Personal Factors That Prevent Therapists from Responding to Clients' Emotions

Just as it is easier for clients to externalize and focus on others rather than themselves, it is also easier for therapists to focus on the client than it is to look within at their own issues and dynamics. If therapists are not willing to work with their own emotional reactions to the material that clients present, however, they will tend to avoid clients' feelings or to engage them on an intellectual level. The purpose of this section is to help therapists manage their own reactions to the evocative material that clients present by learning about their own countertransference propensities.

Little enduring change occurs unless therapists express respect and compassion for the emotional conflicts with which clients are struggling. Realistically, however, it is difficult for therapists to respond to all of their clients' feelings, because every therapist brings his own developmental history and current life stressors to the consulting room. Let's examine how these factors can keep every therapist from responding effectively to clients' emotions at times.

Therapists' Need to Be Liked

Most students who are drawn to clinical practice are generous people, genuinely concerned about others, and ready to give of themselves. Often nurturing individuals by nature, they can readily empathize with and respond to others' hurt and pain. At the same time, many therapists also have strong needs to be liked. As a result, many therapists have difficulty providing a nondefensive response to the challenging relational patterns that some clients will begin to construct or reestablish with the therapist (for example, the client becomes critical, demanding, demeaning, and so forth toward the therapist) (Strupp & Hadley, 1979; Navitts & Strupp, 1994). Other therapists have strong needs to nurture others who are dependent on them. These therapists enter clinical practice, in part, in order to fulfill these needs, and they may have difficulty supporting the client's healthy individuation at times.

Many counselors were caretakers in their families of origin. They are drawn to clinical practice, in part, by the often unrecognized need to rescue their caregiver from alcoholism, an unhappy marriage, chronic depression, and so on. This countertransference issue, in particular, can make it harder for therapists to address clients' painful emotions. Those emotions evoke for therapists the familiar predicament of needing to fix their caregiver and, because in reality this is always an impossible task, threaten them with the prospect of once again feeling inadequate or disappointing their caregiver.

All therapists need to assess their own motivations for becoming clinicians and to examine what needs of their own are being fulfilled by doing therapy. *The issue is not whether therapists have their own needs and countertransference propensities, but how they will deal with them.* All therapists have certain countertransference reactions, and self-examination is one way to manage them effectively.

Beginning therapists may be taken aback when their well-intended attempts to approach the client's affect occasionally meet with negativity: anger, defensiveness, and so forth. Too often, beginning therapists absorb the client's rebuff, comply with the defensive side of the client's feelings, and stop approaching the client's conflicted emotions. This response is most likely from therapists whose needs to nurture or to be liked have been frustrated. To be effective, however, therapists must continue to work with the client's negative feelings that have engendered distance or discouraged others from responding as well. Therapists must find other ways to approach the client's conflicted feelings and change the maladaptive relational patterns that have usually ensued.

To do this, therapists can adopt the strategies for working with the client's resistance that were presented in Chapter 3. Therapists should not press clients to express a feeling they are reluctant to share, but explore what the threat or danger may be for them if they do so.

What does it say or mean about you if you feel _____?
What could happen between us or what could go wrong for you if you let yourself feel _____?
How have people responded to you in the past when you felt _____?

Therapists do not want to win the client's approval by trying to be a "nice" person who does not approach difficult feelings or awkward interpersonal situations. On the other hand, therapists must not demand that clients enter into, or stay with, a certain affect. Instead, they offer invitations and allow clients to enter and leave difficult emotions as they wish. Letting clients have it their own way avoids the issues of coercion and compliance that so many clients had in their families of origin and that people of color often have experienced in the dominant culture. Therapists provide an effective middle ground when they work with the client to understand the resistance by exploring together the threat or danger that once existed. *With understanding comes safety,* and clients soon will be able to explore feelings that needed to be held away in the past.

To sum up, therapists are most likely to have difficulty working with the client's critical, competitive, and controlling feelings toward them. Beginning therapists will need support from supervisors and colleagues to keep from personalizing these negative reactions and responding defensively. Instead, therapists can learn much by considering how this negativity effects clients' relationships others. For many counselors, this is the most challenging task in clinical training.

Therapists' Misperceptions of Their Responsibility

Too often, therapists believe that, if they respond to a client's feelings, they are responsible for causing the feeling or for taking the client's pain away. Assuming responsibility for the client's feelings is unrealistic and causes problems both for the therapist and the client. Let's look at two aspects of this problem.

First, some therapists avoid their clients' feelings because they do not want to see them hurt or feel that they have made their clients feel bad. These therapists mistakenly believe that responding to a painful feeling in the client is the same as causing that feeling to exist. Therapists must keep in mind that they did not cause the client to have this feeling; it was present before the client met the therapist. The therapist has simply responded to the client in a way that reveals the feeling. The misconception that the therapist has caused the client's painful feeling often leads therapists to feel guilty and begin avoiding difficult feelings.

A second problem arises when therapists labor under the impossible burden of having to "fix" the hurt they believe they have "caused" the client to feel. When therapists take responsibility for causing the client's feeling, they also assume responsibility for making the client feel better. This is an unrealistic responsibility; therapists can never take away or fix clients' pain. However, it is empowering when therapists can share clients' experience and respond with understanding and validation.

Therapists who assume responsibility for clients' feelings are apt to avoid those feelings—to tell clients what to do, to reassure them, and to explain and interpret their emotions before they can experience them. In most cases, these approaches prove ineffective.

If therapists should not shoulder this inappropriate responsibility, what should they do for their clients? The therapist's appropriate response to the client's feelings is threefold: to identify, to join, and to affirm. First, the therapist must help clients identify and express more fully whatever feelings they are having. Second, the therapist must be empathic or emotionally responsive to clients, so that clients can share their feelings with a concerned other rather than having to experience them alone, as they often have in the past. Third, the therapist needs to affirm the validity of clients' responses, by helping clients understand their feelings and make sense of why they are experiencing this feeling at a particular moment. If the therapist accepts this challenge but does not assume responsibility for causing or changing clients' feelings, the therapist will be free to respond effectively to whatever feelings clients present.

Even therapists who are well aware of the pitfalls of assuming responsibility for clients' feelings are likely to struggle with this issue when the client's affect becomes intense. For example, a novice therapist was treating a young boy who was in the midst of a potentially life-threatening battle with leukemia. The therapist would sit with the boy in the hospital and, among other topics, they would sometimes talk about the possibility that he could die: "What would be the scariest thing about dying? What are you most afraid of?" When the therapist approached the child's feelings, the boy responded, "I don't want to die, I'm afraid to die!" Seeing the fear in this young child's eyes and feeling totally inadequate, the therapist questioned what she was doing and felt sadistic to be evoking such anguish.

But the child went on to say, "I don't want to be alone, I'll be all alone if I die!" Inviting the child's feeling revealed his concern that he would be all alone. Hearing the boy's separation anxiety and fear of abandonment, the therapist better understood his concern and was able to talk with him in more helpful ways.

More importantly, the therapist was able to help the child's mother understand his basic fear; his mother was able to reassure him that, if he were to die, her love for him would last forever. By talking with the mother and son together about his fear of being left alone, the therapist helped them to become closer during his illness. This closer contact measurably alleviated the child's distress, in general, and his intense separation anxieties and fears about dying, in particular.

When working with such profound emotions, therapists almost inevitably feel to some extent that they are responsible for causing the client's pain. As a result, they tend to avoid these intense feelings. To prevent this, therapists need to consult with a supervisor or colleague to help them manage their own exaggerated feelings of responsibility at times. Otherwise, these countertransference reactions will prevent therapists from responding effectively and giving clients the holding environment they need. In the example of the boy with leukemia, the therapist could not have stayed with the intensity of the boy's fear or the mother's profound sadness without regularly scheduled consultations with a supervisor to help her manage her own emotional reactions.

Family Rules

Therapists may not be able to respond to certain feelings that the client presents because of family rules about emotional expression that they learned in their families of origin. Every family has unspoken rules that govern emotional expression: A rule-bound homeostatic system prescribes which feelings can be expressed and which are unacceptable. This system also prescribes when, how, and to whom certain feelings can be expressed. The permitted intensity or degree of emotional expression is also controlled by homeostatic family mechanisms. For example, if too much conflict, independence, or closeness is expressed in the family, a corrective homeostatic mechanism is activated to bring the level of emotional expression back within acceptable limits. The following example illustrates how homeostatic mechanisms operate to keep family emotional expression within prescribed limits.

Suppose that in the Smith family a moderate degree of conflict and anger can exist between the older sister and the younger brother. A lesser degree of frustration and irritation can be expressed between the father and both children. However, anger or even disagreement is rarely expressed between any family member and the mother, and is never expressed between the father and the mother. What is the homeostatic mechanism that serves to keep this system operating within these boundaries? Whenever a child begins to express angry feelings or have direct conflict with the mother, she stops the child's protest by looking sad and hurt. If the child's anger continues, the father intervenes and says, "Don't talk to your mother that way," or else the mother employs the father as a regulatory influence by saying, "I'm going to tell your father about this when he comes home."

In contrast, if anger or differences of opinion begin to escalate between the mother and father, the children serve as homeostatic regulators and respond in predictable, patterned ways to terminate the parental conflict. For example, the oldest child, 8-year-old Mary, serves as a go-between in the parental relationship. She tries to make peace between her parents by carrying messages back and forth between them. This mechanism usually succeeds in terminating parental conflict, except in periods of high family stress. If the parental conflict continues, the younger son may then provide a diversion to the parental conflict by acting out in some predictable way—by breaking something or having an accident. When the son makes himself a problem that mother and father must jointly address, their rule-breaking level of conflict is terminated, and the family's emotional equilibrium returns to acceptable levels.

Mary, the oldest child in the Smith family who used to serve as the mediator to assuage parental conflict, is now a graduate student in clinical training. The family rules that Mary has learned in her family of origin will influence her ability to respond to her clients' emotions—just as the family rules that all therapists have learned in their families will influence how they respond to their clients.

Ann, a depressed and irritable client that Mary was seeing at her internship site, became frustrated with Mary because of her continuing depression. Just as she accused her husband and others of always letting her down, Ann angrily criticized Mary: "I keep coming here every week, but I'm not getting any better. You make me have these awful feelings and it just doesn't help. I wonder if you know how bad you're making me feel sometimes!"

Mary was taken aback by Ann's angry criticism, even though it was short-lived and Ann became somewhat apologetic. Mary was especially surprised because she thought they had been working together more effectively in recent weeks. Throughout the rest of the session, Mary backed away from Ann's anger by offering reassurances and trying to convince Ann that she would get better if she would continue to come to therapy and work with the difficult feelings they had been sharing. Mary felt terrible that Ann had been critical and anxiously sought help from her supervisor the minute the session was over.

Mary's supervisor was responsive to her distress over Ann's anger and was able to help Mary understand what went on for her during the session. In reviewing a videotape recording of the session, the supervisor helped Mary recognize how she had tried to assuage Ann's anger with reassurances, rather than allowing Ann to feel angry with her or disappointed about her lack of progress. The supervisor also explored with Mary why she had responded defensively, rather than joining Ann in figuring out what wasn't working and exploring together what they could do differently to make treatment more helpful.

With her supervisor's help, Mary quickly realized that she had been unable to tolerate Ann's anger. She had indeed tried to assuage it, just as she had always done with her parents. Mary's role in her family of origin was to take care of the conflict, rather than to allow others to express their dissatisfaction and talk about what they could do to resolve the problem. In addition to clarifying Mary's old response pattern, the supervisor also helped Mary to generate working hypotheses about what this critical response may have meant for Ann. The supervisor observed that Ann had expressed her anger just after sharing her hurt, and hypothesized that the anger may be a part of her affective constellation. In other words, Ann needed the experience of being able to express her anger at someone who would take her concerns seriously and would not punish or abandon her—as she expected. Thus, Ann may have been testing to see whether she could be angry or critical of Mary and still remain in relationship with her—which Ann had not been able to do with her primary caregivers.

Without this helpful consultation, Mary and Ann may have become stuck in their relationship, and therapy might not have progressed. Ann could not move forward and make progress with her relational conflicts as long as Mary could not respond to the anger component of Ann's affective constellation. In their next session, however, Mary was able to recover and invited Ann to discuss her frustration with treatment and anger toward her. Although Mary still felt an anxious compulsion to reassure Ann and move on, Mary was able to contain her own discomfort enough to be able to discuss Ann's concerns more fully, better appreciate what was not working for Ann in their relationship, and suggest some accommodations to make things work better.

Clients will not be able to resolve long-standing conflicts unless the therapist can respond to each component of their affective constellation. *Therapy often stops progressing at the point at which the therapist cannot respond to a particular feeling that the client is experiencing.* With this in mind, all therapists should examine the rules that governed how emotions were dealt with in their own families, and reflect on how they are most likely to bring their own familial roles and rules for emotional expression to their clinical work. For many therapists, their own therapy will be the best way to change ineffective family rules that still govern how they act with their clients.

Situational Problems in Therapists' Own Lives

Situational stressors in therapists' own lives may also prevent them from responding effectively to clients' emotions. For example, when personal problems in therapists' own lives are similar to those their clients are experiencing (perhaps marital conflict, parenting problems, or conflicts over caring for aging parents), therapists may have difficulty approaching their clients' affect or will not be able to allow clients to experience fully the feelings they are struggling with.

Therapy is a very human interchange. Although the therapist's personal availability is the greatest asset in helping clients change, it will also be an obstacle at times. When the therapist repeatedly fails to accurately identify the client's feelings, avoids rather than approaches the client's feelings, or becomes personally invested in the client's choices, the therapist's own countertransference is operating.

Countertransference reactions will occur at times for every therapist; the issue is how they are dealt with. When countertransference reactions occur, therapists need to consult with a colleague or supervisor to better understand and manage their own reactions. When therapists find that they are repeatedly having difficulty with the same type of affect, or when supervision does not free them to respond effectively, therapists should seek therapy for themselves. *Because change is predicated on the relationship that the therapist creates with clients, therapists must make a lifelong commitment to working with their own dynamics.* Without this willingness to work on their own personal reactions to clients, the help therapists can provide will be limited. Furthermore, therapists who are unwilling to work on persistent countertransference reactions are those most likely to have a negative therapeutic impact on their clients. It is a privilege to be able to help people change, and therapists honor that privilege by acknowledging their own limitations and personal involvement in the therapeutic process.

Closing

As we saw in Chapter 4, clients' conflicted emotions will not emerge in therapy unless the therapist focuses clients inward. An externalizing focus will lead therapists into a problem-solving and advice-giving mode that precludes clients' affective exploration. As clients talk about the problem out there with others, the therapist's complementary role is to give advice, reassure, or disclose what the therapist has done to solve similar problems. Although these responses can be helpful at times, limited change will usually occur if they characterize the ongoing course of therapy.

But if the therapist does not tell clients what to do, what does the therapist really have to offer? *An essential component of facilitating change is to help clients resolve their conflicted emotions by integrating feelings that have been too painful, shameful, or unacceptable to tolerate in the past.* Most clients have experienced tragedies in their lives and conflicted emotions will be entwined with the symptoms and problems they present. The therapist helps clients resolve their presenting problems by providing a holding environment for feelings that others have not been able to understand or accept. Perhaps the most important relearning occurs when clients risk sharing certain difficult feelings and find that, this time, they do not receive the same problematic response they have come to expect. Thus, a corrective emotional experience occurs as old expectations and relational patterns are disconfirmed. Yet therapists must also be aware that the biggest obstacle to providing clients with a relationship that can produce change will be their own discomfort with certain feelings. Thus, therapists must make an on-going commitment to working with their own personal issues and how they influence their clinical practice.

In the next three chapters, we will see how the therapist can conceptualize clients' dynamics on the basis of their conflicted emotions, faulty belief systems, and the maladaptive relational patterns they begin to reconstruct with the therapist. Then, in the final two chapters, we will explore how this conceptual formulation can be applied to treatment plans and used to help clients change.

Suggestions for Further Reading

1. Beginning therapists need help identifying their countertransference propensities. A self-assessment scale to help therapists become more aware of their own response tendencies is provided in Chapter 5 of the Student Workbook.

2. For information on basic communication skills to help therapists approach or inquire further about clients' feelings, see Chapters 5 and 6 of G. Egan's *The Skilled Helper* (Pacific Grove, CA: Brooks/Cole, 1994). These two chapters also help counselors to listen without judging and to become more aware of their own cultural filters in listening to clients.

3. One way for therapists to increase their effectiveness with clients is to explore their own motivations for becoming therapists and to examine how their own dynamics are expressed in their clinical work. It is just as important to focus on the personhood of the therapist as it is to focus on client dynamics and intervention strategies. One of the best sources to help therapists explore these issues is J. Guy's *The Personal Life of the Psychotherapist* (New York: John Wiley & Sons, 1987).

Conceptualizing Client Dynamics

CHAPTER SIX

Familial and Developmental Factors

Conceptual Overview

The previous chapters have focused on the interaction that occurs between the therapist and the client. In the next three chapters, we step back from this intensive tracking of the therapeutic process and consider how therapists conceptualize their clients' dynamics. To be effective, therapists need to formulate a case conceptualization that leads to specific treatment plans and intervention strategies. Because novice counselors have had little training or experience, however, it is difficult to formulate the treatment focus that clients need, and write the treatment plans and case reports that third-party insurers and treatment agencies require. Guidelines to meet these professional demands will be provided in the next three chapters, especially learning how to focus on the relational patterns that provide organizing themes to understand clients' symptoms and problems.

How can therapists conceptualize their clients' problems and use this formulation to guide their treatment plans? One way to understand how enduring problems develop is to examine what occurs for offspring in families that function well. Once the basic familial processes that produce healthy offspring are identified, it is easier to understand how normal development can go awry and lead to the enduring problems that clients present. This chapter is intended to help therapists recognize the familial and developmental antecedents of many clients' problems. It introduces family systems concepts that clarify how many problems develop in order to aid therapists in conceptualizing their clients' dynamics. The discussion of developmental issues in this chapter will be integrated with a framework for conceptualizing clients' interpersonal coping strategies in the next chapter.

Chapter Organization

In this chapter, we examine three fundamental dimensions of family life. In the first section, we discuss the structure of family relations and the nature of the parental coalition. In the second section, we examine the separateness-relatedness dialectic that was briefly introduced in Chapter 1. In the third section, we explore childrearing practices. Problems in each of these areas may lead to developmental problems that are central to many of the symptoms that clients present. Understanding how these familial factors shape personality and development will help therapists respond to their clients more effectively. Therapists will be even more effective when they can broaden this perspective and take into account the cultural context in which family relations and childrearing practices are embedded. Cultural values and beliefs circumscribe acceptable behavior and exert considerable influence on family functioning.

Structural Family Relations

Structural family relations comprise the relatively enduring patterns of alliances, coalitions, and loyalties that exist in the family. These relationships are the basis for family organization and define how the family operates as a social system (Minuchin, 1974). Such family patterns are also bounded by cultural values and beliefs. Thus, structural family relations and the culture within which they are delineated shape family communication patterns and the roles that family members adopt. Among the different structural family relationships that exist, the parental coalition is the basic axis of family life and shapes much of how the family functions (Teyber, 1981). In particular, the nature of the parental coalition will be pivotal in determining how well children adjust.

The Parental Coalition

In healthy nuclear families, the marital relationship is the primary two-person relationship in the family. Each spouse has a primary loyalty commitment to the other, and the marital coalition cannot be divided by grandparents, children, friends, employers, or other influences such as alcohol or career. Parents will still have differences or problems between themselves, and they will be committed to others as well, but the marital relationship is a stable alliance. Researchers consistently find that its cohesiveness is a fundamental feature of healthy families (Beavers, 1982). In contrast, two-parent families in which the primary emotional bond or involvement is not between mother and father often produce offspring with enduring behavioral problems. In these more symptomatic families, the primary alliance is between a grandparent and a parent, between a parent and a child, or within some other dyad. Cultures vary in how strongly they emphasize family loyalties, and extended family members may play a more important role in child care responsibilities in some cultures (for example, in Native American cultures). Within varying cultural contexts, however, family researchers find that a cohesive parental coalition is a mainstay in healthy families.

How does a primary parental coalition develop? Both parents must have been able to individuate within their own families of origin. This means they were able to achieve a sense of their own personal identity that is based on the ability to act on their own authentic feelings, interests, and values in early adulthood. In this developmental transition of forming their marital coalition, they must also be able to establish a new balance of family loyalties. This complex process involves maintaining ties and responsibilities toward the previous generation, while shifting their loyalties to a primary commitment to a peer relationship. Thus, as young adults, these offspring have been able to transfer their primary loyalty commitment from their parents to a spouse. This process, accompanied by increased self-efficacy and self-esteem, permits these young adults to define their own sets of values and beliefs. This does not mean losing their sense of connectedness to their own parents, but signals increased flexibility in their relational capacities and involves transferring their primary relational commitment to a peer relationship characterized by mutual empathy and respect.

If young adult offspring have not been able to shift their primary loyalty commitment from their family of origin to their spouse, a primary marital coalition will not develop. The primary loyalty of one or both spouses will remain with their parent. When this occurs, the primary dyadic relationship in the family is a *cross-generational alliance* between a parent and the adult offspring, rather than a primary marital coalition between the husband and the wife. With the birth of children, this cross-generational alliance (for example, maternal grandmother-mother) usually repeats in the next generation and results in a primary parent-child coalition as well (for example, mother-oldest child). Before examining how these cross-generational alliances become so problematic for offspring, we must better understand the important developmental transitions that are occurring while forming a primary parental coalition. To do this, let's examine the wedding ceremony as a cultural rite of passage.

The wedding ceremony is a familiar ritual; its central purpose is to publicly mark a transition in loyalty bonds from the parents and family of origin to the new spouse. How does this occur? In the traditional Christian ceremony, once family, friends, and clergy are all assembled, the father walks the bride down the center aisle to the front of the church. In front of all, the father symbolically gives the bride away by placing her hand in the hand of the waiting groom. The father then leaves the couple by stepping back from the front of the church and sitting down beside his own wife. The bride and groom, who have now been demarcated physically from the parental generation, turn away and step forward to be married. In this ceremony, the bride and groom are shifting their primary loyalty commitments and publicly defining themselves to family and friends as an enduring marital couple.

This shifting of primary loyalties from the previous generation to the new spouse is a major developmental task. According to Erikson (1968), this transition is a normative developmental crisis that arouses difficult feelings of loss and requires a complex redefinition of individual identity, as well as familial roles, membership, and boundaries—even in healthy, well-functioning families. This developmental challenge becomes conflict-laden and may lead to symptom formation when the young adult offspring are trying to emancipate from families of origin that have not had a primary marital coalition. The question of who comes first, parent or spouse, may be painfully enacted in conflicts that surround planning for the wedding day. Offspring embroiled in cross-generational alliances often find that their wedding day, and the preparations for it, are a wrenching battle between conducting the ceremony as the couple being married would wish and doing it the parents' way. Usually, the underlying conflict is over the shift of loyalties from the previous generation to the spouse, but it is played out in arguments over who will be invited to the wedding, where it will be held, how it will be conducted, who is financially responsible, and so on. In contrast, young adults who enjoy their wedding day usually have been allowed to switch their primary loyalty from parents to each other. These couples are not being pulled apart by competing loyalty ties, with the associated feelings of guilt.

How does this model of a primary parental coalition apply to single-parent families? Are they inherently problematic because there is not a primary marital coalition? Absolutely not. However, this structural model can still help us understand why some single-parent families function effectively, whereas others have problems.

In single-parent families that function well, as in nuclear families that function well, there are clearly defined *intergenerational boundaries* that separate adult business from child business. By necessity, children in single-parent families usually need to take on more household duties, and there is often more emotional sharing between parent and child than in two-parent families. Even so, in healthy single-parent families, roles and responsibilities are clearly differentiated between adults and children. This means that the single parent still performs executive ego functions for the family, such as providing an organized household with predictable daily routines for children, making decisions and plans for the family, and setting limits and enforcing rules. Further, although parents have legitimate needs for companionship and support, these needs are met primarily in same-generational peer relationships rather than through the children.

In contrast, single-parent families will have more problems when adult and child roles are not clearly distinguished, when intergenerational boundaries are blurred, and when too many adult needs are met through the children. With this general background in mind, we now examine how the parental coalition influences child development and why primary marital coalitions are more effective than cross-generational alliances.

How the Parental Coalition Influences Child Adjustment

Why is the nature of the parental coalition so crucial for family functioning and child adjustment? To answer this question, we look at four common problems of adjustment among children from families in which the primary coalition is not between mother and father.

1. *Problems with limits and authority.* When mother and father have a stable alliance, children cannot come between their parents and play one parent against the other. For example, if a primary coalition exists between the father and an

8-year-old son, the mother may have trouble setting and enforcing limits with her son. If the mother tries to put her son to bed at 9 P.M. or to have him pick up his room, he can often defy these attempts at discipline by appealing to his father. The son learns that he does not have to do what his mother says because his father is likely to take his side. When children can play one parent against the other in this way, they learn that rules do not apply to them. Having learned that they do not have to conform to adult rules in the home, they will often disobey teachers and try to manipulate other authority figures in their lives. These problems with limits and authority will carry over into adulthood and are especially likely to lead to problems in the work sphere, where adults must be able to conform to rules and cooperate with others. A tragic illustration of this principle is found in Woodward's (1984) biography of the late comedian John Belushi.

When Belushi was an early adolescent, his authoritarian father would work long hours in the restaurant business. Belushi was supposed to stay home and do chores for his mother. By making up humorous but hostile imitations of his father and performing these routines for his mother, he could get her to laugh at his father and manipulate her out of enforcing household duties. With his mother and siblings in stitches, Belushi would escape his chores and head out the door to meet with his friends. Among other problematic messages here, Belushi learned that, if he could successfully play one parent against the other, limits and rules did not apply to him. Despite his talents, his life was a study in compulsively pushing limits and being out of control with food, money, sex, and especially drugs. After years of chronic drug use, he died from an overdose of cocaine and heroin.

Therapists may not see large numbers of clients with this background unless they work in alcohol and substance abuse programs. Clients who learn that they can manipulate controls by dividing the marital coalition are more likely to develop impulsive acting-out symptoms. When therapists do see them in treatment, these clients often will push limits, avoid responsibility, and be demanding of the therapist. For example, they will often arrive late for appointments; expect to be able to stay beyond the scheduled time; be delinquent in or try to avoid payment of fees; be insistent on calling the therapist at home; and so forth. These challenging clients receive a corrective emotional experience when therapists can tolerate the clients' disapproval, set firm limits, and not be manipulated out of their usual treatment parameters. At the same time, however, therapists must do this in a nonpunitive way that also communicates their continuing personal commitment to the client and compassion for the insecurity and loneliness that has been engendered by these developmental experiences. When therapists are able to do this, they will help clients address more directly the real fears and needs that underly this behavior and help them find more effective ways to get those needs met.

2. *Exaggerated self-importance.* Children in families with a primary marital coalition are more likely to gain a realistic sense of their own power and control than are children raised in cross-generational alliances. When children form the primary emotional bond with a parent, they become too important to the parent and can exert too much influence over the parent's well-being. As a result,

these children gain an exaggerated sense of their own importance and a grandiose sense of their own ability to influence others. Children cannot gain a realistic sense of their own limits and capabilities when they are encouraged in the illusion that they can prop up a parent's sagging self-esteem, maintain their parent's emotional equilibrium, or make important decisions for the parent.

It is highly reinforcing for children to feel so powerful vis-à-vis their parent and they will be reluctant to give up their special role. These clients will presume this special status with the therapist and are often effective in reestablishing this relational configuration. In particular, such clients may offer the therapist a subtle but all-too-easily accepted invitation to establish a mutual admiration society. These sophisticated clients are often adept at flattering the therapist and establishing an elite sense of mutual, shared superiority; in this cozy relationship, the client is very responsive to nuances in the therapist's mood, is skilled at making the therapist feel special, and, in turn, expects to be treated as special by the therapist. Once this unspoken deal is struck, however, the therapist feels constrained because challenging or disagreeing with the client, or focusing the client inward on her participation in conflicts with others, feels like a betrayal of this special relationship. Such a reenactment, of course, will keep the therapist and client from addressing the client's real problems.

In parallel with their grandiosity, such clients also have a strong sense of inadequacy. This inadequacy is a pervasive source of anxiety that arises because, as children, they were never actually capable of meeting their parents' needs. These clients often suffer from extreme performance anxieties and feel inadequate to meet the exaggerated demands they now place on themselves. As before, the therapist's task is to remain empathic toward both sides of their dilemma. On the one hand, these clients struggle both with feelings of inadequacy and anxiety over all of the things they must try to control. On the other, if they relinquish the demands they are struggling to meet, the clients surrender their family role and identity as special. In so doing, they dissolve the illusion of safety or secure attachment that this coping strategy has provided. Further, to become healthier and relinquish control also threatens to undo or betray their caregivers, who have convincingly but covertly communicated for decades that they cannot get along without them. Thus, making progress and getting better in treatment arouse two intense conflicts: guilt over leaving their parents to manage their own lives, and anxiety over disrupting the illusion of a secure attachment.

The therapist must not overreact to such clients' compliments, criticisms, or distress. The therapist needs to be emotionally responsive, of course, but (recalling client response specificity) these clients will get worse if the therapist's own personal equilibrium can be too readily influenced. In the past, these clients were able to exert too much control over their parents' well-being, and the therapist does not want to reenact this maladaptive pattern. In other words, these clients feel safer and are reassured when the therapist is responsive but without becoming overreactive—even when they try to elicit overreactions from the therapist.

3. *Emancipation conflicts.* Many children caught in cross-generational alliances have problems with individuation and emancipation. If the mother

and father have a primary marital coalition, children are free to grow up. As young adults, they can leave home successfully while still preserving close ties and involvement; they do not have to emotionally disengage or physically break off contact with the family. This is possible because the parents do not need the child to remain dependent on them in order to fulfill their own lives. In contrast, if the primary coalition is between parent and child, these off-spring may feel separation guilt over leaving their parent alone or forsaking their parent to the unfulfilling relationship with the other spouse. As a result, they are especially likely to feel dissatisfied with their achievements, to be unable to establish fulfilling relationships with others, or to be chronically depressed. Guilt over emancipation frequently underlies academic failure in college, as well as many other symptoms and problems that students present in college counseling centers (Teyber, 1983). Thus, cross-generational alliances impose binding loyalty ties that make young adults feel guilty about leaving home, successfully pursuing their own career interests, and establishing satisfying love relationships. Although emancipation issues are expressed differently in varying cultures and will reflect a different balance of separateness and relatedness, symptomatic guilt and depression often ensue when offspring are not permitted culturally sanctioned avenues of individuation and coupling.

Many clients, like Anna in Chapter 4, struggle with binding separation guilt and survivor guilt. These clients may feel guilty about being happy, succeeding in life, or even getting better in therapy. It is important with all clients, but with these in particular, that *therapists enjoy clients' happiness and express pleasure in their success.* Some trainees are unsure of how to respond when clients begin the session by saying, "I feel really good today; I don't have any problems to talk about right now." When that happens, clients with separation guilt are apt to assume, on the basis of their relational templates, that their happiness, success, or independence has hurt the therapist or evoked envy. The therapist disconfirms such clients' separation guilt and faulty relational expectations—and expands the healthier ways in which they can be connected to others—by enthusiastically responding, "That's great! What's the best thing going on for you today?" Without such affirmation of the successful aspects of clients' lives, therapy will bog down and become stuck as these clients repetitively share sadness, fear, or pain without any progress or resolution.

As we will explore in Chapter 10, conflicts over independence and success will be activated for many clients around the time of terminating therapy. In particular, clients who struggle with separation guilt will feel especially bad about no longer needing the therapist and heading off successfully on their own. Therapists must find multiple ways to give these clients permission not to need them, to leave when they are ready to go, and to enjoy their own successful lives. Guilt over growing up and becoming stronger, succeeding in work and love, or perhaps even surpassing the parent or therapist with more happiness or success is a common issue, although few clients will be able to recognize it as such on their own. Fortunately, it is relatively easy to help many clients dispel these binding guilt-related beliefs by communicating in words and behavior that the therapist enjoys their success, takes pleasure in their competence, and is in no way hurt or threatened by their stronger stance.

Finally, cultural factors play an especially important role in these issues of individuation and family loyalty. Family connectedness or close family ties may be of special concern to some Asian or Latino clients or clients from other strong relational cultures. Entering the subjective worldview of these clients, therapists must learn to appreciate the subtle balance between being loyal to family and culture and possessing their own authentic self. In order to become familiar with the culturally sanctioned avenues for individuation that exist, therapists will find it helpful to consult with others who are knowledgeable about the client's cultural context. With such consultation, therapists can help clients find ways to individuate that are acceptable within their own culture. Although the balance of separateness and relatedness will be different in each familial and cultural context, counselors who are open to these differences can help clients find culturally acceptable outlets for expressing their own individuality and forming love commitments.

4. *Parentification of children.* When the primary coalition is between a parent and a child rather than between the mother and father, the result is frequently the *parentification* of children. A role reversal occurs: Rather than the parent responding to the child's needs, the child takes on the role of meeting the parent's emotional needs. When certain of the parent's emotional needs are not met by his or her spouse, the parent inappropriately turns to one or more of the children to meet his or her own adult needs for affection and intimacy, for approval and reassurance, or for stability and control. Therapists should be alerted to the possibility of such problems when they hear adults describe their children as their "best friend," "lifeline," or "confidant" (Teyber, 1994).

It is problematic for children when they become responsible for, or take care of, the emotional needs of their parent. The problem with this role reversal is that children must give to their parents rather than receive, and the children's age-appropriate dependency needs go unmet. These parentified children grow up to feel overly responsible for others, insecure about depending on others, and guilty about having their own needs met. As adults, parentified offspring often describe themselves as feeling empty or "having a hole inside" as a result of having given rather than received throughout their childhoods. Parentified children initially enjoy having such a special role vis-à-vis their parent. As adults, however, they ultimately come to resent having been deprived of their own childhoods.

Parentification is pervasive in the background of clients and therapists alike. Perhaps as many as one-half of all clients seeking outpatient therapy have been parentified to some extent. Because their basic relational orientation is to take care of others, parentified children select careers as nurses, teachers, and therapists. Although kind and capable, they often are guilty about saying "no," setting limits, and meeting their own needs. Because they do not draw boundaries well, they tend to become overidentified with others' problems and are prone to experience burnout. Since they grew up having to take care of their parent, it is now threatening to relinquish this control in their current lives. For example, it will be hard for them to let others share in meeting obligations, although they may also resent having to do everything themselves. These control issues may also be evidenced in symptoms such as airplane phobias, for example, where these individuals must temporarily relinquish control to the pilot. These clients also have problems in close personal relationships because it is too threatening to relinquish the control necessary to be intimate with someone.

Note, however, that in some families—for economic or other reasons—one child may temporarily assume a parental role with younger siblings. This is not the same as parentification, because it is always clear that the parent is in charge when the parent is home and that this child's temporary role is not to care for the parent's emotional needs but to assist in the functioning of the home in the parent's absence.

Parentified clients will be highly sensitive and responsive to the therapist. It can feel great to work with these clients in the short run, because they astutely discern what they need to be or do so that the therapist will feel competent or secure. If the therapist does not collude in reenacting this relational pattern, however, clients can begin to explore the consequences of having missed their childhood. Simultaneously, it will be relieving yet anxiety-arousing for these clients when the therapist begins this exploration by making process comments.

Therapist:

> Here again, it seems as if you are taking care of me. I really do appreciate your concern for me, but I also see how this puts you back in your old predicament. In fact, it leaves me wondering what happens to your needs—if there is any place where you can ever be responded to.

To illustrate this important concept, let's consider a case study. Carol was a highly regarded psychiatric nurse. An utterly dependable and take-charge person, she could seemingly handle every situation that arose. In a hospital emergency setting where she had to deal with very disturbed patients in crisis, her rapid and accurate assessments, good judgment, and compassion for highly disorganized patients had earned her the respect of the entire staff. Although considered a superstar at work, Carol sought therapy for her recurrent depression and loneliness.

Carol was a bright and engaging client who was adept at getting the therapist to lead, talk about herself, and become involved in collegial discussions about interesting clinical issues. The therapist had formulated working hypotheses about this potential reenactment, however, and was usually effective in recognizing Carol's strong pulls to externalize in these ways. Repeatedly, she focused Carol inward (by saying, for example, "What were you feeling when he . . . ?" "What were you afraid was going to happen if you . . . ?"). Carol began expressing that it was new for her to attend to herself and her inner experience in this way, and that it made her feel uncomfortable. The therapist was genuinely sympathetic to her concern and did not press for this internalizing focus when Carol did not want it, but instead asked Carol on occasion how it was for her to look within at her own experience. It soon became clear to both of them that guilt over being "selfish" and shame over being "the center of attention" were evoked by the internal focus. Thanks to the therapist's affirming responses to these reactions, however, both soon agreed that important new material was emerging.

Four weeks into treatment, Carol disclosed something that she had never told anyone: Her stepfather had sexually assaulted her in her early adolescence. Carol successfully fought him off, although she was scratched, had her blouse torn open, and suffered a bloody nose when she literally pushed him off her. Carol recounted her ordeal in detail but without emotion, and the therapist responded effectively in caring and validating ways. The therapist, herself a mother of two daughters, soon asked what was for her the burning question.

Therapist:
You have kept this awful secret for almost 20 years. I'm glad that you can share it with me now, but I'm sorry that you had to be alone with this for so long. What kept you from telling your mother?

Carol:
I didn't want to put more on her. She couldn't have done much anyway, and I didn't want her to worry.

Carol's poignant response illustrates the plight of the seriously parentified child. As in most aspects of her relationship with her mother, Carol's own profound needs for protection and comfort were set aside in order to meet her parent's need.

To sum up, there are many ways in which children are better adjusted in families with a primary marital coalition or clear boundaries between parent and children than in families with cross-generational alliances. Therapists will see the consequences of these structural family relationships operating with most of their clients. As we will see next, the nature of the parental coalition also influences two other basic dimensions of family life: the separateness-relatedness dialectic and childrearing practices.

The Separateness-Relatedness Dialectic

The second basic dimension of family life that informs therapists about clients' problems is the *separateness-relatedness dialectic*. At times, the separateness side of this dialectic is misunderstood. In this context, autonomy, separateness, or individuation does not imply cut-off, selfish, or uninvolved individuals who are insensitive to their responsibilities to others. Rather, autonomy connotes being connected to our genuine self—our inner voice and our own feelings, needs, and wishes. It is the ability to make choices based on awareness of our own experience within our own relational contexts. Thus, the autonomy side of this dialectic connotes connection to our authentic self and the ability to respond purposively and choose, instead of merely complying with what others want us to do. With this clarification in mind, let's see how this family systems perspective refers, in part, to the family's ability to respond to the child's need for both closeness and autonomy.

Recall the dialectic of separateness-relatedness that was introduced in Chapter 1. As children develop, each family faces the complex task of having to provide both secure attachments and support differentiation—to be close and separate at the same time. Both of these behavioral systems—of attachment, on the one hand, and of differentiation (curiosity, exploration, mastery) on the other—are present at every stage of the lifespan. Although there will be varying expressions and timetables within different cultural contexts, strivings for authentic self-definition and expression figure predominately during adolescence for securely attached offspring. Successful parenting requires great flexibility. Whereas caregivers must respond to young children's basic attachment needs, they must exercise different parental skills with adolescents. For example, they must be able to encourage emancipation by giving adolescents permission to develop meaningful relationships outside the family. Thus, healthy families are characterized by *experiential ranging* between these two poles of intimacy and individuation (Farley, 1979). Effective caregivers have the flexibility to range back and forth along this separateness-relatedness continuum and meet children's changing needs both for closeness and for autonomy.

In contrast, less effective families do not have the interpersonal range to function well at both ends of the continuum. These less effective parents cannot meet the changing needs that children present at different developmental stages. Enmeshed families cannot support older children's developmental growth toward individuation, whereas disengaged families cannot provide a secure base for young children and meet their basic attachment needs.

This conceptualization provides a useful model for therapists, who similarly need a broad interpersonal range to respond flexibly to clients' needs. This interpersonal range is a critical component of client response specificity; depending on their developmental history, clients will need different degrees of separateness-relatedness. Furthermore, these needs will change during the course of treatment. As clients improve in treatment, for example, they will develop an increased sense of competence and need support from the therapist for this new, stronger self. With this flexibility in mind, let's see how the separateness-relatedness continuum relates to the parental coalition. The following example shows how families without a primary marital coalition tend to be more restricted in their experiential range.

Suppose that a young married couple has not made much progress with Erikson's identity and intimacy tasks in late adolescence and early adulthood (Solomon, 1973). As a result, they have not been able to psychologically separate from their families of origin sufficiently to shift their primary loyalty ties from parents and establish a successful marital coalition with each other. Such newlyweds are likely to have a high degree of conflict that they are both unable to resolve. In many cases, the couple will attempt to deal with their relationship problems by having children. As Bowen (1978) illuminates, these less-differentiated parents are more apt to experience the children as extensions of themselves. Without being aware of this overidentification, they tend to project their own unresolved developmental and marital conflicts onto the children—in an attempt to distance and control those conflicts. Typically, such parents unconsciously select a certain child who matches their own relational templates—perhaps an oldest or youngest child, a boy or a girl—to carry or express certain unmet needs, unacceptable feelings, or unwanted aspects of themselves.

Through this "family projection process," described by Murray Bowen, Ivan Boszormenyi-Nagy, and other multigenerational family therapists, children are scripted into roles, and restrictive family rules and myths are established to help parents defend against their own conflicts that have been activated by marriage and childrearing.

To illustrate, if the mother's need for closeness or intimacy cannot be met by the father, she may turn to the child to have her emotional needs met. This solution will work well in the short run, but it establishes the basis for long-term family problems. In many cases, the mother could not emancipate psychologically from her own parents and she is made anxious and/or guilty by individuation. As long as the child remains dependent on her, her need for closeness is met and her anxiety or guilt over individuation is managed. On the other hand, the father in this prototypic scenario is uncomfortable with closeness; it makes him anxious. The primary mother-child coalition allows him to gain greater distance from the anxiety-arousing demands of closeness from both the spouse and the child. In this way, the child's role is to hold the marriage together and, at the same time, to ensure distance between the couple.

This situation eventually poses a dilemma for the child's continuing development, however. The child's innate push for growth will destabilize the marital relationship, and this will arouse each parent's own internal conflicts. Thus, these parents need to undermine the child's increasing independence or competence in order to maintain the marital status quo. At the same time, the parents will express anger and frustration to the child about the child's refusal to grow up and act responsibly. This family's limited experiential range is most likely to come to a head during adolescence, when offspring face the developmental hurdle of beginning to leave the family of origin. The mixed messages that the child has received about growing up and remaining dependent will often erupt in psychological symptoms and acting-out behavior at this time. Furthermore, these underlying conflicts have been internalized by the child and often continue to be expressed in symptoms and problems throughout adulthood.

The separateness-relatedness dialectic is a useful construct to help therapists conceptualize basic familial and developmental processes. We will return to the separateness-relatedness dialectic later in this chapter and see how therapists can use it to better understand the interpersonal process that is transpiring between the therapist and the client.

Childrearing Practices

Three Styles of Parenting

Childrearing practices are the third dimension of family life that helps therapists understand how symptoms and problems develop. Although the following discussion oversimplifies the complexity of parenting, researchers find that most families employ one of three predominant parenting styles: *authoritarian, permissive,* or *authoritative.* Although other parenting styles occur, these three are most common. We will examine these three styles of childrearing and see how each is related to the problems that clients present in therapy. Much of this discussion draws on research by Baumrind (1967, 1971, 1983) on the effects of childrearing practices.

Authoritarian parents are firm disciplinarians. Children are given unambiguous prescriptions for acceptable and unacceptable behavior. Parental rules are clearly explained and the consequences for violating them are consistently enforced. Authoritarian parents also hold high expectations for their children to behave in a responsible and mature manner. Children are expected to perform up to their abilities and to be competent and contributing family members.

Authoritarian parents do not give children reasons or explanations for the rules they set, however. They do not discuss compromises or alternatives with children. Children of authoritarian parents cannot ask why a rule is set; they must simply obey. Furthermore, although authoritarian parents discipline their children and expect much of them, they do not provide their children with much warmth or affection.

In contrast, permissive parents are lax disciplinarians. Children often do not know what behavior is expected of them, and they do not know what consequences will occur if they violate parental norms. Most importantly, permissive parents do not consistently enforce the rules that they do set. As a result, children of permissive parents learn that they do not have to obey because parents will not follow through and enforce the rules they set anyway.

Thus, authoritarian parents exert a high degree of control over their children. However, these limits and rules are not coupled with warmth and affection, with parental explanations for why the rules have been set, or with opportunities for the children to discuss compromises and alternatives with their parents. In contrast, permissive parents may be more expressive and communicative with their children, but their children are not disciplined, nor are they expected to behave in a mature, responsible manner. Before looking at the problems that result from these two ineffective childrearing styles, let us contrast them with the more effective, but less well-known, authoritative style.

The authoritative childrearing style produces the most healthy, well-adjusted children. *Authoritative parents possess a wide range of parenting skills and are able to combine firm discipline with nurturance*. Although authoritative parents set and enforce firm limits, they also communicate with their children. Authoritative parents give reasons and explanations for the rules they set and they will negotiate compromises and discuss alternatives with their children when appropriate. While authoritative parents expect responsible behavior from their children, they also provide physical affection and spoken approval. In short, authoritative parents combine the most effective features of the other two parenting styles. They are nurturing parents who communicate with their children, but they are also firm disciplinarians who place high demands for maturity on children. Research has found that authoritative parents produce children who are independent, self-controlled, successful with peers, and who generally have a positive, happy mood.

We have seen that the authoritarian parent sets limits but does so harshly, whereas the permissive parent cannot take charge and place appropriate controls on children. Most parents are unable to both set rules and do so in a caring way; they believe that they must be either strict or loving. In contrast to this common polarization, authoritative parents are more effective because they have a wider experiential range; they can be both firm and loving. Despite the best of intentions, however, balancing these two domains is hard for most parents. Furthermore, authoritative parents must also be discriminating enough to accurately assess the upper level of their children's abilities. Because growing children are constantly changing, it is not easy to assess what demands for mature behavior will challenge children yet allow them to succeed. Thus, it is understandable—but unfortunate—that only a minority of parents (10–12 percent in many studies) provide such effective authoritative parenting.

As trainees listen to their clients' narratives and learn more about their relational patterns and developmental histories, they will most commonly find that their clients have been reared by authoritarian parents. Offspring from permissive parents also have problems but, as we will see, they are less likely to seek help. What about children of authoritative parents—is this an idealistic family without problems? Parenting and development are always challenging, even in the best of circumstances, and conflict-free families and individuals are a myth. On average, however, children of authoritative parents are better adjusted and are less likely to be seen in longer-term treatment for enduring problems. More likely, they will seek help in crisis situations (such as a child's major illness) or when negotiating developmental hurdles; for example, they may seek premarital counseling. In the following section, we examine more fully how childrearing practices are reflected in the problems that clients present in therapy.

Consequences of Childrearing Practices

In socialization, young children gradually give up their own immediate demands in order to maintain parental approval. It is the threat of parental punishment or disapproval that leads young children to conform to parental standards. If all goes well, children's conflict between the desire to gratify their own impulses and the fear of parental punishment will be successfully resolved. *A successful resolution is one in which children learn to obey, but without sacrificing their own initiative and positive self-regard.* Developmental researchers tell us that healthy children can become self-controlled and self-reliant without inhibiting their own initiative or losing their sense of being prized by their parents (Wenar, 2000).

In contrast, children of authoritarian parents must give up too much of their initiative and positive self-regard to maintain parental approval because authoritarian parents provide too little nurturance and affection. These children are obedient and achieving, but they are also anxious and insecure; they comply with their parents out of fear.

By the time the child is of school age, this conflict with the authoritarian parent has been internalized: What was originally an interpersonal conflict becomes an internal or intrapsychic conflict. These well-behaved, insecure children become harsh, critical, and demanding toward themselves, just as their parents have been with them. Many of these offspring will seek therapy as adults, presenting with symptoms of guilt, depression, unassertiveness, anxiety, and low self-esteem.

Offspring of permissive parents will also have adjustment problems, but their symptoms are different. Although these offspring may enter treatment with anxiety, depression, and other internalizing symptoms, they are prone to develop acting-out or externalizing problems. They may be less likely to seek therapy as adults because they often avoid taking responsibility for themselves and see their problems as external to themselves. Children of permissive parents often learn that they can avoid the consequences of their own behavior by manipulating others. They learn that rules and limits do not apply to them, and expect wishes to be gratified without delay. Research has found these children to be dependent, immature, demanding, and unhappy. They have little self-control, low frustration tolerances, and do poorly with their peers. As adults, these offspring tend to be impulsive and to avoid taking responsibility for their own behavior. As a result, they are more likely to develop acting-out disorders such as alcohol and substance abuse. They also tend to be more self-centered and demanding in relationships, and less capable of making commitments and following through on obligations.

Authoritarian Parenting, Love Withdrawal, and Insecure Attachment

Authoritarian parents typically employ discipline techniques based on love withdrawal. Instead of communicating that they disapprove of the child's behavior, parents respond with anger or rejection and communicate disapproval of the child's basic self. This communication is often nonverbal; it involves inflection, gesture, and facial expression as much as it involves words.

To punish or shape the child's behavior, or to communicate their anger—which may be accompanied by the far more damaging affects of contempt or disgust by more severely authoritarian parents—these parents withdraw their warmth and emotional connection to the child. They may say, for example, "Get out of here! I don't even want to have to look at you." Moderately authoritarian parents may respond in these ways occasionally, when they are tired or upset, and may later try to clarify that it is only the child's behavior that they disapprove of, not the child's personhood. For highly authoritarian parents, in contrast, withdrawal of approval and emotional contact is frequent, and the child's emotional ties to the parents are disrupted on a regular basis. Understandably, the child develops symptoms and defenses to cope with the separation anxieties and the resulting shame-based sense of self that these painful attachment disruptions engender. For these children, shame is evoked because their attachment needs are rejected—leaving them exposed and vulnerable.

A continuum of authoritarian responses   Authoritarian childrearing and love withdrawal techniques both occur on a continuum of severity. In families that are only somewhat authoritarian, disruption of ties from love withdrawal may not be severe, and may occur only when parents are stressed or tired. If love withdrawal does not occur routinely, and if other opportunities for emotional connection are available, ties can soon be restored. Although the anxiety may still be painful, the child can fashion a reconnection through compliance and assuming full responsibility for causing the disruption. As therapists become more familiar with these attachment configurations, they will begin to recognize the coping strategies their clients have learned to employ in their efforts to restore ties to others, and the pervasiveness of *compliance* in the symptoms and problems clients present. For example, the pathogenic beliefs accompanying these coping strategies may sound like these: "I am bad or unlovable, I should just go away" or "It's my fault that they don't want me; if only I could try harder (to be better, achieve more), then they will love me."

In families that are more severely authoritarian, however, the disruption will occur more often and more severely. In daily interactions around discipline and control, children in these families will be exposed regularly to experiences of interpersonal loss and emotional isolation, even though the parent and child remain in physical proximity. As we move further along this authoritarian continuum, ridicule and rejection occur more overtly and may erupt into physical abuse. In this situation, the child experiences his parent's anger and contempt as assaults on his basic sense of self, leaving the child vulnerable and psychologically alone. Concluding that they are to blame for their parents' anger, such children feel ashamed of themselves. The attachment researchers teach us that this sense of shame and feeling blame-worthy are adaptive if these children are to experience some sense of control in their lives. The unwanted alternative would be to view the parents more realistically as rejecting and punitive, which would leave the child feeling even more powerless and unattached. When the child experiences this severely authoritarian style of parenting, it is the genesis of an essentially shame-based sense of self.

Moving even further out on the continuum of authoritarian responses, some parents are so completely removed psychologically that they become dissociated. For a few moments, usually while in the midst of their own shame-rage cycle, these parents lose all vestiges of emotional connectedness to their own experience. More importantly, they lose any awareness of their child's experience and all feeling for the frightening or humiliating impact they are having on their child at that moment. It is terrifying for children when threatening caregivers become dissociated in this way; they may describe their parents in these moments as being "possessed," "not there," or "somebody else." Most clinical trainees will not be familiar with these *dissociative ego states,* and will be understandably upset by the shame and terror these states engender in some of their clients. However, by learning about these highly dysfunctional families, therapists also learn important principles for working with clients from less disturbed backgrounds.

Caregivers may abuse or mistreat children for many different reasons. Some abusive parents were indulged when they were children and experienced no consistent limits or consequences for their behavior. Whenever they did something wrong or got in trouble, their caregivers externalized all responsibility for the problem and blamed the teacher, coach, or other child. Others mistreat their own children in the same ways they were once mistreated. *This multigenerational reenactment is especially likely if the relationship was highly ambivalent—that is, the parent who hurt them also was authentically caring or affirming at times* (Rocklin & Levitt, 1987). Abuse of this sort tends to occur when parents are in dissociated ego states, which often have been facilitated by alcohol or other substance abuse. In those states, parents reenact with their children abusive events from their own childhoods. In many cases, these parents can remember and talk about the abusive events that happened to them, but as we saw in Chapter 5, the fear and shame that accompanied their experience is not available to them and they relay the story without affect. By reenacting the abuse, rather than reexperiencing it, they defend against the painful feelings evoked by their own mistreatment. Typically, *one of their own children comes to represent themselves or the unwanted or disowned aspects of themselves* (the emotional needs, vulnerability, or anger that originally evoked the abuse). Which child is chosen will depend on age, gender, birth order, temperament, and other characteristics.

This disturbing reenactment has received different labels within different theoretical systems: projective identification (object relations theory); identification with the aggressor (psychoanalytic theory); the family projection process (family systems theory); turning passive into active (control/mastery theory). However labeled, it is a primitive attempt to manage trauma and gain control of feelings that once were intolerable. This intergenerational transmission of pathology stops if the abusive parent can not only remember the events, but also tolerate experiencing the fear and shame they originally felt and dissociated from in their own past. This psychological task is not easy to accomplish.

Let's review these complex dynamics again. Object relations theorists and family systems theorists both tell us that the more conflicted or unintegrated we are, the more likely we are to instill unacceptable aspects of ourselves in others—especially in our children. As we discussed in Chapter 4, evoking or instilling our own conflicts in others is a way of seeking an external solution to an internal problem. Thus, in this type of abuse, the abusive parents' reenactment with their own child is a primitive psychological defense—often an attempt to gain some control over the painful feelings associated with their own past abuse. By externally reenacting some version of the trauma over and over again with their children, abusive parents do not have to remember or experience internally their own painful feelings. That is, by evoking the same terrified or otherwise unacceptable feelings in their child that they were once made to feel, abusive parents do not have to experience the frightening trauma and shame-laden defeat of abuse as their own; their fear and shame is expressed vicariously through the child. For children in extremely authoritarian families, such abuse is a persistent threat that organizes their daily experience and, ultimately, their psychological adaption to life.

Effects of severe love withdrawal   Severe love withdrawal as a means of discipline is primarily associated with highly authoritarian parenting, but it occurs in other families as well. For example, variations of this dynamic can be seen in parents who act like martyrs. These parents communicate their hurt and disappointment nonverbally, by turning away and withdrawing emotionally or by means of a sigh or long-suffering look. Other parents also demand perfection from the child, exert excessive control over the child in order to obtain it, and withdraw their love when the child does not fulfill their perfectionistic demands. In these moments of parental love withdrawal, however they occur, the child in effect loses the parent. The child's emotional connectedness to the parent is temporarily broken and the child's attachment ties are situationally disrupted. This engenders separation anxieties and shame until the child can find a way to comply and restore the tie. Of course, all children's ties to their parents will be threatened at times. Significant problems occur when disruption of such ties is so intense and frequent as to characterize the relationship, and when these rupturing interactions are disavowed by the parent rather than acknowledged (for example, Parent: "Daddy just lost his temper. I'm sorry I scared you, that was my fault, not yours.").

A constellation of significant emotional reactions occurs when authoritarian parents cut off their emotional connection to children in anger or disgust. Even though the caregiver is physically present, the child is psychologically alone and suffers painful separation anxieties. This withdrawal also gives rise to helplessness and hopelessness, because the child cannot win the parent's love, and eventually to dysthymia and a shame-based sense of self as ineffectual. Thus, this child is made to feel bad and alone, and desperately wants to restore the relationship and renew emotional ties, but is ultimately helpless to do so until the parent reengages. Moreover, the child's inefficacy is further exacerbated by being made to feel responsible for and deserving of the parent's disapproval or withdrawal. In this immobilizing double-bind, children believe that their behavior (spilling a glass of milk, crying, feeling angry, needing help, and so forth) has caused the parent to go away, which is the response the child fears most because it threatens already insecure ties.

The child is angry at being abandoned and wishes to protest, of course, but this would only elicit further threats from the authoritarian parent. Through power assertion, the authoritarian parent usually does not allow the child to disagree, let alone find appropriate means of expressing anger. (The child may be told, for example, "I'm your father. You are never angry at me. Do you understand that? Look at me and say, 'Yes, sir.'") Thus, the child cannot protest interpersonally, or even experience anger internally, because such reactions will further threaten already tenuous ties to the parent. As a result, the child often turns the anger inward; it is expressed through self-deprecation, dysthymia, and having a shame-prone self. This tendency toward self-blame is exacerbated as the child comes to identify with the parent and adopts the same critical or contemptuous attitude toward himself or herself that the parent originally communicated. Just as this type of parent loses touch with the child's feelings or experience in these angry moments, the child in turn loses the clarity and authenticity of his or her own internal experience; such children lose touch with important aspects of themselves. Denial, splitting defenses, and disassociative tendencies result when these processes have been intense and pervasive. Poignantly, when they enter treatment as adults, such clients often describe their parents in idealized, conflict-free terms.

Low self-esteem, internalized anger, inefficacy, and loss of the caregiver's love constitute a prescription for depression. They also engender anxiety symptoms and the control problems found in eating disorders. In addition, these offspring develop identity conflicts because their feelings have been so pervasively invalidated. Because highly authoritarian parents are so rigidly demanding of conformity and obedience, children soon lose touch with their own internal experience. They may not know what they like and dislike; they may even be confused about what does or does not feel good to them. Their own experience can be so completely overridden that they have no basis for later developing their own belief systems, differentiating their own values, or formulating occupational interests in late adolescence and young adulthood. In other words, they haven't been allowed to "have their own mind."

Trainees will often find themselves working with the emotional conflicts and family dynamics described above. With these clients, the therapist's initial intervention strategies are (1) to validate their subjective experience which has been so pervasively invalidated; (2) to encourage their initiative which has been undermined—for example, by following their lead in therapy whenever possible; and (3) to provide a treatment focus by helping them clarify and act on their own feelings and interests whenever possible. To better understand these dynamics, let's look further at the developmental experiences of children in highly authoritarian families when emotional ties are disrupted by severe love withdrawal.

The need for constancy of parental relatedness   There is already too little affection in authoritarian families, and these thin threads of connection are repeatedly disrupted when angry or demanding parents emotionally disconnect from children. When young children are unable to maintain parental affection, they will not grow up to feel love-worthy. These children have missed the essential experience of constancy in their attachment bonds. This developmental deficit has two aspects.

First, these children missed the experience of someone actively reaching out and choosing them. They do not feel loved, cherished, or even wanted in many cases. If the parent's responsiveness is dependent on the child's efforts, and the child feels responsible for eliciting the caretaker's love, the child will not develop a secure attachment. In authoritarian families, children must find ways to cope with the anxiety generated by their insecure ties. To establish more secure bonds, they attempt to control or manipulate parental feeling for them; they learn to employ coping strategies aimed at winning approval—for example, by being compliant and pleasing, by compulsively striving for achievement, or by perfectionistically trying to be good. As we will examine in Chapter 7, this attempt to win approval and love often becomes a pervasive coping style that is integral to clients' symptoms and problems. Because of their childhood experiences, these clients often believe they have to elicit, or be responsible for, the therapist's interest in them.

Second, whereas secure children do not feel they have the power to break the parent-child tie, insecurely attached children are often induced to believe that their 'bad' behavior disrupts the parent's loving commitment to them. Children in authoritarian families learn that their anger, their tears, or even their questions can provoke their parent to disrupt the parent-child tie with threats such as these:

> Stop those tears right now or I'll give you something to really cry about!
> Wipe that angry look off your face. You will do what I say, and like it, or else!
> Never ask me why. Just do what I say, when I say it.
> Now you've really done it. This time I've had it with you for good.

Even as adults, the sons and daughters of authoritarian parents may be subjected to the same threats: If they do not do what parents demand, they will (seemingly) destroy the relationship.

> If you marry him, we won't come to the wedding or visit you anymore.
> If you get a divorce, we are going to disown you and take you out of the will.
> If you do that, no one in this family will ever speak to you again.

The important distinction here is that authoritarian parents and others who use severe love withdrawal techniques to control their children are not just setting limits on unacceptable behavior; they are threatening to cut off fundamental relational ties. As attachment-seeking children try to cope with the intense anxiety this arouses, compliance becomes a generalized trait, pervasive personality constriction and inhibition occur, and obsessive/compulsive symptoms often develop.

A diversity of attachment configurations   In their initial caseloads, most beginning therapists will work with some clients from authoritarian backgrounds. For counselors who have had better developmental experiences, it may be hard to appreciate the emotional severity of highly authoritarian parenting. These therapists often wonder how the ostensibly normal and, in many other ways, decent parents of these clients can be so rejecting on occasion and can have caused such profound insecurities or even self-hatred in their clients. Other trainees may have difficulty with these client dynamics because they evoke the therapists' feelings about their own authoritarian backgrounds. To help therapists better understand this sometimes challenging material, let's look at a range of secure and insecure attachment configurations.

While growing up, Molly generally experienced secure relational ties without fear of love withdrawal. She recalled an incident when her mother was angry about something Molly had done when she was 7 years old. Her mother made strong eye contact with her, reached out and touched her on the shoulder, and said in a calm but firm voice, "I love you, but I don't like it when you act this way and I don't wish to be around you." Molly, feeling confused, protested, "But you have to want to be with me. You're my mom!" Her mother went on to explain that it was Molly's behavior that she did not like, and Molly recalled that she began to understand that concept.

Her mother was angry, and she got her point across that she did not like what Molly was doing, but Molly also felt secure in her mother's love. Looking back, Molly thought that her sense of secure attachment in this conflict was maintained primarily by the nonverbal messages that accompanied her mother's restrictions and explanations. Because of these secure relational ties, Molly was able to internalize her mother's loving feelings for her and to develop object constancy, as discussed in Chapter 1. As an adult, Molly is able to establish friendships in which she feels respected and affirmed and to have a marriage in which aspects of this same loving affect are present.

Often, children will be secure in their emotional ties with one parent but struggle with a lack of constancy with the other parent. For example, Ellen recalled that she always felt secure with her mother—even when her mother was mad or disapproving of something she had done. Her father was highly inconsistent, however; at times he was affectionate and available, but at other times he was overtly rejecting. Ellen recalled happy memories of her father patiently painting ladybugs on her roller skates, as well as painful memories of her father angrily yelling, "Get out of here. I can't stand having you around!"

Ellen's secure base with her mother allowed her to cope with this intensely ambivalent relationship with her father. With the constancy of emotional contact provided by her mother, she could learn to anticipate her father's moods and stay away from him when necessary. Her feelings about herself were not based on the unstable fluctuations of her father's moods. As an adult, Ellen is especially perceptive and aware; personality strengths often develop from such conflict-driven demands for coping. Ellen's brother, however, was not so fortunate. Ellen's mother seemed to enjoy raising girls more than boys, and her brother did not receive the secure emotional base with his mother that Ellen enjoyed. As a result, he was left to bob about unconnected at the mercy of his father's stormy emotional seas. Ellen describes her brother as always feeling very bad about himself and being depressed a lot. Now in his late 20s, he has been unable to make commitments to relationships or a career and cannot find a life for himself. Ellen says she worries about him a lot.

Let's look more closely at insecure attachments. With certain clients, therapists may wish to ask directly about the constancy of attachment ties: "How did your parents respond to you when they were angry or disciplining you, and how did that make you feel about yourself?" or "Tell me, as specifically as you can, what each parent said, did, and felt toward you when they were upset with you."

In response to such queries, many clients will describe disciplinary techniques of love withdrawal that threatened their ties to their parents. Often, the only way for these clients to cope was to accept their familial role of being bad, too demanding, too sensitive, and so on. The following are some sample comments directed at clients by their authoritarian and/or rejecting parents, along with the clients' statements about the impact of these messages. Beginning therapists will find that they commonly hear comments like these from their clients. Oftentimes clients will be able to remember and talk about such interactions, but the painful feelings accompanying them are unavailable.

Parental Comment or Action	Impact on Child
"Stop it, or I'll send you to live with your father!"	The client, whose father was an alcoholic, said this threat made her feel that her mother didn't really want her or care about her. As a child, the client was good all the time and stayed away from her mother to avoid this threat; because of her *vigilance*, it was not often voiced.
"How could you do this to me? Can't you think of anybody but yourself? What's wrong with you?"	The client said she always felt guilty and tried to "earn love" by figuring out what her parent wanted and trying to provide it.
"I've had it with you! Just you wait until your dad gets home!"	After this comment, the client's mother would withdraw and not speak to her for the rest of the day. The client said it made her feel "very alone."
"Look what you did! Get away from me! I don't want anything to do with you!"	The client said he had always felt that he was a "terrible person."
"You did it wrong again; you always do it wrong. You're ruining my life!"	The client said she "hated" herself.
A father would wordlessly slap his son's face for "eating wrong."	The client said, "I don't remember having any feelings about it."

Highly authoritarian parents may threaten their children, break off emotional contact with them, and—in words and, especially, through tone—may communicate contempt or disgust for them. Children from such families typically struggle with anxiety and depression throughout their adult lives, and often report feeling guilt, loneliness, and low self esteem when they enter treatment but without understanding why. These clients have also developed elaborate interpersonal strategies to cope with this trauma, such as being good, taking care of the parent, or being quiet and "going away inside." As we will examine in the next chapter, *the therapist needs to highlight these interpersonal coping strategies and focus on them with clients, or else they are likely to be reenacted in the therapeutic relationship along the process dimension.* These interpersonal defenses originally helped to protect the client from experiencing the painful feelings that resulted from repetitive hurtful interactions. If therapists acknowledge or inquire about these coping strategies when they see them occurring, this accompanying pain will soon emerge. When therapists respond with affirmation and compassion, however, clients can tolerate reexperiencing these feelings, benefit greatly from the therapist's validating response, and begin integrating these sequestered feelings. Only then can these loyal clients stop protecting their caregivers at the expense of their own symptoms, stop denying their own feelings, and stop maintaining family myths of happiness and togetherness that mystify and disempower.

Clinical implications for working with disrupted ties   Parenting is probably the most challenging task in life. The influential family therapist Salvadore Minuchin says forgivingly that parenting has always been more or less impossible. Almost all parents are trying hard to do the best they can for their children. Even many of the highly authoritarian parents we have been discussing, who indeed caused significant lifelong problems for their children, are not cruel or ill-intended people in most cases. In parenting, as in other aspects of personality, people are uneven in their development. Most of these parents do other things well for their children, live by certain moral standards, believe they are trying to do what is best for their children much of the time, and usually are treating their children as well as or even better than they themselves were treated. Although children certainly are hurt by such childrearing practices, almost universally they still love their parents and seek their approval.

Therapists will not do well when they fail to appreciate adult clients' willingness to give parents another chance or their lifelong efforts to improve or repair these flawed but all-important relationships. Therapists need to help clients assess whether caregivers have changed and to what extent they are capable of responding better now than before, and to help these current relationships between parents and adult offspring become as good as they can be. It is important that therapists encourage clients to express the full range of their positive and negative feelings toward their caregivers. Therapists who foreclose on this process by quickly echoing, and remaining stuck on, the negative aspects of the caregivers will rob their clients of the very important experience of integrating both the good and the bad parts of their caregivers. Over the long term, clients who cannot integrate the good and bad aspects of their parents will also have difficulty accepting and integrating the good and the bad parts of themselves (and of their children), which they need to do if they are to live a more authentic life.

In sum, the therapist's role is not to foster splitting defenses by bashing parents who have been hurtful, by making them bad, by encouraging clients to reject them or break off contact, or by replacing the parents with idealizations of the therapist. Nor is it to deny or in any way to minimize the real impact of the hurtful interactions that have occurred. Instead, the therapist helps clients to realistically come to terms with the good news and the bad news in their family of origin, to change their own responses to problematic others in current relationships, and to facilitate clients' current attempts to establish new relationships that are more affirming and responsive.

Therapists provide a corrective emotional experience when they can sustain a working alliance and remain emotionally available to their clients. They provide many clients with the consistent, secure base that they missed developmentally. *Such consistent emotional availability, week in and week out over the course of treatment, usually has more effect on client change than do more dramatic but isolated incidents of compelling insight, important self-disclosure, or other significant therapeutic interventions.* As we will see, however, providing this consistent presence in the face of the client's maladaptive relational patterns is often difficult. The inevitable mistakes, misunderstandings, and reenactments that occur will "rupture" the therapeutic alliance at times. The issue is not whether such ruptures will occur, but the therapist's willingness to acknowledge the conflict in the therapeutic relationship and engage the client in working together to restore the alliance (Safran & Muran, 1998). Let's examine this more closely.

The clients we have been discussing share the experience that caregivers withdrew from them in one way or another or were not emotionally available to them at important times of need. This developmental deficit—and the maladaptive relational patterns that result from it—can readily but subtly be recapitulated in the therapeutic relationship. For example, some therapists may have trouble being emotionally available to clients' pain and vulnerability; they may not be able to remain present with certain affects, such as the client's raw shame or intense sadness. The client's individuation may make other therapists uncomfortable because success or independence was not supported in their own development or because the client's success evokes the lack of fulfillment in their own work or marriage. Some therapists may also feel sad about losing the caretaking role they were scripted to assume in their family of origin. When such common countertransference reactions occur, clients are again left alone and unconnected in their experience. Clients' pathogenic beliefs that being sad, mad, or successful causes others to be hurt or angry, or to disengage emotionally, are confirmed rather than resolved by this interpersonal process. When such reenactments occur without being addressed and rectified, they generate further insecurity for clients as old templates and problematic relational expectations are confirmed.

In addition to countertransference issues, certain factors in the client also make it difficult for therapists to remain consistently available. For example, many clients will report that they like and trust the therapist, but they may also believe that, if the therapist really knew them, she would not respect or care about them. These clients believe they have deceived the therapist or manipulated the therapist's positive feeling for them. This is why it is necessary for the therapist to help clients adopt an internal focus and to draw out the full range of clients' feelings toward the therapist—including the client's perceptions of the therapist's reactions to the client (for example, Therapist: "What do you think I am feeling toward you as you tell me this?").

Unfortunately, clients often hold the pathogenic belief that their despised, weak, or otherwise unacceptable emotions constitute their real selves. This far-reaching problem is resolved when the therapist sees the vulnerable, dependent, shameful, or other parts of themselves, which the clients believe are the damning proof or confirming evidence of their basic unacceptability, and still remains affirming and emotionally connected. *The clients' encounter with compassion and understanding, rather than the emotional withdrawal, rejection, and so forth that they have learned to expect, is one of the most powerful relearning experiences that therapists can provide.* Rogers

(1951) emphasized that clients do not change until they accept themselves, and that they begin to accept themselves when they feel such acceptance from the therapist. Thus, by providing a safe holding environment that contains these core conflicted feelings, as clients are experiencing them, the therapist allows clients to resolve them.

This pivotal experience not only permits the integration of those warded-off feelings, but also opens up a variety of therapeutic opportunities. For example, in the next minute or two, clients often disclose important new material or make significant links to other related problems they have been working on. Clients may risk trying out previously threatening new behaviors with the therapist; they may be more assertive or risk bringing up problems between them that had been too threatening to broach. In the week that follows, clients may act in these stronger ways with significant others in their lives. Clients may also express more self-acceptance, resolve some ambivalence, or make an important life decision they had been unable to act on. Further, therapists can often successfully re-frame problems or help clients see a situation in a different way—a new perspective that did not seem realistic or relevant previously. Because they feel safer now, clients may be able to hear the therapist challenge a pathogenic belief about themselves or identify a recurrent relational pattern. In short, clients often can make use of a wide variety of therapeutic and educational interventions at this point that they could not utilize before.

Finally, *some clients are adept at eliciting the same problematic responses from the therapist that they have received in the past.* As we have already emphasized, a basic tenet of the interpersonal process approach is that most clients' conflicts will be temporarily reenacted with the therapist along the process dimension and need to be identified and reworked. It is easy to be empathic, warm, and genuine with clients who are usually cooperative, friendly, and respectful. However, it is far more challenging for therapists to remain committed to a working alliance with difficult clients because their interpersonal coping strategies—for example, being hostile, dominating, passive, and so forth—eventually alienate, intimidate, or frustrate the therapist. With these clients, it is especially important for therapists to keep process notes and formulate working hypotheses about the clients' maladaptive relational patterns—*to hypothesize what the clients have tended to elicit from others and what they are likely to evoke in the therapist.* When therapists prepare such conceptual formulations, they increase their chances of providing a corrective experience, rather than responding automatically to the client and merely reenacting the problematic scenario with which the client is familiar (see Appendixes A and B). The following case study illustrates how the therapist was able to resist reenacting the client's maladaptive relational scenario and, instead, provide a corrective relational experience by remaining emotionally connected to him during a challenging disagreement.

John, an 11-year-old client, was being treated in a strict, physically punishing manner by his authoritarian stepfather. His mother passively accepted her husband's harsh corporal punishment and derision of the boy; she often walked away and left the room when the stepfather was angrily deriding him.

In treatment, John defied his female therapist, responded contemptuously toward her, and repeatedly tried to push her away. Near the end of an especially frustrating session, in which he had repeatedly pushed every limit, John was unwilling to speak. Trying to find some way to salvage something of a working alliance with him, the therapist agreed that they would not have to talk together if he did not want to and that he could just throw the ball as he wished. Trying to join him in his chosen activity, she asked if she could silently play catch with him. He agreed but, of course, began throwing the ball too hard. This process continued throughout the session: John tested, and the therapist set limits by saying, "You have to throw the ball below my shoulders," while still working hard to try and find some way to connect with him.

Near the end of the hour, the therapist invoked their standing rule and asked him to help her pick up the room. John refused; the therapist insisted. Exasperated, she held him firmly by the arms, whereupon he recoiled. At this moment, the same conflicted emotions that John struggled with at home were evoked toward the therapist and were about to be reenacted in the session. The therapist had been sorely pushed by John for weeks; but even though she was frustrated with John, she was still trying to find another way to relate. She did not withdraw in resignation as his mother did, although she felt like it. She also realized that she did not want to physically dominate him as she was beginning to do, and merely reenact a milder version of what his stepfather did. Instead, she met his eyes and slowly said, "You're a good boy. I still like you. I want us to work this out together." Even though she was upset, the therapist could still communicate that she felt for John and was remaining emotionally connected to him. This was a turning point in treatment: John was not able to reenact with the therapist the same painful scenario of domination by stepfather, withdrawal by mother, and angry but helpless isolation for John that had been played out so many times at home. John grudgingly helped the therapist pick up the room; at their next session, however, John was more respectful of and communicative with the therapist.

In closing, clients will present with an extraordinary diversity of developmental experiences, family dynamics, and interpersonal coping strategies. Many clients' dynamics will be unrelated to the issues discussed here. However, authoritarian childrearing practices, insecure attachments, and love withdrawal techniques will be among the most common experiences clients report in therapy.

Relating the Three Dimensions of Family Life

Let's now examine some relationships among the three basic dimensions of family life: the nature of the parental coalition, the separateness-relatedness dialectic, and childrearing practices. Authoritative parents have the emotional flexibility and range of parenting skills necessary to meet children's needs for both closeness and individuation. The experiential range of authoritarian and permissive parents is more limited—specifically, to opposing halves of the separateness-relatedness continuum. Thus, the range of authoritarian parents does not include intimacy, nurturance, and mutuality. These parents tend not to satisfy young children's attachment needs and older children's age-appropriate needs for warmth and understanding, although their emphasis on mature, responsible behavior often allows children to achieve. In contrast, some permissive parents may meet children's emotional needs better, but they do not foster their children's individuation by expecting responsible behavior from them. Moreover, permissive parents are more likely to embroil children in cross-generational parent-child alliances, which further stifle their individuation. Of the three parenting styles, permissive parents are least likely to establish a primary marital coalition, most likely to embroil children in cross-generational alliances, and most likely to parentify children.

The family dynamics presented in this chapter offer a parallel to the relationship between the therapist and the client. Therapists, like authoritative parents, must develop their own capacity to respond at both ends of the separateness-relatedness continuum. On the one hand, effective therapists are those who have a broad experiential range that allows them to care and emotionally respond to clients. On the other hand, therapists must also be able to set limits with clients, question their actions, and tolerate clients' anger or disapproval at times. More specifically, therapists need to be able to respond to the emotional deprivation of clients who have been reared by highly authoritarian parents, and they must also be able to set limits with the provocative, demanding, or testing behavior of clients who have been reared by permissive or inconsistent parents. Thus, the personal challenge for all therapists is to examine their own experiential range, which has been shaped in their families of origin. Therapists must be willing to explore the interpersonal spheres that are anxiety-arousing for them, and broaden their own interpersonal range so that they can respond to the diversity of problems that clients present. In order to achieve this, therapists must do their own family-of-origin work.

Such work is an integral part of training for those who specialize in family therapy. In particular, Murray Bowen (1978) has written extensively about how clinical trainees can systematically study, and sometimes change, maladaptive relational patterns in their own families of origin. *By achieving greater understanding of their own family backgrounds, and perhaps greater empathy for all family members, therapists can make a significant contribution to their own clinical skills.* Readers are encouraged to read and embark on the great adventure of doing family-of-origin work (Boszormenyi-Nagy & Spark, 1973; Bowen, 1978). In particular, therapists can prepare a family genogram—a three-generational map of family rules, roles, myths, and structural family relationships (Hartman, 1978; Minuchin, 1974). By learning about their families of origin in this way, therapists learn much about their own countertransference propensities, become more compassionate toward their clients' struggles to change, and better appreciate the profound influence of familial experience on adult life.

Closing

In this chapter, we have examined three basic dimensions of family functioning: structural family relations and the parental coalition; the separateness-relatedness dialectic; and childrearing practices. These three aspects of family life will help therapists better understand many of the problems that clients present. For some trainees, it is disruptive to read this material because it violates familial rules and cultural prescriptions to question or examine what transpired in their families of origin. Also, many readers are parents or grandparents themselves. As they learn more about family functioning, and parenting practices in particular, guilt may be evoked as they recognize the limits of their own parenting skills and the problematic consequences that have resulted for their own children. Such readers often need to relinquish unrealistic expectations of themselves, and forgive themselves for not knowing more about childrearing than they did at the time.

Suggestions for Further Reading

1. Clients make more progress in treatment when therapists can help them come to terms with the good news and the bad news in formative relationships with caretakers. However, it is far more challenging for both therapists and clients to do this when abuse has occurred. To help with this complex issue, readers are encouraged to read "Ambivalent Feelings: Case Study of Sheila" in Chapter 6 of the Student Workbook.

2. *Becoming Attached*, Robert Karen's highly readable overview of attachment, is an illuminating book that will help therapists better understand many of their clients' symptoms and problems (New York: Oxford University Press, 1998).

3. Basic information on child development and child psychopathology will help therapists understand their adult clients' personalities and problems. One outstanding text is *Psychopathology from Infancy through Adolescence: A Developmental Approach* (4th ed.) by C. Wenar (New York: Random House, 2000); see especially Chapters 4–8.

4. Parentification is an essential concept to understand because so many clients and therapists have been scripted into this familial role. Further information about this important concept can be found in Chapter 9 of E. Teyber's *Helping Children Cope with Divorce* (San Francisco: Jossey-Bass, 1994).

5. Structural family relations provide an illuminating road map for understanding family functioning. A seminal work in this area is Salvadore Minuchin's *Families and Family Therapy* (Cambridge, MA: Harvard University Press, 1974); see especially Chapters 3 and 5.

6. Even healthy, well-functioning families will often struggle with symptoms and problems when offspring reach junior high and high school. Mary Pipher's important work, *Reviving Ophelia,* stresses how young people are overwhelmed by our culture's extreme emphasis on appearance, thinness, and peer approval; by media exposure to violence; and by social pressure to engage in sexual activity (*Reviving Ophelia,* New York: Ballantine, 1994).

CHAPTER SEVEN

Inflexible Interpersonal Coping Strategies

Conceptual Overview

This chapter provides a model for helping therapists further conceptualize their clients. This model is based on clients' *generic conflict* and their interpersonal strategies for coping with it. Generic conflicts are the central problems that pervade the client's life and link together the different issues that the client presents. The generic conflict has arisen out of the client's repetitive interactions with significant others, especially parents or primary caregivers, and often represents a basic unmet need such as an insecure attachment. Clients' problematic interactions with caregivers often give rise to pathogenic beliefs about themselves and others, and to faulty relational templates that shape their views of themselves, their expectations of others, and what will occur in close relationships. To defend against the painful feelings that accompany these maladaptive relational patterns, clients develop a repetitive or "characterological" interpersonal style. This coping strategy—such as always pleasing others or being perfect; repeatedly taking charge or being in control; routinely taking care of others and meeting their needs; consistently remaining invisible or avoiding all conflict—originally helped them to cope. However, it is no longer necessary or adaptive in most current relationships and actually causes many of the clients' presenting symptoms and problems. Therapists will become far more effective when they can (1) identify the formative maladaptive relational patterns and resulting painful feelings that comprise a particular client's generic conflict; (2) recognize the interpersonal coping strategy that the client has adopted to defend against this conflict; and (3) understand how the generic conflict and interpersonal coping strategy are expressed in current problems and relationships, especially in the current interaction with the therapist. As beginning therapists formulate the client's generic conflict and interpersonal adaptation to it, this conceptualization will lend structure and organization to the disparate material that the client presents. In turn, this will provide a treatment focus to guide therapists' intervention and treatment strategy.

As noted in Chapter 2, client conceptualization is an ongoing process that begins with the initial client contact. General hypotheses about client dynamics

179

are formulated early in treatment, and these tentative working hypotheses are further refined or discarded as the therapist learns more about each client. It is not until the client's conflicted emotions emerge, however, that the two or three core conflicts that link together the client's varying problems become clear. Having initially focused the client inward, the therapist can use the client's emotions that begin to emerge as guideposts to signal generic conflicts, which can now be addressed effectively.

Chapter Organization

In the first section of this chapter, we look at a model for conceptualizing the client's generic conflict—how it originally developed, the client's interpersonal adaptation for coping with it, and how it is being expressed in current symptoms and problems. This model for conceptualizing the client's conflicts includes five components:

1. The client's unmet developmental needs
2. The original environmental block that created the conflicts
3. The client's intrapsychic defenses against her own generic conflict
4. The client's interpersonal strategy to rise above or overcome the generic conflict
5. An interpersonal resolution of the generic conflict

Adapted from Horney's (1970) interpersonal theory, these five components provide a useful schema for conceptualizing client dynamics (see Figure 7.1). This schema is an integrated system composed of several, complex psychological processes. After discussing each component of this model, we illustrate it by considering an extended case study in the second section and two case summaries in the third section. After reading this chapter, therapists can turn to Appendix B for a structured outline they can use to write treatment plans and formulate intervention strategies.

A Conceptual Model

In this section, we examine an interpersonal model for further conceptualizing how client conflicts are expressed, maintained, and resolved. Our discussion does not emphasize how therapists can intervene with clients but instead gives therapists further guidelines for conceptualizing clients' conflicts and adaptations to them. Much of the discussion that follows is derived from the interpersonal theory of Karen Horney (1966, 1970). The five components of this model are schematized in Figure 7.1.* Let's begin by examining how en-

---

*The author wishes to thank Robert Hilton and Karen Humphrey for their help with this schematization.

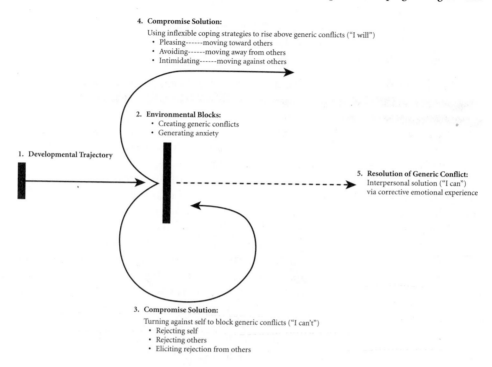

4. Compromise Solution:

    Using inflexible coping strategies to rise above generic conflicts ("I will")
- Pleasing------moving toward others
- Avoiding------moving away from others
- Intimidating------moving against others

2. Environmental Blocks:
- Creating generic conflicts
- Generating anxiety

1. Developmental Trajectory

5. Resolution of Generic Conflict:
    Interpersonal solution ("I can")
    via corrective emotional experience

3. Compromise Solution:
    Turning against self to block generic conflicts ("I can't")
- Rejecting self
- Rejecting others
- Eliciting rejection from others

Figure 7.1   Adaptation of Horney's interpersonal model

during problems result when one or more of a child's basic developmental needs go unmet.

Clients' Developmental Needs

This conceptual model begins with children's basic needs for secure relational ties. Component 1 in Figure 7.1 refers in the broadest sense to young children's dependency needs on their parents and, in particular, to children's need for a secure attachment. If parenting figures are consistently responsive to the children's bids for attention and affection, children will be able to freely express this need. Of course, as children develop, they benefit from learning to accept limitations on parents' ability to respond and to tolerate delays in parental response. If parents are punitive or unresponsive, however, anxiety will soon become associated with the child's needs. Children are then forced to fashion a solution; the *compromise solution* they construct often alienates them from themselves and their authentic experience.

Consider children of preschool age whose need for warmth and affection goes unmet from authoritarian parents. When such children approach their caregiver for comfort or emotional refueling, their need is often rebuffed in a characteristic way:

What do you want now? Why are you always needing something!
Big boys don't want to be held all the time. That's sissy stuff.
Leave me alone, honey. Can't you see I'm busy?

If this type of interchange occurs repeatedly, children will soon learn to anticipate rejection and feel anxious whenever emotional needs for affection or reassurance are aroused. Children are biologically organized to continue experiencing attachment needs, but their direct expression is blocked by the parents' cold response (component 2 of Figure 7.1). The child is then compelled to find some way to cope with this painful set of circumstances.

Although clients will present with a wide range of problems that result from varied developmental backgrounds and environmental blocks, enduring psychological problems often begin when basic childhood needs for secure relational ties are not met. Most generic conflicts also involve a failure to provide children with nurturance, clear communication, and consistent emotional access, on the one hand, and a predictable environment with consistent limits and discipline, on the other. Although clients vary in their developmental needs, experience different types of blocks, and develop their own individualized coping strategies, most clients experience anxiety when their needs are blocked. As a result, they begin to construct a compromise solution designed to manage the continued anxiety engendered by these unmet needs. At the same time, however, they also find indirect ways to express and partially gratify this need. Components 3 and 4 of our model elaborate on how this compromise solution works to block or minimize the felt need that has now become anxiety-arousing and, simultaneously, to implement different coping strategies to rise above the need and indirectly gratify it. We will now discuss the complex psychological maneuvers that individuals adopt to cope with these generic conflicts and examine how these relational templates and inflexible coping styles are replayed in current relationships with the therapist and others.

Clients' Compromise Solutions

In Chapter 2, we proposed three domains for conceptualizing clients' conflicts: maladaptive relational patterns, faulty beliefs about self and unrealistic expectations of others, and core conflicted affects. Although all three domains are intertwined, the environmental block in component 2 of Figure 7.1 principally evokes the core conflicted affect. As we will see, the faulty beliefs that arise from these generic conflicts and the interpersonal strategies used to cope with them are highlighted in components 3 and 4, respectively.

Clients employ two psychological mechanisms in order to defend against the anxiety associated with expressing, or even experiencing, their blocked needs. First, we will examine three ways in which clients defend against their generic conflicts by responding to themselves in the same way that others originally responded to them (component 3 of Figure 7.1). These can be thought of as *cognitive schemas* that clients develop; that is, beliefs about themselves and others that, once adopted, regulate their behavior. Second, we will examine three ways in which clients adopt a general *interpersonal coping style* to try and rise above their unmet need (component 4 in Figure 7.1). In therapy, this style will often quickly become apparent in the way the client interacts with the therapist. Taken together, these two mechanisms of *blocking* and *rising above*, which represent attempts to minimize anxiety and provide a modicum of control, are clients' compromise solutions to their generic conflict. Let's see how this complex defensive system works.

Blocking the unmet need   We have already described insecurely attached children who try urgently to get their parents to respond to their emotional needs, with the limited skills and means available to them, but are unable to elicit the support they need (for example, lack of parental response; responsiveness on the parent's timetable rather than when the child initiates contact; punitive or inconsistent response). In order to gain some active mastery over their helplessness and to ward off the anxiety aroused, *such children begin to block their own need in the same way that the environment originally blocked it.* For example, children may deny it as insignificant and regard themselves as unimportant (for example, "I'm stupid") or may reject it and view themselves contemptuously as demanding (for example, "needy"). Such clients adopt the same hurtful responses toward themselves that others originally expressed toward them; that is, they behave toward themselves in ways that make the original environmental blocks seem reasonable. In turning against themselves, children are able to ward off the unwanted anxiety that would be engendered if they felt that their true personhood was being seen and that their developmental needs were reasonable and legitimate but still viewed as unacceptable to those they need and love most—their caregivers. As well as preserving attachment ties, this adaptation provides a sense of control and allows them to continue to view the world and their parents as reasonable and just. This then becomes their cognitive schema for viewing themselves and others; by acting accordingly, they elicit responses consistent with the template. This adaptation has been referred to as identification or modeling; some clinicians speak colloquially of "using the devil's tools to fight the devil." Component 3 in Figure 7.1 reflects three ways in which children begin to do to themselves what was originally done to them.

First, clients block their own need intrapsychically and respond to themselves in the same hurtful ways that others have responded to them. When the children of nonresponsive parents feel a need for reassurance or affection, they will block their need by feeling critical or contemptuous of themselves and their own need. As adults, their own emotional needs are still unacceptable, and they will often say to themselves the same critical things they heard from others years ago—often adopting the exact words and tone—but without recognizing the source of these internalized templates or tapes. Furthermore, *the judgmental, punitive, or rejecting affect that clients feel toward themselves as they replay these internalized tapes is the same affect that parenting figures originally expressed toward them years before* (for example, "What's *wrong* with you!").

Second, clients block their anxiety-arousing need by reenacting in current relationships the same conflict that was originally experienced in earlier, formative relationships. Thus, on an *interpersonal* level, clients will say and do to others what was originally done to them. For example, the child of strictly authoritarian parents may grow up to be a parent who similarly disparages the same emotional needs in his or her own children—especially the child who most resembles the parent in gender, birth order, temperament, and so forth. Especially in the arena of parenting, clients who have not integrated or come to terms with their own conflicted feelings are prone to reenact the same maladaptive relational patterns that they experienced in their own childhoods. As parents, many clients feel helplessly dismayed as they watch themselves respond to their children in the same hurtful ways that their parents responded to them. Despite sincere pledges that "I will never do to my children what my parents did to me," variations of the old themes occur with clocklike regularity. Here again, in the interpersonal sphere, adult offspring gain some control over their own unwanted feelings by reenacting what was originally done to them.

Third, clients block the experience and expression of their own conflicted need by *eliciting* the same unsatisfying response from others in current relationships that they received in their childhood. For example, unless the children with problematic developmental experiences have other reparative relationships (with friends, teachers, mentors) to help them resolve these concerns, they are likely, as adults, to select a marital partner who cannot respond well to their emotional needs. In the event that their spouse is capable of being affectionate or intimate, they may not be able to accept this emotional responsiveness. Why? The possibility of having a response to their old, unmet needs—even a well-intended, benevolent response—will often arouse shame or anxiety. That is, the spouse's affection is likely to arouse the pain of the original deprivation. Following their relational templates, such clients also will anticipate that their spouse will reject their needs as others have done in the past—even though the spouse has never responded in this hurtful way. Operating on schemas or internal working models that fit past relationships better than current ones, such clients may simultaneously elicit and reject nurturance from their spouse. This will certainly frustrate and confuse the spouse, especially if these mixed messages activate the spouse's own problematic relational templates. Unresolvable marital conflict that results in emotional gridlock occurs when the spouses' templates dovetail and each presents the other with a version of their old relational pattern. In such cases, the likely outcome is enduring conflict and perhaps marital counseling or divorce.

To sum up, clients defend against their unmet needs by (1) intrapsychically responding to themselves in the same rejecting ways that parental figures did; (2) identifying with the rejecting parent, usually in an attempt to overcome their powerlessness, and then responding to others in the same hurtful way; and (3) repeatedly becoming involved with others who provide the same hurtful response to them that they originally received as children. Although one of these three modes may be predominant for a particular client, most clients with longstanding problems will employ all three mechanisms at different times to defend against their generic conflicts.

At the same time as clients are blocking their needs on each of these three levels, they are simultaneously trying to get them met in an indirect way. Let's examine the other side of clients' compromise solution: their compensatory attempts to rise above their conflict and be special.

Rising above the unmet need   Perhaps the most important component of this model for therapists to understand is clients' attempts to *overcome* their conflict by being perfect or superior. Before discussing clients' attempts to rise above their conflict in this compensatory way, let's review. As we have seen, clients employ three strategies to defend against their conflicts. In all three maneuvers, clients do to themselves and others what caretakers originally did to them. In other words, they turn a passive experience that originally happened to them into an active experience over which they now have some control. Therapists can focus on how clients take up the same hurtful affect that was originally expressed toward them in childhood and actively turn it against themselves, or others, to block their own unmet need. This turning against the self in component 3 of Figure 7.1 (often expressed as variations of "I can't") is only half of the conflict, however. Paradoxically, while defending against the anxiety of reexperiencing the unmet need, clients simultaneously try to rise above it and have it met indirectly. Component 4 of Figure 7.1 shows the other side of clients' compromise solution—clients' attempts to rise above the need and indirectly gratify it.

If they do not find reparative experiences with significant others, children's unmet developmental needs do not dissipate or go away on their own as they grow up. These needs may be denied or sequestered away but they continue to seek expression. As a result, clients do not give up trying to have the needs met. At the same time, however, the unmet needs still remain too anxiety-arousing to be expressed directly. So clients try to rise above the unmet needs and indirectly fulfill them by adopting various interpersonal coping styles. Horney (1970), Beck and Freeman (1990), Strupp and Binder (1984), and others have utilized other terms to describe the interpersonal coping styles that clients employ. To illustrate these interpersonal coping strategies, we will focus on Horney's three lifestyle adaptations: moving toward, moving away, and moving against others.

Moving toward, moving away, and moving against   Clients adopt a fixed interpersonal style both to reduce the anxiety associated with an unmet need and to find some partial, indirect gratification of it. For example, in their families of origin some clients learned to cope with their unmet needs and problematic familial interactions by *moving toward* or pleasing people. These clients could earn some needed approval and diminish the threat of further rejection by complying with their parents and being unfailingly nice or good. Generalizing from their formative relationships with caregivers, they develop a primary mode of relating to others in which they please others and rarely assert their own wishes. Thus, these clients have learned (1) to defend against the anxiety engendered by their legitimate needs, (2) to win some approval, and (3) to indirectly gratify these needs, by consistently moving toward others in a pleasing, servile way.

Other clients have learned that aggressiveness and resistance to parental wishes, if pursued long enough, will ward off the pain of not having their needs affirmed. They may have needed caretakers to act more strongly or take charge of the family effectively, which would have provided the safety they needed to feel secure without having to take control. These individuals cope by *moving against* others and expansively seeking to be in control of themselves and, especially, others. They approach relationships with the orienting attitude that they must win what they deserve, and they often assert themselves aggressively for this purpose. Thus, these clients adopt a repetitive interpersonal style of moving against others to protect themselves from feeling the unwanted emotions evoked by their generic conflicts.

In Horney's third interpersonal style, clients have learned that the best way to reduce their anxiety and defend against their unmet needs is by *moving away* from others through physical avoidance, emotional withdrawal, and self-sufficiency. Although clients may employ all three of these interpersonal coping styles at different times, one style usually becomes predominant or characterological.

To review, first we saw that clients use three ways to cope with problematic developmental experiences. These turning-against-self defenses repeat the ways in which clients' original needs were blocked by caregivers in the environment (component 3 in Figure 7.1). We then discussed how clients try to rise above their blocked needs and indirectly meet them (component 4 in Figure 7.1). These rising above strategies include the interpersonal styles of moving toward, against, or away from others. *These coping strategies are maladaptive because they are employed inflexibly in all types of situations.* We are now ready to examine further how clients use their interpersonal coping style to try to rise above their generic conflicts. As we will see, however, even though clients derive self-esteem and an identity from their interpersonal coping style, it remains a defensive adaptation that avoids their basic conflicts and, ultimately, exacerbates rather than resolves their problems.

Being special to rise above their generic conflicts   When the developmental problems were pervasive or severe, clients often adopt one of the three interpersonal styles—moving toward, moving against, or moving away. This inflexible coping style is a habitual way of relating to others that is also reflected in vocational and other life choices. Although there are many exceptions, individuals who move toward may find that careers in nursing or counseling are a good match for their caretaking skills and needs. Individuals who move against may find that careers in law, medicine, or money management are congruent with their interpersonal style of taking charge and needing to be in control. Individuals who move away may feel comfortable as researchers, artists, or in a meditative lifestyle. Many clients will find some way to turn this defensive coping style into a virtue and use it to feel special and rise above their generic conflict. Thus, clients do not experience their interpersonal strategy as a defensive coping system to avoid conflicts, but as a virtuous way of relating to others that may make them feel superior. However, this unrealistic sense of being special is compensatory and actually reflects their low self-esteem or shame-based sense of self. *Their self-worth comes* *from their ability to rise above their generic conflicts by pleasing others, by achieving and succeeding, or by withdrawing and feeling cynically superior.* Thus, clients convert their defensive interpersonal coping style into a virtue and adopt an aggrandized *false self.* Although moving-toward clients are compliant and submissive, they do not experience themselves as servile. Instead, they often feel special or secretly better than others. They accomplish this by perceiving themselves as selflessly loving; always altruistic and sensitive to the needs of others, and committed to the ideals of peace and harmony. In this way, the defensive interpersonal style becomes a source of self-esteem—a virtuous quality that wins needed approval from others and wards off anxiety and shame.

Similarly, moving-against clients do not see themselves as angry, competitive, self-centered, or demanding, although they frequently behave in these ways. Instead, they see themselves as heroes or heroines—strong leaders entitled to direct others, who often are seen contemptuously as weak or incompetent. Finally, moving-away clients glorify their aloofness. They are often cynical and may feel better than the ordinary masses, whose involvement in everyday happenings is seen as mundane and small-minded.

Thus, to lessen their anxiety, clients inflexibly employ an interpersonal coping style that allows them to feel special; gain a sense of identity and self-esteem; and rise above their unmet needs, unacceptable feelings, and pathogenic beliefs. Their <u>ability to turn a defensive coping style into a virtue or talent that testifies</u> <u>to their specialness is what Horney</u> terms a *neurotic pride system.* For these clients, however, their self-esteem is brittle and vulnerable. It is based on being special, controlling others, and rising above, rather than learning to value and understand their authentic feelings.

We now have the two sides of clients' conflict. First, clients block their own needs and respond to themselves in the same critical or rejecting ways that adult caregivers originally responded to them. Clients come to feel toward themselves the same affect that the caregiver originally communicated toward them (for example, disappointment, resentment, or indifference). This affect will usually be evidenced as low self-esteem when clients enter therapy, and clients are often seeking treatment to be rid of this intrapunitive feeling. Usually, it is relatively easy for counselors to recognize how clients are turning against themselves and to respond in more affirming ways.

Second, clients rise above their generic conflicts and find ways to interpret their defensive strategy as evidence that they are special. This aspect of their compromise solution is usually more covert and is harder for therapists to address because the clients' neurotic pride or false self is typically bound up with shame. Let's examine some other reasons why clients are reluctant to let go of their attempts to rise above their conflict.

Clients' interpersonal coping style has indeed granted some relief from the anxiety of unmet needs and unacceptable feelings by providing a sense of control over them. It has also allowed clients to earn some attention from others and respect for themselves. In this way, *clients' interpersonal coping style has become the primary source of their self-esteem and their principal avenue to succeeding in life.* Further, clients have often developed impressive abilities in the service of their coping strategies. For example, many moving-toward clients have not only been able to make others like them, but have also become highly skilled at caring for others. Likewise, moving-against clients have often been able to achieve a great deal of success and power, and may develop significant technical or other abilities along the way. Similarly, although moving-away clients have often been able to become safely aloof, they may also have developed a rich inner life or been able to develop creative or artistic abilities.

One of the most important interventions therapists can make is to help clients recognize the significant emotional price they pay for their interpersonal coping style. Therapists do this by highlighting, as concretely as possible, how this adaptation is creating stress and limiting their lives now. At the same time, however, therapists need to appreciate the genuine strengths that clients have developed through these coping strategies. Later in this chapter we will see how clients can retain the skills and strengths they have developed in these compensatory modes, but without utilizing them in the same inflexible or unidimensional ways as in the past. As before, therapists need to help clients honor their interpersonal coping styles and appreciate rather than pathologize them. Although they were absolutely necessary and represented the best possible adaption they could have made at that point in their development, they are no longer adaptive in most current situations. *An important treatment goal for many clients is to learn that more flexible or varied interpersonal coping styles will serve them better.*

To sum up, by turning their defensive interpersonal adaptation into a sense of being special, clients often gain significant secondary benefits and genuine personality strengths. It's no wonder, then, that some clients initially may be reluctant to explore this adaptation with the therapist. It has become the basis of their identity, the primary source of their self-esteem, and the only way they have had of defending against their seemingly unresolvable conflicts. Once therapists fully grasp all that this defense has provided for clients, and help clients to appreciate what it has accomplished, both therapists and clients will be less judgmental or impatient with the clients' symptoms. Both will have greater empathy for the clients' struggles to change, and will be able to work together collaboratively to develop more flexible coping strategies that will be more adaptive. Understandably, clients will not want to risk giving up what they have without having something better to replace it. As we will see, however, clients never really succeed in rising above their generic conflicts through these compensatory adaptations. By acknowledging conflicts and addressing them more directly, clients can develop a wider experiential range that will be an important part of resolving many relational problems.

Shoulds for the self, entitlement from others   Clients both gain and lose something by their compromise solutions. As a result of feeling special and virtuous, clients gain what Horney terms *false pride* and a sense of *entitlement* from others. For example, the self-effacing, moving-toward client expects unwavering approval from others and often demands constant reassurances. The expansive, moving-against client claims leadership and demands deference from others. And, although it may look as if detached, moving-away clients have no expectations of others, they demand that others never criticize them or make demands upon them. These neurotic claims leave clients exposed to continuing frustration, however, as others frequently do not accept such unrealistic demands and unspoken expectations.

Accompanying clients' excessive demands on others is a harsh and uncompromising set of "shoulds" that they place on themselves; these recurrent cognitive messages reflect their impaired self-esteem and impoverished sense of self. For example, moving-toward clients suffer under the self-imposed demands that they should be the perfect lover, teacher, spouse, and so forth. They must always be caring and responsive to the needs of others and never feel angry, compete, or advocate on their own behalf. Moving-against clients demand of themselves that they should be able to quickly overcome all difficulties. They must win and be on top, control their feelings at all times, and overcome insecurities simply by an act of will. Moving-away clients pay for their specialness by believing that they should be able to work tirelessly and always be productive. They demand that they should be able to endure anything without becoming upset and that they should never need help from anyone. As we will see, such individuals will not always be able to meet these demands they place on themselves. When the inevitable failures occur (for example, they cannot win someone's love, do not succeed in an expansive new venture, or cannot remain aloof but are critically evaluated or forced to compete in some way), they often feel defeated and may suffer great shame. Anxiety attacks, suicidal feelings, and depressions all may result when their coping strategies fail, and some will enter treatment in crisis.

Clearly, the *tyranny of the shoulds* is an exacting price to pay for the client's sense of specialness. One of the most comforting interventions the therapist can offer is to be able to effectively and emphatically articulate how this coping strategy is core to clients' sense of self, the basis of whatever self-worth they may possess, and the organizing theme in their life. Therapists also need to help clients appreciate how these interpersonal coping strategies were once necessary and adaptive, while clarifying the unnecessary price in symptoms and pain that the client is paying.

Resolving the Generic Conflict

Component 5 in Figure 7.1 illustrates how the client's generic conflict can be resolved. Ultimately, clients' compromise solutions to their generic conflicts will never succeed. The client can block the anxiety-arousing need and try to rise above the generic conflict in the ways described here. However, these compensatory maneuvers will not release clients from the original conflict and will only expose them to unending interpersonal problems. Let's examine three reasons why the client's compromise solution will not succeed, and then explore what can be done to help the client instead.

First, the tyranny of the shoulds places unrelenting demands on the client. These can be heard, for example, when clients enter therapy exclaiming, "I can't do all of this anymore! I can't be a perfect parent, a loving spouse, and a successful professional. I can't be everything to everybody anymore." An individual's quality of life is diminished by these harsh and fatiguing expectations on self, even if these demands do not provoke a crisis that brings him or her to therapy.

Second, clients' sense of entitlement and being special will usually fail as well. Self-effacing clients will not be able to win the love or approval of everyone they need; expansive clients will not always succeed or be able to make others defer to their demands; and detached clients will encounter criticism from others or situational pressures that force them to compete. Inevitably, clients' attempts to become special and rise above their conflict will lead to frustration and interpersonal problems. Typically, clients enter therapy when situational life stressors, developmental transitions in adulthood, or aging have caused their interpersonal coping strategies to fail.

Therapists must grasp the pain that clients suffer when their attempts to be special and rise above their conflict fail; Kohut (1977) writes eloquently of the *narcissistic wound* that is so painful to clients' core sense of self. Even though the demands that clients place on themselves and others are unrealistic, it still can be excruciating when their interpersonal adaptation fails. Tragically, some individuals will even attempt suicide in response to such a failure, and many others will contemplate it. Why are some clients driven to such an extreme response? The interpersonal coping style that many clients have adopted is the basis of their identity and their primary source of self-worth, but it is a brittle substitute for the genuine self-esteem they lack. From these clients' point of view, they *are* their ability to please, to achieve, or to remain superior and aloof. Such clients overreact to seemingly insignificant events (for example, gaining 3 pounds over the holidays; losing their patience with a trying 2-year-old) because their entire sense of self is threatened when their interpersonal coping strategy fails—and, following their internal working models, their shame-based sense of self is revealed by their mistakes.

When such students get an A minus, rather than the A they demand of themselves, they may feel completely undone or defeated. Too often, the therapist's initial response to this disappointment is to question the overreaction to the grade and provide a more realistic appraisal of the situation. This well-intended, reality-based response misses the deeper meaning for the client. Such a reality check is certainly useful, but a more productive initial response would be a bid for an empathic connection. The therapist does this by reflecting that, from the client's point of view, this grade seems to indicate once again that the client is imperfect and cannot get it right—despite Herculean efforts. In other words, the therapist's goal is to communicate that he grasps that this experience really is devastating for the client, and that he feels compassion toward the burden of perfectionism that the client has suffered under for so long. Such empathic reflections behaviorally demonstrate that the therapist grasps the core meaning of this prototypic experience that occurs over and over again for the client. Therapists who can provide this understanding will become far more important to their clients at this point, will intensify the collaborative alliance, and will provide the safety clients need to explore this issue further. By appreciating the full impact of this failure—an impact that has always seemed so irrational to clients themselves and insignificant to others—therapists understand clients in a way that most clients have not experienced before. This deep understanding is a corrective emotional experience and helps clients free themselves from the tyranny of the shoulds and their sense of entitlement from others. In this way, therapists provide the holding environment that allows clients to enter into their disappointment and shame, to grieve their deprivation, and to come to terms with the various parts (lonely, angry, and so on) of who they are.

It can be shattering for some clients when their interpersonal coping style fails because it arouses the helplessness, vulnerability, shame, and other feelings that were originally evoked by their problematic developmental experiences. Eventually, the therapist must respond to the emotional deprivation beneath clients' compensatory attempt to obtain security by moving toward, moving against, or moving away. As noted before, therapists can do this only after they have first responded to the conflicted emotions aroused by clients' pathogenic beliefs that—if they are no longer able to successfully please, achieve, or remain aloof—they are not love-worthy, cannot protect themselves, are inadequate, and so on.

Third, the most important reason the client's compromise solution cannot succeed is that the client can never rise above the original conflict. No matter how saintlike or loving, how successful or accomplished, or how self-contained and independent, the client is still defending against the original conflict, rather than resolving it. All these interpersonal maneuvers are attempts to avoid the problem by blocking it, or to manage the problem by trying to rise above it, rather than address it directly. This is clearly illustrated when the client's increasing success in life is not met by satisfaction, but by increasing depression or cynicism.

Let's review this complex sequence (Figure 7.1). When the child's basic developmental needs are blocked by the environment, the individual will adopt a compromise solution that temporarily relieves the anxiety of having basic emotional needs unmet but does not resolve the underlying conflict. On the one hand, clients try to block their own needs, just as they were originally blocked by significant others. Component 3 in Figure 7.1 illustrates how clients turn against themselves and develop self-punitive feelings, and lose their self-efficacy to feelings of helplessness and hopelessness ("I can't").

On the other hand, clients try to rise above their generic conflict by becoming special, turning their defensive lifestyle adaptation into a virtue, and obtaining some compensatory self-esteem. Clients will be highly invested in trying to overcome and rise above their generic conflict; typically, to have to face it will feel like a shameful defeat. Thus, component 4 in Figure 7.1 indicates clients' exaggerated sense of being special and defiant attempts to rise above the conflict. (They might vow to themselves, for example, "I will never need anything from anybody ever again.") This rigid, defensive stance reflects an act of will to overcome these basic developmental needs and rise above them, rather than trying to resolve them by addressing them directly.

In order to resolve these generic conflicts, clients must relinquish their compromise solutions. Attempts to block the original need and to rise above it are fragile solutions that cannot be sustained; failure experiences that precipitate crises and symptoms are inevitable. Thus, instead of avoiding the conflict in these two ways, clients will have to approach it. They do this by allowing themselves to experience their generic conflict in a more supportive relationship that does not meet the original need, but does give clients the understanding and validation they need to live through the feelings that were intolerable before. The interpersonal solution provided by the therapist's more effective response also allows clients to integrate other reactive feelings, helps clients to disconfirm pathogenic beliefs about themselves, and alters clients' faulty expectations about what may occur in relationships with others. Now, as adults, clients are able to understand and resolve the original conflict by reexperiencing the hurt or pain with the therapist. In this way, what was once dreaded can now find compassion and validation, and clients can integrate these difficult feelings rather than continuing to defend against them. When clients allow the therapist to help contain the conflicted feelings, old relational patterns are broken and new, more flexible templates become possible. Therapists' compassion and understanding disconfirms these faulty beliefs and, in turn, allows clients to be more self-accepting and forgiving of themselves. Thus, clients can now integrate the previously unacceptable parts of themselves, and can better accept both good and bad in significant others (component 5 in Figure 7.1). In other words, once the original hurt or need begins to emerge in therapy, the therapist's task is to provide a more affirming response than clients have received in the past. Therapists do this, in part, by holding the client's pain in their compassionate understanding, as discussed in Chapter 5. We will examine more fully how this corrective emotional experience leads to meaningful change.

If clients receive a more effective response from the therapist than they have received from others in the past, they will not need to block their generic conflict or attempt to rise above it anymore. The therapist's more accepting and affirming response to the client's original predicament alters the maladaptive relational pattern and alleviates the anxiety that was originally evoked. As this occurs, clients become capable of generalizing this experiential relearning with the therapist to others. That is, clients begin responding in more adaptive ways that improve their relationships with others and expand the narrow coping style that they originally adopted to cope with their dilemma. In Chapter 9, we will examine more specifically how the therapeutic relationship can be utilized to help clients resolve their generic conflict, and how this emotional relearning with the therapist can be generalized to other relationships in clients' lives. For now, an extended case study of a moving-toward client, and two case summaries of the other coping styles, will help therapists apply this conceptual model to therapeutic interventions.

Case Study of Peter: Moving Toward Others

Developmental History and Precipitating Crisis

Peter was an insecurely attached child and his authoritarian parents left his age-appropriate dependency needs profoundly unmet. In addition, Peter's parents divorced when he was 6 years old, and his father did not take an active parenting role after the divorce. Peter's infrequent contacts with his father were superficial and inconsequential.

Peter's mother was overburdened by the demands of raising three children on her own, working full time, and trying to make some kind of personal life for herself. She was involved with several men in the years following the divorce but was never able to establish an enduring love relationship. Frustrated by the many demands and few pleasures in her life, she often made Peter the target of her resentment. She was irritable toward Peter, criticized him when things went wrong in her life, and felt resentful of his needs. Although she tried her best to be fair to the children and give them a good home, she could not be responsive or affirming toward Peter.

After his father left, Peter quickly learned that taking care of his mother was the best way to ward off her disapproval and to win whatever affection he could. By the age of 10, Peter had adopted a pervasive interpersonal style of moving toward people. His teachers described him as an especially responsible and well-behaved boy who was a pleasure to have in class. Now let's turn the clock forward 15 years and see how these developmental conflicts are expressed in Peter's early adulthood.

At the age of 25, Peter is a graduate student in psychology. Becoming a therapist felt like a perfect career choice to Peter. He prided himself on his sensitivity and concern for others and took pleasure in being able to help those in need. And now that he was carrying his own client caseload, it was great to find that he enjoyed being a therapist as much as he thought he would. Best of all, Peter was finding that he was often able to help his new clients. At least, all of his clients seemed to like him and they kept returning to their therapy sessions each week. As his second-semester practicum got underway, Peter felt that he was on his way.

Later that semester, though, Peter had a setback. After presenting a videotaped recording of one of his therapy sessions in group supervision, Peter received some unexpected feedback. The practicum instructor told Peter that he was being "too nice" to his clients and that he seemed to need his clients' approval too much. The instructor went on to say that Peter seemed to be afraid to challenge his clients, and tended to avoid conflicts in the therapist-client relationship that needed to be addressed.

Peter was stunned. Although he had some awareness of his aversion to conflict, he did not truly understand what the instructor was talking about, and he felt hurt and confused by the criticism. It was important to Peter that his instructor like him and approve of his clinical work. Peter tried carefully to explain that the instructor did not understand the close relationship that Peter was developing with his clients or recognize all of the important issues that his clients had been revealing. The instructor responded that this was probably true, but Peter was missing the point, and he repeated that Peter needed to think about his aversion to conflict and consider its impact on his therapeutic style. To make matters worse, two students in the practicum group chimed in and agreed with the instructor's comments. With that, Peter's anxiety became so high that he could no longer defend or explain himself, let alone try to understand or learn from their comments. Peter finally stopped arguing with them and quietly nodded agreement throughout the rest of the supervision session.

The next few days, Peter felt sick. He was dismayed by the criticism and his thoughts and feelings were racing. For a while, he thought that he should drop out of the practicum group, but then he decided that, if he only tried hard enough, he could make the instructor see that his critical observations were inaccurate. Peter's mind kept searching for a way to discount the feedback and stop the anxiety that was churning inside.

A week later, Peter found out that his girlfriend was having an affair with another student in the program. Although he tried to be understanding at first, he felt shocked and betrayed. He alternated between withdrawing and announcing that their relationship was over and desperately trying to win her back. Peter felt shattered. He became so anxious that he was unable to eat or sleep, let alone study. It felt as if a motor were racing inside of him—accelerating out of control. Peter began hyperventilating and having anxiety attacks. By the end of the week, he was so anxious that he couldn't drive his car.

To make matters worse, Peter tried to keep all of this to himself. He thought he should remain calm and together, and he was afraid that his supervisors would not want him to see clients if he was "so messed up" that he was having anxiety attacks himself. But despite his attempts to cover up his distress, his individual supervisor soon asked him what was wrong.

Although he could not ask for it, Peter desperately wanted his individual supervisor's support, and he was greatly relieved to receive it. Peter explained how his practicum group was becoming one of the worst failure experiences of his life and how hard it was to accept what his girlfriend had done. The supervisor was supportive, but also found a tactful way to say that the practicum instructor's comments fit with some of his own observations. Because he knew that his individual supervisor liked him, Peter was able to consider the feedback this time. The supervisor suggested that these were important issues for Peter to work with, but that they could be dealt with better in his own therapy rather than in supervision. Peter agreed and began seeing his own therapist.

It helped that Peter's supervisor thought he could be a fine therapist, in spite of his anxiety attacks. However, Peter's anxiety remained paralyzing as he began his own treatment. Fortunately, Peter was assigned to a skilled and experienced therapist who soon formulated the predisposing vulnerability, interpersonal coping strategy, and situational stressors that precipitated his crisis. Peter's coping strategy of moving toward people had worked well for him up to this point. However, both of the crises that Peter had just experienced ran headlong into the heart of his generic conflict and coping style. As Peter's ability to rise above his generic conflict by pleasing others failed, the anxiety associated with his unmet developmental needs broke through, and anxiety attacks resulted. Let's examine more closely why Peter developed these symptoms.

Precipitating Crises, Maladaptive Relational Templates, and Symptom Development

Many people would have coped with the two stressful events that Peter experienced without developing such significant symptoms. As we have seen, however, we must apply the concept of client response specificity. We will use Peter to illustrate how symptoms develop when situational stressors reenact maladaptive relational patterns, confirm pathogenic beliefs, or cause interpersonal coping strategies to fail.

The first stressor for Peter was his practicum instructor's critical feedback. Criticism from a respected authority figure would be unsettling for most people, but generally they could cope with it. For someone like Peter, however, such disapproval carries far more weight. Because of Peter's unmet needs for approval, his history of receiving excessive criticism, and his lifestyle adaptation of trying to win approval by being helpful and nice, the practicum instructor's criticism was devastating.

Second, the practicum instructor specifically challenged Peter's interpersonal coping style of moving toward others. This confrontation not only aroused the anxiety of receiving criticism, but also undermined Peter's primary means of warding off this anxiety (pleasing others and accommodating himself to their needs). Thus, *the instructor's feedback intensified Peter's generic conflict and, simultaneously, weakened his defenses against it.* In other words, Peter's developmental experiences made him acutely sensitive to criticism. He adopted the attitude toward himself communicated by his early caretakers—that he was unimportant and not worthy of approval. He also expected others to be critical of him. His attempt to defend against the anxiety evoked by this set of circumstances was to try to be perfect—to please others and meet their needs, so that this self-critical schema could be refuted by their appreciation. This coping style was, of course, a formula for failure: It was impossible for Peter to be perfect and appreciated at all times by everyone he met. He experienced even constructive criticism, especially from an authority figure, as total rejection of his self-worth, and he became flooded by feelings of shame for not being able to win everyone's approval.

If not for the subsequent crisis with his girlfriend, Peter probably could have recovered from the first setback without developing incapacitating symptoms. Most likely, he would have reconstituted his interpersonal coping style and been somewhat successful in winning the approval he needed from others in his life—from instructors, friends, and perhaps clients. As we will see, however, the subsequent stressor with his girlfriend also struck at the same generic conflict. At that point, his moving-toward coping style failed, and the strong anxiety associated with his insecure attachment history was evoked. More specifically, this was shame-anxiety—anxiety over the threat of having his unlovable and unworthy self exposed. This shame-anxiety became too intense to be blocked; it broke through in anxiety attacks and symptom formation.

A partner's infidelity will be highly stressful for almost everyone. Here again, though, this particular stressor held far greater significance for Peter when viewed within the personal context of his developmental history and subjective worldview. Peter had to cope with far more than just the loss of trust with his girlfriend; he also suffered a blow to his identity and basic sense of self-worth. In Horney's neurotic pride system, Peter's coping style of moving toward entitled him to be special, so that others would love and prize him inordinately. Although Peter was largely unaware of these unrealistic claims, he expected his girlfriend to idealize him as the most sensitive and loving man there could possibly be. It was impossible for Peter to believe that his girlfriend could actually be interested in someone else. Peter's neurotic pride system was painfully shattered when he found that he was not the commanding center of his girlfriend's life that he felt entitled to be.

Multiple stressors occurred for Peter in a short period of time. In and of themselves, these would precipitate a crisis for many people and lead them to therapy. If these situational stressors do not tap into preexisting templates, faulty relational expectations, and pathogenic beliefs, however, they will not usually provoke such strong symptoms as anxiety attacks. The client will often be able to recover in a relatively short time with crisis intervention or short-term supportive therapy. In contrast, when life stresses tap squarely into a client's generic conflict, a client such as Peter has to cope with far more than just the demands of the current situational stressors.

Thus, when Peter's girlfriend became sexually involved with another man, he had to cope with much more than just the pain of this betrayal. He also had to cope with his long-standing attachment conflicts, with relational templates of criticism and rejection, and with the wound to his self-esteem when his sense of being special was shattered. Clients such as Peter do not have a secure sense of self-esteem to fall back on in times of crisis. As unrealistic as it is, the sense of being special—yet vigilant for slights—and the secondary gains that clients can earn by rising above their need is the only self-worth that some clients have been able to achieve. It is understandable, then, that Peter was overwhelmed by anxiety attacks. Both of the situational stressors intensified Peter's original conflicts, and they both took away his coping strategy for defending against them. In this light, *we see that clients' presenting symptoms are not irrational.* They make sense once they are conceptualized in the broader context of the client's developmental history and the relational templates, pathogenic beliefs, and inflexible coping styles that result from this history.

The Course of Treatment

Fortunately, Peter's therapist was knowledgeable and kind. He was genuinely empathic to the pain that these situational crises brought on for Peter but also grasped the predisposing templates that made Peter so vulnerable to them. The therapist recognized that the current crisis provided an opportunity to resolve the more important developmental conflicts that left Peter prone to reexperience crises such as these whenever others disapproved of him. In the months that followed, Peter was able to resolve the precipitating crisis. More significantly, he made progress in coming to terms with the impact of his developmental history and the pathogenic beliefs and maladaptive relational patterns that followed from it. As a result, Peter was able to begin expanding his rigid moving-toward coping style and adopt a more flexible interpersonal repertoire. Here and in the next chapter, we will review the course of therapeutic events that allowed Peter to initiate these far-reaching changes.

Peter entered therapy in crisis, and the therapist was able to respond effectively to his distress. From the first session, Peter felt understood and cared for by the therapist. As they developed a strong working alliance, Peter's anxiety attacks stopped. The intense anxiety and profound distress that had been so disorganizing for him steadily subsided as the therapist continued to provide an effective holding environment. As soon as Peter felt better, however, he began to be resistant to the feelings he had been sharing with the therapist and wondered about ending treatment.

The therapist observed that, as Peter felt understood and affirmed by him, Peter's unmet needs for emotional support were assuaged and the attendant anxiety diminished. As a result, Peter felt safer, his symptoms abated, and he began to function better. At the same time, however, he also began to reconstitute his interpersonal coping style of moving toward others. Rather than terminating therapy at that point—which might need to occur in crisis intervention or time-limited treatment—the therapist was able to help Peter go further and address the developmental conflicts that made him so vulnerable to rejection and criticism. One way the therapist did this was to help Peter recognize his habitual coping style of pleasing; the therapist *named it*, so that they could talk about it together. Peter also learned to focus inward in order to identify what he was feeling, and what was occurring with others, at the instant when he felt compelled to employ this coping pattern. Usually, at such moments, he was expecting others to be disapproving of him or to disengage from him in some way.

Additionally, as he listened to Peter's narratives, the therapist would repeatedly highlight instances of Peter's key relational themes: (1) pleasing others; (2) avoiding anger and other interpersonal conflicts; and (3) expecting others to be critical and rejecting. Most importantly, the therapist used process comments to point out these three repetitive relational themes whenever he saw them occurring in their interaction. This provided direction to the treatment process, and served as the basis for developing mutually held treatment goals. For example, the therapist encouraged Peter to risk not pleasing him, and focus instead on just saying and doing what he wanted in their relationship. Not having to be vigilant about what the therapist liked or wanted gave Peter the opportunity to live "inside his own skin" and have the experience of being cared about even when he was not trying to please or take care of the therapist.

The therapist continued to focus Peter inward. In particular, they explored the feelings that were evoked for Peter when instead of employing his usual coping style of pleasing others, he risked new responses he had rehearsed with the therapist. Gradually, by focusing on what he was experiencing just before he employed his usual coping patterns, Peter began to clarify the emotional deprivation he had suffered as a child but had always been too ashamed to reveal to others or even acknowledge to himself. Relatively quickly, the therapist's accepting presence allowed Peter to overcome the shame of having his pain revealed. Now Peter could be fully seen by someone with all his strengths and flaws and continue to be accepted and cared about. Peter began to understand how he had coped with his childhood predicament by taking care of and pleasing his mother, and then others as well. As this significant progress occurred, however, Peter still continued to struggle with feelings of anxiety, shame, and sadness.

This triad of contradictory feelings continued to wax and wane for Peter over the next few months. Peter was increasingly recognizing how unwanted and alone he had felt as a child and how ashamed of himself he had always been. (He recalled that the animal he felt most similar to during elementary school was a worm.) Accompanying these feelings, Peter realized that he had always believed that he was not loved because he was somehow unworthy and did not matter enough to be important to others. As Peter continued to feel emotionally held by the therapist's understanding, his long-withheld feelings of sadness over this deprivation and invalidation could be expressed for the first time. Peter moved through the old block as he received a far-reaching, corrective emotional experience. The therapist's kind response was a powerful disconfirmation of Peter's old relational expectations and was deeply comforting, offering him the affirmation for which he had always longed. This response also provided him with a sense of being seen or known for the first time and accepted for who he really was (component 5 in Figure 7.1).

As Peter began to mourn his childhood loss, two reactive feelings followed closely behind. Peter found himself feeling furious that his father could value him so little that he could walk away so easily, and angry at his mother for making him feel responsible for trying to earn her love and ward off her disapproval. As soon as this anger emerged, however, the third feeling in Peter's affective constellation was aroused. Following his relational templates, Peter also became exceedingly anxious—afraid of being left helplessly on his own by the therapist, his girlfriend, his father, and anyone else with whom he would ever risk becoming close. Each of these three feelings in Peter's affective constellation—the shame surrounding his unmet emotional needs; his anger over being rejected and dismissed so readily; and the anxiety of being left or emotionally cut off if he protested—was repeatedly aroused. With the therapist's help, Peter was gradually able to integrate or resolve these feelings by, first, being able to fully experience them, then, to risk sharing or expressing them to the therapist, and finally discovering that he could live through or contain them with the therapist's support. This was not a smooth or simple process, however. As the primary feeling of sadness or deprivation emerged, Peter's own shame and internal blocking defenses were activated (component 3 in Figure 7.1). Let's examine how the blocking side of Peter's compromise solution was expressed in therapy.

Peter's situational crises disabled his rising-above defense. He could no longer be special and win the approval he needed from either his practicum instructor or his girlfriend. As his interpersonal coping strategy failed, it exposed his shamefully unacceptable ("weak") attachment needs and evoked intense anxiety. However, the other side of Peter's compromise solution was still operating: To defend against the painful feelings being evoked, Peter turned the same type of block that was originally imposed by the environment against himself.

In treatment, and throughout his life, Peter had tried resisting each feeling in his affective constellation. He did this by blocking both the subjective experience and the interpersonal expression of his generic conflict. For example, as the sadness of his childhood deprivation emerged, Peter thought the therapist would be critical and rejecting. He was afraid that the therapist would want him to "stop crying, grow up, and act like a man." This *transference reaction* again served to block his sadness until the therapist was able to draw it out more fully and clarify this misperception by talking it through with Peter. More importantly, Peter was able to experience the therapist's continuing respect for him in the face of emotional needs. Peter learned in this relationship that he did not have to be perfect or take care of the therapist to be cared about. Although his relational templates led him to expect that the therapist would be irritated by his emotional needs and resent them, as his mother had done, Peter experienced a reparative new response.

Next, as Peter's anger toward his girlfriend, and then, both of his parents, emerged, Peter felt guilty. He felt that he should be understanding of them, as his coping style demanded. He also felt ashamed of his inability to sustain this forgiving response. This intrapsychic defense reflected his moving-toward adaptation and served to block expression of his profound need for more secure relational ties. With the therapist's skillful balance between affirming the validity of Peter's anger and acknowledging that Peter still loved his mother and had also received many good things from her, Peter gradually felt safe enough to risk experiencing both the longing and the anger that had always been present but too anxiety-arousing to acknowledge. Why was it so threatening for Peter to experience his hurt and anger or share it with the therapist? While growing up in his family, Peter had learned that relationships were not resilient. His needs and his angry or sad feelings were unacceptable. He knew that, because of his father's unavailability and his mother's strong propensity toward love withdrawal, he could easily lose the little emotional support that he had. Aware that many different feelings could rupture his attachments, Peter had convinced himself that his needs were unimportant. In particular, his anger aroused strong separation anxieties. Realistically, he feared that he would be emotionally cut off if he expressed even disappointment, let alone the outrage he felt. Peter then tried to block these threatening feelings by telling himself that he should be calm and accepting—especially if he were going to be a therapist and help other people. Fortunately, the therapist was able to approach each of these feelings and help Peter understand and contain them.

To sum up, the therapist provided a skillful balancing of the good news and the bad news in Peter's development. He was able to acknowledge the reality that Peter loved his mother and still felt loyal to her despite all that occurred, and that some of the qualities he valued most about himself were qualities derived from her. However, these reality-based strengths were accompanied by the painful realities of Peter's emotional deprivation and invalidation. By exploring with Peter and then affirming what was legitimately good in his relationship with his mother, rather than simply blaming or rejecting her, the therapist helped Peter preserve important aspects of this tie. This gave Peter the attachment security he needed to risk Bowlby's *protest*—his anger at both of his parents for what was seriously wrong in their relationship. Being able to protest what had been so unfair and hurtful was to say these things did happen, were painful, but that he didn't deserve them. Rather than deny the reality of his circumstances any longer, being able to have the validity of his own feelings and perceptions for the first time led Peter to a stronger stance that he was able to generalize to other relationships. Peter became more confident and assertive; developed firmer boundaries and more clarity about who he was; and began to disconfirm the pathogenic belief that he would be criticized or rejected if he wasn't taking care of others. Peter came to understand that his mother's rejection was unwarranted, had absolutely nothing to do with him, and was just the projection of her own hopelessness about ever being loved. Significantly, Peter was able to feel compassion toward her and appreciate the predicament she was in, but without denying the reality of the painful consequences this engendered for him.

As a result of these changes, Peter no longer needed to defend so strongly against his attachment need and its attendant anxiety. He felt less compelled to make others like him, more comfortable with interpersonal conflict, and more self-contained. Finally, Peter was also able to keep the best parts of his moving-toward style: He remained a caring and responsive person, but this was no longer the unidimensional response pattern that it had been in the past. Peter had grown through this crisis and become a more resilient person. Therapy helped to immunize him against his predisposing vulnerability to rejection, criticism, and loss. By experiencing a reparative response from the therapist and finding that some relationships can be different from those he had experienced and expected, Peter expanded his internal working models of what can occur in relationships. In this way, the therapist helped Peter to stop carrying emotional baggage from his past into his present relationships and from reenacting the same maladaptive relational patterns and inflexible coping strategies. Peter will certainly have further crises in his life, especially when circumstances again tap into his "old wound," but the changes he has made will help him respond to future problems in a more empowered, present-centered manner.

Two Case Summaries

In the case study of Peter, we have seen how our model for conceptualizing clients may be utilized with a moving-toward client. Let's look at two additional case summaries that highlight salient features in the functioning and treatment of a moving-against client and a moving-away client.

Carlos: Moving against Others

Carlos had been riding high, on his way to becoming a real estate "king," when his business failed. Buoyed by his initial success in a booming real estate market, Carlos had been woefully overextended when interest rates rose unexpectedly. His business went bankrupt, his new home and sports car were repossessed, and friends deserted him as creditors and the IRS aggressively pursued him. When his dreams of wealth and power were dashed, Carlos became seriously depressed and reluctantly entered therapy.

After six months in treatment, however, his depression had greatly improved. Carlos recognized how helpful the therapist had been and had recently told a friend that he might not be alive if not for the therapist. On this particular day, however, Carlos was ready to quit. He was frustrated that he had to come to therapy and he did not want to need the therapist, or anyone else, anymore. On this particular day, he was fed up with the therapist's fees. As Carlos often did with others when he was distressed, he began the session by taunting the therapist:

Carlos:
Who do you think you are to charge so much? Do you really think you're worth all that?!

The therapist acknowledged the feeling in Carlos's provocation:

Therapist:
You're angry about my fees. Let's talk about them.

Carlos:
(*continuing his assault*) What are you doing—trying to prove that you can gouge as much money as the psychiatrist next door? Did your mother want you to be a "real" doctor and make a lot of money or what?

Fortunately, the therapist was able to remain nondefensive and stay emotionally connected to Carlos during this critical incident. The therapist did not retreat or counterattack, which is what Carlos's provocations usually elicited from others. Instead of acting on his own reflexive response to justify and explain, the therapist tried to find a more neutral or effective middle ground of involvement:

Therapist:
Yes, of course you're angry with me. I do charge you for my time, and it's hard for you. And there are still problems in your life that haven't changed yet.

The therapist's accepting response only infuriated Carlos further:

Carlos:
I hate your understanding, I hate coming here, I don't want to do this anymore! What do you do—be nice to people when they're falling apart so they become dependent on you? Then they can't leave you and you can keep taking all of their money. There's something wrong with you; you're sick!

Therapist:
It seems like I am trying to make you dependent on me so that I can take advantage of you financially?

Carlos:
(*loudly*) Yes, you idiot, I want to leave and get out of here, but I can't. I'll just get depressed again if I leave. I need you and you know it, and you use that to take the little money I have left. I'm trapped by you. You've got me and I hate it. I hate you!

The intensity of Carlos's reactions told the therapist that, in addition to whatever reality-based issues needed to be addressed regarding fees, Carlos's generic conflict was being activated by this issue. On the basis of other hypotheses he had already formulated, the therapist reasoned that, for Carlos, having a need of someone meant being used or exploited by them. Trying to articulate what Carlos was experiencing in their relationship right then, the therapist responded:

Therapist:
It's as if you don't experience the decision to remain in therapy as your own choice, Carlos. You feel trapped by your own need to be helped or understood. You have learned that to let yourself need somebody puts them in control of you and leaves you subject to their own selfish needs. Your only choice then is to resist by leaving the help you still want or to stay in treatment with me but feel exploited. You may have experienced something like this with others, Carlos, but I want you and me to have a different kind of relationship. I want to respond to your need for help, not to use your trust and vulnerability for my own gain.

Unable to provoke the therapist, Carlos gradually calmed down and was somewhat able to accept what the therapist was saying. Carlos's transference reaction—that the therapist had manipulated Carlos's dependency in order to meet his own financial need—encapsulated Carlos's life story. Carlos had grown up in an orphanage, emotionally deprived throughout his early childhood. When he was placed with foster parents during second grade, they took advantage of his need to belong.

By the time he was 9 years old, Carlos had become an exploited laborer on his foster parents' farm. In order to earn their approval and affection, he worked sunup to sundown on the days he wasn't in school. By the time he was 18 and ready to leave home, Carlos had long since realized that his foster parents were taking advantage of him and were not to be trusted. At some point during his early adolescence, Carlos's anxious efforts to please his foster parents had evolved into an angry defiance toward them and others. He had vowed to himself that he was never going to need anybody again. Attempting to disconfirm the shame of having been used, Carlos resolved that he was going to be rich and powerful—strong enough that no one would ever be able to take advantage of him again. Carlos's feelings of exploitation and betrayal were further fueled by his racial and ethnic history. A Latino, Carlos was the child of undocumented migrant workers. According to the story he had been told, his baby-sitter had been instructed to leave him at the orphanage if his parents were to be suddenly deported. Carlos was angry at their exploitation by farmers but also angry at his parents for never returning to claim him. These factors all contributed to Carlos's determination to never need again. He was going to be rich so that no one would ever use or betray him; he would simply buy whatever he needed.

The dramatic confrontation with his therapist proved to be a corrective emotional experience for Carlos. Although his fear of being used was aroused in relation to the therapist, as it was whenever Carlos began to get close to anyone, his coping strategy of anger and intimidation did not succeed in pushing the therapist away. The therapist was able to tolerate Carlos's provocative attacks and remain concerned about him. The therapist validated the betrayal in Carlos's original and adoptive experiences, and how this was compounded by seeing the impact of racial exploitation in the society.

As the therapist repeatedly provided a different response to his moving-against style than Carlos expected, the next feeling in his affective constellation emerged: His primary feelings of hurt and betrayal became the predominant issue in therapy. Just as the therapist was able to tolerate Carlos's anger, he was also able to provide a holding environment to contain these unwanted feelings. Before long, they, too, became less threatening. Carlos's accompanying shame about being vulnerable or used continued to be activated in the therapeutic relationship, however. Thus, Carlos experienced the same fears and concerns over and over again with the therapist. Each time these issues were manifested in a conflict between them, however, Carlos received a genuinely empathic and understanding response from the therapist that was different from what he expected.

In addition, it was helpful to Carlos when the therapist validated his perception that he had indeed been taken advantage of as a child, that people of his culture were often devalued and exploited, and that elements of devaluation and exploitation were present in some of his current relationships. However, the therapist patiently but repeatedly questioned whether others were always trying to exploit him in his current relationships. Honoring his interpersonal coping strategy, the therapist also observed that, as a child, Carlos had made the best adaptation that he could to his unsolvable predicament. The therapist pointed out that there was strength in Carlos's defensive attempt to rise above his conflict by adopting an aggressive and defiant interpersonal style. Although this defensive attempt to rise above his conflict was no longer necessary or effective in many situations, it had indeed once served a necessary purpose—to protect him from the vulnerability and exploitation that he experienced as a child and was determined never to experience again. Eventually, Carlos understood what the therapist meant when he suggested that it took just as much courage for Carlos to face these feelings now, as an adult, as it originally did to defend against them as a child by compulsively striving to achieve power and control.

With the therapist's support, Carlos was gradually able to reexperience his old feelings of desperate need and simultaneous outrage at being used by his foster parents and, later, at being left by his biological parents. As he gradually let go of his combative and competitive stance with the therapist, and let himself undergo those painful feelings of his childhood that had always seemed disgusting and weak, Carlos began to change. First, his provocative, argumentative style gave way to a friendlier comradery with the therapist. Carlos began to disclose more easily and felt less competitive with the therapist. In particular, Carlos became less concerned about needing the therapist or being exploited by him.

These changes with the therapist began to carry over to other relationships as well. Carlos began to establish male friendships for the first time; his competitiveness had always precluded such friendships in the past. Carlos's relationships with women had always been short-lived, superficial sexual contacts, which often were manipulative. During counseling, Carlos realized that he felt less threatened by these short-term relationships. As Carlos stopped defending against his shame, and began to feel compassion instead for his own childhood predicament, his pattern of using women diminished.

However, he was still unable to establish an intimate, egalitarian love relationship at this time in his life.

Other evidence of change came from Carlos's business associates, who told him he seemed less driven than in the past. His attempts to rise above his conflicts through achieving wealth and power were modulated and gave way to an ambitious work ethic. When an established banker in town offered Carlos a good job as a loan officer and appraiser in his real estate division, Carlos felt ready to terminate therapy. All of his problems were certainly not solved, but Carlos now felt capable of successfully moving on in life on his own. It was important to Carlos to be reassured that he could return to therapy if he needed to in the future, and that doing this would not signify failure but rather strength in recognizing when he needed help. Assured of the therapist's continuing availability, Carlos terminated and did not recontact the therapist.

Maggie: Moving away from Others

Four months after her adolescent daughter had been raped, Maggie was still in crisis. Nightmares disrupted her sleep; tension headaches tormented her by day. Helpless to comfort her daughter and enraged at the casual indifference of the police and courts, Maggie was afraid her life was spinning out of control. At work, her supervisor evaluated her as irritable and sullen and suggested that she seek counseling. Although she had always been reluctant to ask for help with anything, Maggie contacted a therapist when she realized that her job was in jeopardy.

The therapist was responsive to Maggie's feelings of rage, guilt, and helplessness evoked by her daughter's tragedy. To her surprise, Maggie felt understood by the therapist and began looking forward to their meetings. She gradually came to trust the therapist and increasingly invested herself in their relationship. During one session, Maggie reported a dream from the previous night. In the dream, Maggie was alone in a vast desert night. No other people existed in this great, silent space. The desert night was black; no light shone from stars or moon. As she walked across the endless sand, a cool, dry wind began to move lightly across her face. Maggie laid down on the sand, closed her eyes, and silently slipped away into the darkness.

Maggie said that variations of this dream had recurred throughout her life. Because this was a recurrent dream, the therapist knew that it held much meaning and probably encapsulated her generic conflict. Hoping that the dream could also provide an avenue for joining her central feeling of aloneness, the therapist tried to bridge Maggie's moving-away orientation:

Therapist:
  Can you close your eyes and see the dream again?
Maggie:
  Yes. (*settles back and closes her eyes*)

Therapist:
Describe what you see to me.
Maggie:
I'm walking; it's quiet and dark. I'm alone. I can feel the breeze. Now I'm lying down on the sand, and I close my eyes, like going to sleep.
Therapist:
I don't want you to be there alone. Will you let me join you?
Maggie:
(*pause, then slowly*) Yes, you can.
Therapist:
Keep your eyes closed, and hold that same, familiar image. But you're not alone this time. I am walking toward you, and I reach my right hand out toward yours. Will you take it?
Maggie:
OK.
Therapist:
I'm now in the dream with you. We're holding hands and walking together through the desert night. Can you see that?
Maggie:
Yes, we're walking toward the lights of a distant city. (*opens her eyes and looks kindly at the therapist*) I'm not alone. It's good to have you with me.

This joining experience was a turning point in therapy. Maggie had grown up in a silent void, much like the setting in her dream. She had never known her father and her mother was often "away." By 10 years of age, Maggie was regularly spending most of the weekend alone, fixing her own meals, and putting herself to bed while her mother was out. The dream reflected both the emptiness of her childhood and her lifestyle adaptation of moving away from others. However, the crisis with her daughter overwhelmed her coping strategy to withdraw and be self-sufficient and aloof. In letting the therapist join her in her dark, empty space, Maggie took the enormous personal risk of accepting the human contact that she longed for but had long since learned to hold away. This single experience did not resolve her generic conflict or adaptation to it, of course, but incidents of sharing continued in her relationship with the therapist. Maggie's sense of aloneness was heightened by her consciousness of being "different" because she was bicultural, as her therapist, herself a bicultural woman, was well aware. The therapist's support and bold articulation of all the factors, familial and cultural, that contributed to Maggie's sense of isolation helped her begin to change in three ways.

First, Maggie's conflicted affective constellation emerged. The sadness resulting from her childhood neglect became accessible. To ensure that the lack of response that Maggie had experienced as a child was not recapitulated, the therapist was careful to let Maggie know that her feelings were being heard. The therapist was skilled in communicating to Maggie that she was not in a dark, silent void anymore, but in a caring and responsive relationship. Maggie became more comfortable sharing her loneliness with the therapist. However, the intense anger at her mother for virtually abandoning her, and the painful separation anxieties aroused by this protest, were worked through more slowly.

Second, as the therapist continued to affirm other difficult feelings, Maggie began to change in her relationship with the therapist. Although Maggie had become attached to the therapist during her crisis, she was still deeply reluctant to accept help from anyone and held a part of herself back. Interestingly, the therapist's bicultural identity had given her an *ascribed credibility*, based on perceived similarities of race and gender, which helped Maggie remain in treatment initially when she was so unsure of this process. Soon, however, the therapist's *achieved credibility*—her skillfulness and the effectiveness of her responses—took center stage. As their working alliance continued, Maggie incrementally relinquished her coping style and gradually allowed the therapist to help her. Maggie began to act in a less-reserved manner toward the therapist and became more expressive than she had been with others. She talked more freely about herself and found herself being curious about the therapist's own personal life. The therapist saw this as positive for this particular client and responded willingly at times. Maggie's nightmares and headaches already had been alleviated, but now a sense of humor and graciousness were emerging as well.

Third, as Maggie talked about feeling "fuller" inside, the therapist observed that changes were occurring in her other relationships as well. At work, her supervisor was pleased to observe that Maggie was less irritable and it was easier for others in the office to work with her. When tensions surfaced with others, Maggie's initial reaction still was to withdraw and "go away inside" since this felt safe to her. However, Maggie became better able to discuss her problems and she and her therapist worked on generating new ways for her to respond. Together, they would role-play different responses that would keep her involved in the situation, rather than to remain aloof as she had done in the past. Maggie also became more accessible to her two adolescent children. For the first time, she talked more about her own interests and personal history with them. The nearly grown children welcomed this contact and, in turn, they began to disclose more to her as well. Maggie was feeling closer to her children and somewhat more capable of helping her daughter cope with the aftershocks of her assault.

Finally, Maggie began to change how she responded to her boyfriend. This had been a superficial relationship, like most others in her life had been. Even though they had known each other for 2 years, Maggie now invited more personal involvement and asked for more commitment from him. She asked him to talk more about himself and to spend more time with her, which he was able to do. It was exciting but anxiety-arousing to take each of these steps forward. Her old coping style of moving away was activated each time she took a step toward her boyfriend or others, especially when others did not respond in completely positive ways. Still, Maggie readily defended herself against the disappointment that relationships had originally held for her. As the therapist repeatedly helped her work through her concerns in each of these arenas, however, Maggie tentatively suggested that it might be time to end therapy.

Although things had been going well, Maggie became depressed as soon as they talked about ending. Even though she had brought up the topic of termination, she missed their next session and arrived late to another. As before, the therapist continued to focus Maggie inward, and her profound feelings of loneliness emerged again. The therapist suggested that they put off setting a termination date for a while and work further with this sadness. With more vividness and detail than before, Maggie recalled her childhood depression. Painful memories returned—for example, the memory of being 8 years old, sitting alone on the living-room couch, and listening to the clock tick the empty afternoon away. Maggie sobbed as she recalled her aching wish for someone to come home to, which was only reenacted in her marital choice. Following her relational templates, Maggie had married a salesman whose job took him away from home for extended periods—and he had been emotionally unresponsive to her when he was home as well. Thus, in her first significant love relationship, Maggie chose someone who repeated, rather than resolved, her history of aloneness. However, touching the pain of her childhood neglect so directly, and sharing so fully how it was echoed in the disappointment of her marriage, relieved the depression that the suggestion of terminating had precipitated. Before long, Maggie again felt ready to terminate. This time, with the therapist's reassurance that she could contact her again if necessary, Maggie successfully terminated.

Closing

In this chapter, we have studied an interpersonal model for conceptualizing clients. In an extended case study of Peter, a moving-toward client, we saw how this model can be applied to treatment. We also looked at two case summaries that illustrate its application to moving-against and moving-away clients. At this point, however, two issues remain incomplete.

First, we need further guidelines for conceptualizing clients and using this conceptualization to shape treatment plans and intervention strategies. Information will be provided in Chapter 8 about clients' current interpersonal functioning that will help therapists further conceptualize their clients. At this point, however, readers are prepared to utilize the structured guidelines for keeping process notes provided in Appendix A and for writing case conceptualizations in Appendix B. These guidelines will help therapists to formulate treatment plans by clarifying the maladaptive relational patterns that provide a focus for treatment. These guidelines will also help therapists recognize the process dimension and intervene more effectively by linking the current interaction with the therapist to the problems that clients are having with others.

Second, trainees need further guidelines for negotiating the crucial stage of therapy in which the compromise solution fails (for example, a crisis is provoked when the client's attempts to rise above fail in some way) and the client's generic conflict emerges. The therapeutic relationship is the most important tool therapists can utilize at this critical juncture to help clients resolve problems and change. In Chapters 9 and 10, we will discuss more fully how the therapist can best respond when the client's generic conflict emerges, provide a corrective emotional experience, and generalize this experience of change with the therapist to relationships outside of therapy.

Suggestions for Further Reading

1. In the current managed care milieu, it has become essential to write effective case formulations with clear treatment goals and programmatic intervention strategies. In Chapter 7 of the Student Workbook, an extensive case formulation is provided that illustrates the format for writing case formulations provided in Appendix B.

2. Useful guidelines for conceptualizing clients are found in Chapters 5 and 7 of *Psychotherapy in a New Key: A Guide to Time-Limited Dynamic Psychotherapy* by H. Strupp and J. Binder (New York: Basic Books, 1984). Readers are also encouraged to examine D. Wiger's practical book for all aspects of psychotherapy documentation: *The Psychotherapy Documentation Primer* (New York: Wiley, 1999).

3. *Coping with Conflict: Supervising Counselors and Psychotherapists* by W. Mueller and B. Kell (New York: Appleton-Century-Crofts, 1972) is unsurpassed in helping trainees utilize their own reactions and the process dimension to conceptualize client dynamics and effect change.

4. Karen Horney's work is essential reading for interpersonally oriented therapists; it has been widely incorporated by psychodynamic and cognitive behavioral theorists alike. Readers may be especially interested in *Neurosis and Human Growth* (New York: Norton, 1970) or *Our Inner Conflicts* (New York: Norton, 1966). *Cognitive Therapy of the Personality Disorders* by A. Beck and A. Freeman (New York: Guilford, 1990) effectively links Horney's interpersonal coping styles to cognitively based interventions for clients with difficult-to-treat personality disorders.

CHAPTER EIGHT

Current Interpersonal Factors

Conceptual Overview

We now turn to the third component of our conceptual formulation. As we have seen, the first component comprises the three basic dimensions of family life introduced in Chapter 6. Most clients who enter therapy have experienced problems along each of these dimensions. They have usually come from families without a primary parental coalition, are often bound in cross-generational alliances and loyalty conflicts, and/or have not had their developmental needs for secure attachments and responsible discipline met. When development goes awry along these dimensions, children are likely to develop enduring symptoms and problems.

The second component of our conceptualization is the schematic model for understanding how client problems are expressed, maintained, and resolved (Chapter 7). Central to this conceptual model are clients' characteristic interpersonal styles for coping with their generic conflicts. In this chapter, we consider the third component of this conceptual formulation—clients' current interpersonal patterns. These maladaptive relational patterns provide important information for conceptualizing clients and understanding what is occurring in the therapeutic relationship. In particular, we will see how therapists' own reactions toward clients can be the best source of data for understanding clients' conflicts and interpersonal coping strategies. Thus, in this chapter, we examine how therapists can utilize clients' current interpersonal functioning to formulate treatment plans and guide their interventions.

Chapter Organization

In the first section, we examine three basic ways in which client conflicts are expressed in the therapeutic relationship: eliciting maneuvers, testing behavior, and transference reactions. Next, we consider the degree of separateness versus relatedness in the therapist-client relationship. Using the process dimension, therapists can try to maintain an optimal degree of interpersonal relatedness with clients to facilitate change. Finally, we look at the ambivalent or two-sided nature of clients' conflicts. Applying the guidelines for writing case formula-tions in Appendix B, therapists consider each aspect of clients' current interper-sonal functioning in formulating treatment plans and intervention strategies.

How Clients Bring Their Conflicts into the Therapeutic Relationship

The symptoms and problems that clients strive to resolve in treatment origi-nated from, and are now being expressed in, close personal relationships. Be-fore long, the same relational patterns that are causing problems with others emerge in the therapeutic relationship. As we have seen, clients do not merely talk with counselors about their problems with others. In a far more personal and immediate way, they recreate the same maladaptive relational patterns with the therapist. The challenge for the therapist, then, is to provide a new, more satisfying solution to this familiar but problematic scenario in their real-life re-lationship. Thus, in this section, we will examine three similar but distinct ways in which clients bring their conflicts with others into the therapeutic relation-ship: eliciting maneuvers, testing behavior, and transference reactions.

Clients often employ *eliciting maneuvers* (Sullivan, 1968) in order to avoid anxiety and defend against their conflicts. At other times, clients will employ a second interpersonal strategy, testing behavior (Weiss, 1993), to approach and try to resolve their conflicts. *Testing behavior* reflects a healthy, growth-oriented attempt to resolve problems. Clients assess whether the therapist will respond in the familiar but problematic ways they expect, or in the more helpful ways they actually need. Working with these complex and sometimes contradictory maneuvers can be one of the most interesting aspects of clinical work.

The third (and better known) way that clients bring their conflicts into the therapeutic relationship is through *transference reactions*—the clients' system-atic misperceptions of the therapist. As we will see, much confusion arises from two different definitions of the term *transference*. We will clarify important dis-tinctions between the traditional usage of this term and contemporary defini-tions that are more useful. These three interrelated concepts reflect the most fundamental aspects of interpersonal psychotherapy.

Eliciting Maneuvers

Interpersonal theorists have described how clients develop fixed interpersonal styles to avoid anxiety and defend against their conflicts (for example, Beck & Freeman, 1990; Fromm, 1982; Horney, 1970). Most clients are inaccurate in their perceptions of others, restricted in their range of emotions, and respond to others inflexibly, using, for example, Horney's fixed coping styles of moving toward, moving away, and moving against. In part, clients systematically employ these interpersonal styles to (1) elicit certain desired responses that will avoid their conflicts and diminish anxiety and (2) preclude responses from others that will arouse their generic conflicts. An *eliciting maneuver* is a behavior that wards off anxiety and brings about certain desired, safe responses. To illustrate how eliciting maneuvers serve to defend against conflicts, we will return to the case study of Peter, the moving-toward client in Chapter 7.

Peter's moving-toward style tended to elicit approval and kindness from others. Peter was helpful; he sympathized, agreed, and cooperated with people. For example, Peter tried to understand his girlfriend's infidelity and to win his instructor's approval. When Peter began to see a therapist, he employed the same interpersonal style. He tried to elicit his therapist's approval by being a good client who was quickly getting better—at least until the therapist began using process comments to question this compliance.

In contrast, Peter was never appropriately forthright, assertive, angry, or skeptical. These responses, which a well-functioning person needs to employ at times, were not part of Peter's interpersonal repertoire. For example, he did not communicate how angry he was with his girlfriend or even experience how angry he was with both of his parents. Similarly, he did not set limits with his practicum instructor or classmates regarding how much critical feedback he could incorporate at one time. *His pleasing interpersonal coping style was designed to discourage angry or critical responses from others that would arouse anxiety for him.*

If Peter's therapist had merely responded automatically, without first considering what Peter's interpersonal style tended to elicit from others, Peter would not have changed much in therapy. That is, if the therapist had merely supported Peter and met his need for approval, Peter would have only reenacted in therapy the same rising-above defense that he had used throughout his life. Shoring up these defensive coping styles is often the goal in some time-limited or supportive therapies, and helps many clients in crisis to regain their equilibrium. The therapist wanted to do more than this with Peter, however. Having the opportunity to seek more enduring changes, and not merely symptom relief, the therapist used process comments to make Peter's moving-toward style overt and a focus for discussion. Thus, instead of automatically responding to what Peter elicited, the therapist focused on it as a part of Peter's problems that needed to be addressed in therapy. How did this intervention occur?

Over the course of several months, the therapist repeatedly pointed out, in a supportive and noncritical way, instances where Peter's coping style of moving toward others was operating. Working together in a collaborative alliance, Peter and his therapist considered both how this was serving to protect him from his conflicts on the one hand, and how Peter suffered as a result of this coping style on the other. Once they named this pattern, Peter began observing himself using this moving-toward style with others, and recognized how it elicited from others the approval and support he had missed as a child. In addition, he began to recognize how this style discouraged the critical or rejecting responses from others that were so anxiety-arousing for him. As Peter made progress in treatment, he gradually became less intent on winning others' approval and was able to expand his interpersonal repertoire. He felt more resilient, and began to be appropriately assertive and direct for the first time in his life.

The eliciting maneuvers used by moving-toward clients like Peter are familiar, and may not be especially difficult for beginning therapists to work with. In contrast, moving-against clients employ eliciting maneuvers that are often challenging for beginning therapists. In the first few minutes of the first session, these clients often do something to take command of the relationship. The moving-against client seems to quickly find some way to intimidate or disempower therapists, making them feel insecure. For example, clients may insist on sitting in the therapist's chair, questioning the therapist's adequacy, or criticizing one of the therapist's early responses. This provocative presentation elicits anxiety in many beginning therapists. Many therapists will respond by withdrawing emotionally, others will respond in kind and become competitive, a few will counter with their own hostility. Treatment will not progress when moving-against clients succeed in eliciting one of these defensive responses from the therapist. By reenacting this maladaptive relational pattern with the therapist, these clients have successfully neutralized the therapist's effectiveness. In this way, they have defended against their anxieties associated with having problems, asking for help, relinquishing some control, and so forth. Thus, clients' eliciting maneuvers protect them from their generic conflicts, but at the price of change.

Unfortunately, therapists tend to respond "reflexively" or automatically to the client's eliciting maneuvers or *relational enactments* with their own countertransference propensities—with their own characteristic tendencies toward flight or fight. What can therapists do instead? First, they can try to retain legitimate control of themselves internally by trying to understand what is occurring right now in their relationship with the client. They can begin formulating working hypotheses about (1) the impact this behavior might have on others in the client's life and (2) what conflicts the client might be avoiding by these maneuvers. Second, on an interpersonal level, therapists can try to engage the client in some way in which the maladaptive relational patterns of domination, intimidation, or competition are not enacted. There are several possibilities.

1. One possibility is for the therapist simply to withstand the client's criticism, without avoiding the challenge.

Therapist:
Your previous therapist was a more experienced psychiatrist, and you're not sure that a younger social worker such as myself can help you. Let's talk about that. Tell me more about your concern.

2. Another alternative is for the therapist to make a process comment about the current interaction.

Therapist:
You're speaking in a loud voice right now. You sound angry. Are you aware of speaking loudly, and do you feel angry?

3. With some clients, the therapist may wish to take this approach one step further and begin to explore the effects of this interpersonal presentation on others.

Therapist:

I often feel criticized by you and I'm wondering if that is characteristic of just our relationship or if others in your life feel this way as well.

Many of these clients will be unaware of the impact they have on others, and this interpersonal feedback can provide a productive new starting point. However, the purpose here is not to win the battle for control and get one up with the client. On a process level, this would only continue the same maladaptive relational pattern that recurs with others. Rather, this approach is intended to help the therapist find some way to engage with the client other than by reflexively responding with anger or compliance to the eliciting maneuvers. It will also help clients develop increased awareness of when they employ their coping style and its impact on others.

4. Still another possibility is for the therapist to focus on how the client's eliciting maneuvers affect the client.

Therapist:

You have insisted on sitting in my chair, even though I asked you not to. Now that you are there, how does it feel?

Routinely, clients become aware of feeling alone, empty, or anxious at these moments, and the hollowness of their victory provides a new shared point of departure for the therapist and the client.

Particular types of responses will work well with one client and not at all with another. Therapists' goals in responding to such challenging clients are as follows:

1. To attend to their own subjective reactions and observe what the client is eliciting in them
2. To formulate working hypotheses about how others would typically respond and consider how this may serve to help the client avoid his conflicts
3. To try to find another way to respond (for example, via process comments) that does not go down the old familiar path and reenact the same scenario that the client usually elicits from others

Although moving-against clients may intimidate beginning therapists, this usually does not last very long. With a little more experience and a supportive supervisor, beginning therapists learn not to charge so readily at a waving red flag. With more exposure to moving-against clients, therapists learn to appreciate that the intensity and rigidity of clients' eliciting maneuvers, no matter how alienating initially, are commensurate with clients' degree of anxiety and conflict. As therapists better understand the purpose of these distancing maneuvers, they may begin to find some compassion for the predicament such clients are living out, which is usually the best way to step out of the old relational pattern being elicited. By flexibly trying out different approaches, therapists can usually find ways to engage these clients that do not reenact the intimidated, competitive, or hostile patterns that they elicited from others.

As we have seen, the confrontational eliciting maneuvers of moving-against clients—such as Carlos in Chapter 7—place considerable demands on beginning therapists. However, the meeker, moving-away clients—such as Maggie in Chapter 7—actually pose a greater challenge. Most people who select counseling careers have strong needs for close personal relationships. When moving-away clients continue to maintain emotional distance, many therapists are discouraged and ultimately give up—disengaging as many others in the clients' lives have done before. When this occurs, the clients' eliciting behaviors have successfully defended them against a relationship with the therapist, and thereby allowed them to avoid the threats entailed by closer involvement. Clients' conflicts will not be activated, but the therapist's only real vehicle for resolving these conflicts is closed as well. Rather than responding reflexively to these clients' aloofness by turning off, the therapist may use process comments to try to find other avenues for establishing a relationship. Beginning therapists must be patient with themselves and these clients, however, appreciating that it will be threatening initially for these clients to relate more personally.

To sum up, the therapist can attend to what clients systematically elicit in others and hypothesize how this interpersonal strategy protects them from their generic conflicts. In doing this, the therapist must attend to the clients' developmental history, current life situation, and to the cultural context in which these are embedded. More importantly, *therapists must also attend to their own feelings that are elicited by clients.* This self-awareness is one of the therapist's most important sources of information about the client. If the therapist can be nondefensive enough to observe what clients elicit, it will provide information about the clients' conflicts, coping strategies for defending against them, and how the relational themes being played out in the therapeutic relationship may be effecting other relationships.

We have already seen that the therapist can generate working hypotheses to understand why clients elicit certain responses, such as reassurance, disinterest, competition, and so forth. The therapist's task is to keep from responding reflexively to what clients elicit and, if possible, to try to respond in a way that does not reenact the clients' maladaptive relational patterns. Over time, the therapist can begin to help clients identify their eliciting mechanisms, recognize the situations or types of interactions where they tend to employ these measures, and understand how these maneuvers have served to create safety in the past. It does take courage, however, for beginning therapists to tolerate their own unwanted or unacceptable feelings long enough to consider how they may be used to understand the client. The key element here is for therapists to remain nondefensive toward their own reactions so that they can reflect upon and learn from them.

Finally, significant problems may arise when clients' eliciting behaviors tap into the therapist's own dynamics. With this new development, we enter the arena of countertransference. For example, suppose that the client acts helpless and escalates his distress, and the therapist tries to help but fails at each attempt. Ideally, the therapist would remain involved and keep exploring other ways to respond—for example, by making the process comment that whatever the therapist tries seems to disappoint the client. However, if the therapist's own personal concerns regarding performance demands or criticism are activated, the therapist may lose her effectiveness. For example, the therapist may feel inadequate and withdraw, become impatient with the client and criticize, and in so doing, recapitulate maladaptive relational patterns.

Scenarios in which the clients' eliciting maneuvers tap into the therapist's countertransference propensities are commonplace. When this happens, clients often switch to the other side of their ambivalence and try to reengage the therapist, perhaps by reassuring the therapist about how helpful the sessions are. Clients do not really want to succeed in immobilizing the therapist or ending the relationship, and will often try to get the disengaged or discouraged therapist involved in the relationship again. Premature and unnecessary terminations routinely occur when the therapist (and perhaps the supervisor) fails to hear clients' renewed bids for relationship because the disengaged therapist has given up on the relationship. Especially with more troubled clients, therapists must keep both of these disengaging and subsequent engaging maneuvers in mind.

It is essential for therapists to distinguish between their client's eliciting maneuvers and their own personal or countertransference reactions toward the client. This is referred to as *client-induced versus therapist-induced countertransference* (Springmann, 1986). How do counselors know if the client's eliciting maneuvers (i.e. client-induced) or their own personal issues (i.e. therapist-induced) are evoking their strong reactions? Guidelines for making these critical distinctions will be given later in the chapter.

Testing Behavior

We have seen how clients employ their inflexible coping style to systematically elicit responses from others that may seem self-defeating at first, but have served to effectively avoid anxiety in the past. These eliciting maneuvers—or security operations as Sullivan (1968) sometimes called them—are interpersonal defenses.

Although eliciting maneuvers are used defensively to avoid anxiety, at other times clients use different interpersonal strategies to approach and try to resolve problems. While this may sound contradictory, it reflects clients' ambivalent feelings about treatment and what change might entail. Clients often carry out the approach side of their ambivalence by means of *testing behavior.* Clients systematically reenact their conflicts with the therapist, in some direct or metaphorical way, to test whether the therapist will respond in the familiar but problematic way that others have in the past, or whether the therapist can respond in a new way that allows them to make progress on their problems. *Therapists should not underestimate how vitally interested clients are in assessing whether the therapist is going to reenact or resolve maladaptive relational patterns.* Many clients have histories of exploitation, denigration, and rejection; their trust in others has been betrayed. Their testing behaviors often reflect their intense wishes for healthier, safer relationships, as well as their fear of again being disappointed or hurt. As we will see, the change process is set in motion when therapists respond effectively and "pass" important tests. Immediately afterwards, clients often make visible progress in treatment. They may feel more confidence in themselves and in the therapist, symptoms such as anxiety or depression may diminish, clients may risk bringing up threatening new material, and so forth. Much of this discussion on testing behavior follows from the work of Weiss (1993) and Weiss and Sampson (1986).

In a healthy attempt to resolve problems, clients often actively (but not consciously) elicit the same types of problems with the therapist that they are having with others in their lives. Routinely, clients are testing to see if they can change these maladaptive relational patterns and disconfirm the pathogenic beliefs that stem from them. However, it is difficult for clients to find disconfirming evidence because repeated confirmation of the beliefs has made clients deaf to such evidence. *It is the power of immediate, in-vivo relearning with the therapist that allows clients to relinquish deeply held but faulty beliefs about themselves and others.* The therapist can fail the client's test by responding in a way that reenacts maladaptive relational patterns, or the therapist can pass the test and provide a corrective emotional experience by behaviorally demonstrating that relationships can be different. Suppose that the client's pathogenic belief is: "If I become stronger and do what I want with my life, my parent will be hurt and withdraw, and I deserve to feel guilty." In this case, the therapist would fail the test and recapitulate the maladaptive relational pattern if she expressed a worry or concern about the result of a new venture that this overly inhibited client is considering. In contrast, the therapist would pass this test—and embolden the client—by expressing interest in the venture and providing support for the client's expansiveness.

Such testing behavior pervades all counseling relationships. Testing is especially likely to occur in the initial sessions, in situations where strong feelings are being evoked, and when the current interaction between therapist and client parallels relational conflicts that the client is having with others. Many of these tests are easy to recognize. With only a little experience, therapists can readily pass them, allowing clients to progress in treatment and move on to other issues. For example, suppose a compliant, dependent woman says to her male therapist, "Where should we start today?" The therapist will fail the test and reenact a maladaptive relational pattern if he says, "Tell me about _____." Here the therapist has confirmed the client's pathogenic belief that she must assume the subservient role, that she ought not to assert herself or take the lead. This belief may reflect her own sense of inadequacy—that she is incapable of knowing what is best for herself. In contrast, the therapist will pass the test if he says, "I'd like to hear about what is most important to you. Let's begin with something that you would like to tell me about." This response demonstrates that the therapist is capable of engaging with the client in a more egalitarian way, and this response also acknowledges the client as capable of deciding what she needs to deal with. Repeated interactions of this sort will be reparative; they will begin to bolster the client's self-esteem and self-efficacy.

Tests that involve issues of power and control are more likely to prove challenging to beginning therapists. Clients with different developmental experiences may need to test the strength or resolve of the therapist. Before they feel safe expressing strong feelings or disclosing threatening issues, clients usually need to assess whether the therapist can manage what they present. For example, when clients push limits with the therapist, they often are testing whether they can control or manipulate the therapist (as they could their parent). Or, is the therapist able to tolerate their disapproval and set limits to provide the safety they did not receive before. Let's look at a case example.

Graduate students in Carl's training program were supposed to videotape all therapy sessions. Early in treatment, however, Carl's client Lucy complained that the videotaping made her self-conscious. She said it was already hard for her to talk about what was really important and that the videotaping made an already difficult situation impossible. She continued to stress this concern for several weeks, and finally told Carl that he was not helping her because there were too many things that she could not talk about when the tape was on.

Carl was torn and did not know what to do. He knew the clinic rules, and that his supervisor would abide by them, but he also felt that he should be flexible enough to do what his client needed. Without consulting his supervisor, he reached a compromise with Lucy and agreed to turn the videotape off for the last five minutes of each session. Lucy was appreciative and did indeed disclose new material to Carl. Carl felt badly about doing this without his supervisor's permission, but also thought it was the only way to help his client and keep her in treatment.

Soon, however, Carl observed that treatment was not going well. Lucy became scattered and started jumping from topic to topic. Carl tried to pin her down and focus on her escalating feelings and scattered thoughts, but she slid away each time. Lucy complained that she was very upset and needed more guidance from Carl, but whatever he did failed to help. Realizing that things were getting worse rather than better, Carl decided to talk with his supervisor.

First, Carl and his supervisor spent some time honestly discussing what Carl's decision to turn off the tape meant for their relationship, and then his supervisor began helping Carl hypothesize what this test might mean for Lucy. Assisted by some new ways of thinking about what was going on in the client-therapist relationship, Carl accepted his supervisor's firm limit that the videotape must remain on throughout every session. When Carl reinstated this limit at the beginning of the next session, Lucy protested strongly. She accused Carl of being untrustworthy because he broke their agreement, informed him that she could no longer talk about certain important things, and told him that she would have to think about whether she wanted to remain in treatment with him.

Carl felt badly about her threats, but he tolerated his discomfort and stuck to the rules. Although she continued to blame and complain, Lucy did return. In fact, despite her protests, she became calmer and began to focus more productively in therapy. Over the next few weeks, Lucy went on to recall how disheartening it had been for her to be able to "boss my father around."

It is adaptive for clients like Lucy to play out their maladaptive relational patterns with the therapist and test whether the therapist will reenact or resolve them. Clients cannot progress to more vulnerable material if the current therapeutic relationship is not safer than past relationships have been. However, there is very limited value in verbally reassuring clients. Moreover, most clients are not consciously aware of testing therapists in these ways, and usually it is not very helpful to point out testing behavior to them. Instead, as in all human relations, *it is what the therapist does that counts, rather than what the therapist says*. Therapists pass tests and provide a corrective emotional experience when they consistently respond in ways that behaviorally disconfirm pathogenic beliefs and maladaptive relational patterns. Fortunately, even though all therapists will fail tests at times, as Carl did, they will have many opportunities to recover from failed tests. Therapists can recover from these inevitable reenactments and ruptures if they stay emotionally present in the relationship, attend closely to the process dimension, and continually assess whether the client is progressing. Therapists can recover from mistakes and respond more effectively the next time this issue or theme comes up—often later in the same session or in the next session.

When the counselor passes a client's test and responds more effectively than significant others have in the past, the client experiences a new degree of safety in the therapeutic relationship. Exciting to witness, therapists will often observe clients responding to this greater sense of felt safety by beginning to remember more clearly, feel more deeply, and share more fully.

How can therapists pinpoint the specific interpersonal response that this particular client needs to pass a test and provide a reparative experience? Therapists must (1) consider client response specificity and (2) learn how to *assess* the client's positive or negative reactions to the different responses and interventions that the therapist employs. Let's consider both carefully.

Client response specificity   As discussed in Chapter 1, therapists need the flexibility to be able to respond differently to different clients. The same therapeutic response that passes a test for one client may fail the test and reenact conflicts for a client with different templates and expectations. For example, in response to pressure from her client, Beverly agreed to reduce her fee substantially. Almost immediately, her client began wearing more expensive jewelry, delighted in telling Beverly about shopping trips, and described how wonderful it was to fly first class on her vacation. Realizing that the client was taking advantage of her, just as the client had done with her doting mother and indulgent husband, Beverly renegotiated the fee. Later in the session, the client brought up important new material about feeling lonely when she felt she had too much control over others, and together they entered a productive new phase of treatment. Beverly concluded that, by being firm, she had passed the client's test and provided her with the corrective response she needed. Although unaware of her own testing behavior and unable to articulate it in any way, the client did not feel safe enough to broach this sensitive issue until she was behaviorally convinced that she could not manipulate Beverly as she could her mother, her husband, and previous therapists.

In contrast, lowering the fee can be a helpful response for another client. Chris was coping with a diminished income following a recent divorce and was having trouble making ends meet. Knowing that she grew up with demanding, withholding parents, the therapist realized that offering Chris a more flexible fee arrangement would be a significant overture. This client would never ask the therapist for a reduced fee and was initially unwilling to accept any of the alternative fee arrangements that the therapist suggested. When the therapist asked about her reluctance to even consider his offer, Chris began to cry. As their discussion soon revealed, the therapist's generous response disconfirmed her belief that she "didn't matter," and evoked the longing she had felt for her parents to reach out to her lovingly. Changing the payment schedule with this client disconfirmed pathogenic beliefs about herself and faulty expectations about others. Within this session, treatment progressed into more intensive work about her emotional deprivation and her pathogenic belief that she did not matter much to others. As these two examples illustrate, the same response by the therapist will often have very different effects on different clients.

Clients' ethnic backgrounds will further complicate response specificity. For example, people of color who have experienced the feeling that they don't matter in a broader social context will have an additional layer of issues to sort through. Moreover, racial, gender, religious, or other differences between therapist and client will affect how the client responds to the therapist's offer.

Client response specificity puts substantial demands on the therapist. It takes away the security of having a rule-based approach to therapy, and it requires the therapist to remain close to the unique and personal experience of each client. However, within this highly idiographic approach, guidelines are available to help the therapist respond more accurately to the specific needs of differing clients. The therapist can learn to assess the specific relational experiences that each individual client needs by identifying the maladaptive relational patterns that recur in (1) the current client-therapist interaction; (2) the current interpersonal relationships with significant others; and (3) the client's family of origin that engendered these generic conflicts. Therapists also learn how to pass tests and provide clients with the specific relational experiences they need by assessing how clients react to different interventions.

Assessing client reactions   How do therapists know if they have passed or failed the client's test? Are there guidelines to assess whether the therapists' response has been helpful or reenacted some aspect of the client's conflict? If clinical trainees rely only on their supervisor's advice, they will remain dependent and insecure. Further, they will have a false sense of security, and risk becoming rigid and dogmatic, if they foreclose on a particular theoretical approach to avoid the anxieties and ambiguities inherent in finding their own identity as a therapist. What better alternatives exist? Although supervisors and theories certainly are of help, beginning therapists develop more autonomy by learning how to assess client responses in order to determine the effectiveness of their interventions.

Clients inform the observant therapist of the responses they need in treatment. They do not usually verbalize this, and may not be conscious of it, but they predictably guide the therapist by their behavioral responses to the therapist's interventions. How does this happen?

When the therapist responds effectively to a client's test and disconfirms pathogenic beliefs about self and faulty expectations of others, the client often feels safer and begins to behave in new ways. For example, in the minutes after the therapist passes a test, clients may stop talking about other people and begin talking more directly about their own thoughts, feelings, and actions. Perhaps the client will relate a vignette that is clearer or more meaningful than the material they have produced before. The therapeutic process may acquire more immediacy and become less intellectualized or repetitive. Clients will often initiate a stronger, more direct interpersonal contact with the therapist, perhaps by expressing warm feelings toward the therapist or bringing up a problem they are having with the therapist. Further, the client's affect will often emerge or intensify. The client's feelings about whatever has happened—happy, sad, or angry—become clearer or are shared more openly with the therapist.

Therapists also know that they have responded effectively and passed the test when clients bring up new material and enter new conflict areas. In such client-initiated moments, clients may also make useful connections between the current behavior and past relationships without any guidance from the therapist. The insights that they generate in these circumstances are enlivening and hold enduring meaning. As a rule of thumb, *it is not productive to make historical interpretations or to lead clients back to familial connections. However, significant behavioral changes result when clients spontaneously make these links on their own.*

The therapist has passed the client's test when the client responds to the therapist's intervention by acting stronger with the therapist, becoming more committed to the working alliance, or risking trying out new behavior with others. For example, when Carl turned the videotape on for the full session, Lucy complained and even threatened to terminate. But by maintaining these appropriate professional boundaries, Carl demonstrated that he was different from Lucy's father and was therefore safer. Carl could tolerate Lucy's disapproval and do what she actually needed rather than what she demanded. As a result, Lucy began to work more productively and introduced relevant new material.

If therapists identify what they have done to pass (or fail) the client's tests, they will be able to successfully formulate treatment plans. These conceptualizations can then be used to shape subsequent interventions and better provide clients with the responses they need in order to change. For example, at other anxiety-arousing points in treatment, Lucy will again push the limits with Carl. If Carl has learned from their previous interaction, however, he will be better prepared to provide Lucy with the clear limits she needs.

In their behavioral responses to therapeutic interventions, clients also tell therapists when they have failed. On having failed a test, the therapist will often observe that clients become weaker—more anxious, dependent, confused, or distant. These reactions are often evident almost immediately (within the next minute or two) and may continue into subsequent sessions. Therapists should formulate working hypotheses about how their recent interaction may have reenacted the clients' conflict when (1) clients begin talking about others rather than themselves; (2) the therapeutic process becomes repetitive or intellectualized; or (3) clients become compliant, lose their initiative, or cannot find meaningful material to talk about.

We have just seen that when therapists pass tests (for example, take pleasure in the client's success rather than feeling envious or threatened as others have been), clients usually acknowledge the therapist's achievement by acting stronger and progressing in treatment. In addition, when a therapist repeatedly fails a specific test (for example, by brushing aside the potential validity of a client's criticism of the therapist, just as the client's parent always needed to be right), the client may inform the therapist that something is awry by recalling other relationships in which the same conflict was enacted. That is, the client may begin to tell the therapist vignettes about other people the client has known, or perhaps characters in books or movies, who are enacting the same maladaptive relational pattern. *Therapists always want to consider the possibility that, whenever the client is talking about another relationship, the client may be using this as a metaphor or encoded reference to describe what is going on between the therapist and the client as well* (Kahn, 1997).

Therapists are listening intently for the relational themes that characterize the vignettes clients choose to relay. If clients repetitively share stories in which, for example, trust is betrayed, control battles are being enacted, and so forth, they are often expressing how these same patterns are currently being enacted in the therapist-client relationship. When this happens, the therapist can use a process comment to make this relational theme overt and bring it back into the immediacy of the therapeutic relationship—where something can be done to address and resolve it.

Therapist:
Both of these people you have been telling me about failed to hear something important that others were trying to tell them. I'm wondering if something like that could be going on between you and me right now. Maybe you told me something earlier that was important to you that I didn't hear very well. Do any possibilities come to mind?

To sum up, just as clients use eliciting behaviors to avoid and defend against their conflicts, they also use testing behavior to try to resolve conflicts in the therapeutic relationship. By tracking the therapeutic process in the ways suggested here, therapists can assess whether they are providing clients with the corrective responses they need. This approach refines therapists' conceptualizations of clients and gives therapists the flexibility they need to respond to the varying issues that clients present.

Transference Reactions

The third interpersonal factor for conceptualizing clients is *transference reactions*. Transference is one of the most important but misunderstood concepts in therapy. Problematically, the field has used the same term to convey different meanings, and, over the years, two competing definitions have evolved. Traditionally, transference has referred to the thoughts, feelings, and perceptions that the client has toward the therapist. This psychoanalytic model of transference is based on the notion of the therapist as a blank screen. Since the therapist is neutral, whatever emotional reactions the client has toward the therapist are *distortions* that have been unrealistically transferred to the therapist from historical relationships. Especially during vulnerable or affect-laden moments, most clients do indeed have predisposi-tions to misperceive the therapist and others along old familiar lines. These persistent distortions of the other (as wonderful, withholding, contemptuous, and so forth) are an enduring characteristic of the clients' inner life. They re-peatedly disrupt their close relationships and are almost certain to be evoked toward the therapist at times. When these types of traditional transference reactions occur, they systematically distort clients' perceptions of, and emo-tional responses to, the therapist.

In contrast, the contemporary definition of transference is broader and more useful. It refers to *all* of the feelings, perceptions, and reactions that the client has toward the therapist—both realistic and distorted. These reactions, which may be positive or negative, range on a continuum from fully accurate and reality-based perceptions of the therapist to highly distorted responses that are based on past relationships. Thus, they do not necessarily represent pathology. Although transference distortions in the traditional sense certainly occur, the contempo-rary definition is emphasizing that many of the client's reactions toward the ther-apist are reality-based and, most importantly, *almost all hold at least a kernel of truth.* For example, the client's attraction to the counselor may be a genuine re-sponse to the very real caring the client feels in this relationship. Similarly, the client's anger or competitiveness may be based on the reality that the therapist has been late for sessions, emotionally unavailable, unable to grasp what is most important to the client, or invested in the client making certain decisions regard-ing divorce, abortion, and so forth. Therapists make a significant error when they too readily attribute the client's positive or negative feelings toward them to tra-ditional transference distortions. Before doing so, they should seriously examine the potential or partial accuracy of those reactions and consider how the client may have understandably misperceived the therapist's behavior.

As we will see, therapists can use transference reactions both to conceptual-ize their clients better and to help them resolve their problems within the thera-peutic relationship. More specifically, contemporary short-term dynamic thera-pies have encouraged an important change in how therapists work with transference reactions. In decades past, transference reactions were used as a springboard to discuss historical relationships and the genesis of conflicts with parents and early caregivers. Now they are used instead to understand and change the current interaction between the therapist and the client (Mills, Bauer, & Miars, 1989). Although therapists will often recognize important con-nections between current problems and historical relationships, few clients will find these links meaningful. It will be far more productive to address and re-solve these distortions or conflicts first in the real-life relationship with the ther-apist. In this more contemporary approach, the therapist is working with trans-ference in the immediacy of the current relationship, rather than through the more distant or intellectualized approach of trying to interpret or lead the client back to historical relationships. The goal is to: (1) help the client recognize re-peated misperceptions of the therapist; (2) clarify how this repetitive pattern disrupts the therapeutic relationship and leads to problems with others; and (3) work together to replace this repetitive pattern with a more flexible and reality-based relationship.

In most cases, clients' reactions toward the therapist accurately capture *aspects* of the interaction with the therapist. It is essential for the therapist to acknowledge the potential validity of the client's perceptions. Therapists want to enter into a genuine dialogue with the client to clarify what each was thinking, feeling, and intending during any misunderstanding or rupture of the working alliance. If therapists are willing to explore *nondefensively* what they may have done that activated the client's relational templates, the client sees that the therapist doesn't always have to be right but is willing to risk a genuine relationship. Clients feel empowered in this egalitarian relationship, and often are able to be more authentic with the therapist and others.

In the interpersonal process approach, the therapist affirms the validity of many client reactions and tries to resolve actual distortions in the current relationship, rather than focusing back on the historical genesis of conflicts. Ironically, after the therapist and client have addressed and resolved a significant misperception or misunderstanding between them, clients will often go on to make meaningful connections between formative and current relationships on their own. Before going further with this clinical focus, let's become better acquainted with the everyday expressions of transference.

Transference in everyday life   Transference reactions are commonplace. We all systematically misperceive others as a result of overgeneralizing previous learning to the present situation. For example, imagine a male university professor who stands in front of a large lecture hall about to convene the first class meeting of Psychology 100. Before he has even begun to lecture, many students will be transferring expectations from significant male authority figures in their own lives onto the professor. Thus, in the first row of the lecture hall, 18-year-old Mary looks up to the professor admiringly. Just as she still somewhat idealizes her father, Mary respects the middle-aged professor, even though she actually knows little about him.

In contrast, Joe looks down on the professor disparagingly from high up in the last row of seats. Joe anticipates that the professor does not know as much as he thinks he does and that he will not be very well-prepared for class. Joe expects that, just like his father who always had to be right and on top, this professor will probably try to get away with a lot in class. Joe has already decided that this know-it-all professor is not going to get away with much if Joe can help it.

Let's hope that the professor is aware that people tend to transfer expectations onto authority figures who hold power and onto those who have high public visibility (movie stars, sports heroes, and political leaders). If the professor is not sensitized to transference reactions, his self-confidence as a teacher will take a roller-coaster ride. One day, Mary will follow him out of class and tell him what an interesting lecture he gave; the next day, Joe will drop his books loudly when the professor starts to speak and demand that he defend half of what he says.

Transference reactions such as Mary's and Joe's are everyday occurrences. The more conflicted or disturbed one is, however, the more one distorts the current reality by inaccurately displacing reactions from the past onto the present. For example, Joe and Mary have strong initial reactions to the professor that are based on little real-life experience with him. If Joe and Mary function fairly well psychologically, their internal working models will be flexible and varied, and they will become more realistic in their assessment of the professor as they have more exposure to his teaching. In contrast, the more thoroughly Mary idealizes her father and the more pervasively Joe disparages his father, the more likely it is that they will not be able to perceive the realistic strengths and limitations of the professor. Instead, their relational templates will be more fixed and self-fulfilling, and they will maintain their initial expectations despite the fact that they are inaccurate or overdetermined.

Transference reactions are also more likely in emotion-laden situations. For example, transference reactions are especially pronounced when two people are falling in love. Many married couples recall their courtship as the happiest time in their lives. During this time, each member systematically misperceives the partner as someone who can fulfill unmet developmental needs and make one feel love-worthy or capable when one has not felt this way before. This romantic period is usually short-lived, however. The transference projections soon break down—they both realize that the other person cannot fulfill their unmet needs and their old templates are reinstated. If the unmet needs and accompanying transference distortions are strong, the result is often dashed dreams or angry feelings of betrayal. In contrast, if the transference distortions of the partner have not been extreme, they can readily be shed, and a more realistic and enduring relationship can develop. Relinquishing these transference distortions of the new spouse is one of the major psychological tasks of early marriage.

In treatment, the therapist can systematically utilize the client's transference reactions to better understand the client's conflicts. One of the most effective ways to discern the client's internal working models—and their interpersonal expression in maladaptive relational patterns—is to note how the client perceives and reacts to the therapist. Almost without exception, aspects of the client's maladaptive relational patterns that cause problems with others will be reenacted in the therapeutic relationship. If the therapist tracks these transference reactions, they provide an invaluable tool for conceptualizing these conflicts, and for changing the problematic interpersonal process that is being reenacted.

Using transference reactions to conceptualize clients   Some therapists are uninterested in the client's transference reactions because they do not see these reactions occurring in therapy. Therapists usually miss the client's transference reactions, and the invaluable information they provide, unless they actively attempt to draw them out. How can the therapist highlight the client's transference reactions? Let's look at two ways of making the client's transference distortions overt in such a manner that the client can recognize them nondefensively and join with the therapist in a collaborative effort to understand them.

First, the therapist should find tactful but direct ways to inquire about the client's feelings and reactions toward the therapist. As we saw when discussing resistance, it is important for both therapist and client to learn about the client's reactions to being a client and seeking help. Now further along in treatment, it is just as important to learn about the client's reactions toward the therapist—especially in the event of strong emotions or conflicts between the therapist and the client. If well-timed, the sample questions that follow can provide a wealth of new information:

> As you were driving to our session today, how did you feel about coming to see me?
> When you find yourself thinking about therapy; what kind of thoughts do you have about me and our work together?
> What are you feeling toward me right now as we talk about this sensitive issue?

Second, the therapist should also explore what the client *projects* onto the therapist—what the client believes the therapist is thinking, expecting, or feeling toward the client at that moment:

> What do you think I am feeling toward you as you tell me that?
> How do you think I am going to respond if you do that?
> What do you think I expect you to do in that situation?

Oftentimes, it is uncomfortable for the beginning therapist to break ingrained familial and cultural norms and invite the client's feelings and perceptions so forthrightly. Although it may not feel this way in the beginning, therapists can be direct while still remaining sensitive and respectful. Therapists can be more forthright than they may have been in the past without crossing the line and becoming demanding or intrusive. If therapists take the risk to pose such questions, clients' transference reactions will be made overt. The issues that will be revealed are rich and often surprise both the therapist and the client:

> I think you are feeling disappointed in me.
> I think you'll act nice, but you really wish I would stop coming.
> I think that you believe I should get divorced.

As these misperceptions are made overt, therapists can correct them and clarify their actual responses to the client, which in itself often provides the client with a reparative experience. When transference reactions are revealed in this way, therapists have the opportunity to respond to the client with an immediacy and authenticity that is not often found in other relationships.

Most beginning therapists find it difficult to believe that their client has significant feelings toward them that are not based on the reality of how they have responded to the client. Seasoned therapists have become more familiar with clients' distortions and often are prepared to respond to them nondefensively. In contrast, beginning counselors usually find it hard to accept that their clients have deeply held but unspoken beliefs that the therapist really doesn't like them, feels overwhelmed or bored by them, believes they are childish or foolish, and so forth. Even though the therapist may never have responded to the client in any of these problematic ways, the discerning therapist still works under the assumption that such client reactions are likely to be present but covert. If these relational templates are not made overt and the corresponding distortions rectified, the potential for change is limited. Perhaps this is the critical point: The beginning therapist can accept intellectually that the client may hold such faulty expectations of others, but it is too threatening (evokes too much guilt and fear of having made a mistake) to accept that the client also has such reactions toward the therapist.

Exploring clients' subjective reactions toward the therapist reveals three important types of information. First, just as adopting an internal focus will reveal issues that would not have become accessible otherwise (Chapter 4), exploring clients' reactions toward the therapist will also reveal new issues—central to the clients' presenting problems—that need to be addressed in treatment. Detailed examples of this important intervention will be given in Chapters 9 and 10. Second, exploring transference will also show how aspects of the same problems that clients are having with others—and that they originally experienced in formative family relationships—are being reenacted with the therapist. Unless the clients' reactions toward the therapist are directly solicited, these reenactments are far less likely to be identified and resolved. Thus, transference reactions reveal clients' generic conflicts and, as we will explore further in the next chapter, also provide an opportunity to resolve them in the therapeutic relationship.

Third, transference reactions provide evidence that the clients' feelings are trustworthy. Because clients' emotional reactions may not fit the situation in which they are aroused, they may not make sense at first to the clients or others. As a result, many clients experience their emotional responses toward spouses, children, employers, and others as irrational. Consequently, they do not trust the validity of their own experience and, as a result, their self-efficacy is undermined. This also occurs in the therapeutic relationship when clients have emotional reactions toward the therapist that seem inappropriate or exaggerated. However, clients' emotional reactions are not irrational. Though they may not fully fit the present circumstances, they do make sense when they can be placed within their original context. Real-life events have legitimately caused clients to feel the way they do. If the therapist helps clients trace the feelings back to their source (by asking, for example, "When is the first time you can remember feeling this way? Who were you with and what was happening?"), *they will prove to be an appropriate reaction to someone else in another time and place.* Once the therapist and client have been able to clarify the transference distortion in their current interaction, this developmental link can often be made readily. In contrast, the developmental context will not come alive or be meaningful to the client if it is not first resolved in the current interaction with the therapist. In this way, transference reactions ultimately reveal that the client's feelings make sense.

To sum up, clients do transfer emotional reactions from past formative relationships onto the current relationship with the therapist. However, therapists begin by clarifying and resolving these misperceptions in the current, here-and-now relationship with the therapist:

Therapist:
No, I'm not thinking you're selfish and demanding if you ask him—or me, for that matter—for what you want.

OR

Therapist:
> No, my tears are not because I'm burdened by your needs. It's because I'm touched by how much you've been hurt.

This clarification gives clients a real-life experience of change; they learn that their inaccurate expectations or problematic working models do not fit in this relationship. As they experience another way to be in a relationship, clients are often able to make meaningful connections to past experiences. By themselves, without historical interpretations from the therapist, clients often begin to understand the source of their emotional reactions and find that their inner experience does make sense and has validity. This powerfully affirming sequence leaves clients feeling in control of themselves and of their emotional reactions—perhaps for the first time in their lives. When clients find that their feelings do make sense, even though in another time and place, they are empowered. By resolving the transference distortion in the real-life relationship with the therapist, and then making meaningful connections between the past and the present, clients are able to leave the past behind and find new ways of responding.

Optimum Interpersonal Balance

In understanding clients' interpersonal functioning in the therapeutic relationship, we need to apply the separateness-relatedness dialectic (Chapter 6). In order to establish a relationship that is capable of producing change, therapists must have an effective degree of involvement with their clients. If the therapist is overly invested, this will be a reenactment for many clients and they will fear being controlled and will not be able to take the risk of letting the therapist influence them. Similarly, if the therapist is too distant or removed, a corrective emotional experience cannot occur because the relationship is too insignificant to effect change. Throughout treatment, therapists must monitor the balance of separateness versus relatedness in the therapeutic relationship. At times, all therapists will have difficulty maintaining an optimum interpersonal balance with some clients. The therapist must be both a genuine participant in the relationship and an objective observer of it. Sullivan (1968) captures this double role in his enduring term *participant-observer,* while more contemporary discussions speak of *empathy* and therapist-client *boundaries* (Jordan et al., 1991). Combining these emotional and cognitive components is no easy task for most beginning therapists, however. With another year or two of experience, it will become easier to maintain this self-reflective capacity.

Enmeshment

At times, all therapists will become overidentified—enmeshed—with certain clients who tap into the therapists' own issues. Moreover, certain therapists, because of their own personal dynamics, will consistently become overinvolved with clients. How can therapists recognize when this traditional countertransference is occurring? Whenever therapists find that they are angry or critical of a client for not changing, they are overly invested in the client. Usually, therapists are also enmeshed when they have dreams about a client, often think about a client outside of the therapy session, feel depressed when a client is not changing, see the client as being just like themselves, or become envious or elated over positive changes in the client's life. This overinvolvement is different from genuine care and concern for the client, which is appropriate and necessary. When enmeshment occurs, therapists usually need the client to change in order to meet their own needs—for example, to shore up their own feelings of adequacy or to manage conflicts of their own that resemble the client's.

Substantial problems occur when therapists become overinvolved with a client. When they do so, they lose sight of the fact that clients have their own subjective worldview, shaped by many factors that actually differ from those of the therapist. They tend to become controlling of the client and to project their own problems and solutions onto the client. This is likely to recapitulate the client's generic conflict in the therapeutic relationship. For example, many clients have grown up in families with overly controlling or intrusive parents. When the therapist's own experiential range is too limited to encourage clients' own autonomy, the therapist is responding in the same problematic way as the parents. This reenactment will prevent clients from being able to resolve their conflicts around intimacy and control. Consequently, the most important reason for paying attention to traditional countertransference is to avoid treating clients in the same ways that originally caused their problems.

Further, when therapists become too close or overidentified, they lose their ability to think objectively about the client and their heightened emotional reactivity to the client will cloud their understanding. Instead of accurate empathy, therapists will offer a global, undifferentiated sympathy. Most importantly, therapists will fail to discriminate the real differences between themselves and the client that do exist, and will begin to see the client as just like themselves. Although this overidentification may feel like closeness or caring and may be reassuring to clients at first, little enduring change occurs in an enmeshed relationship.

Disengagement

All therapists will hold themselves too far away from certain clients at times— usually when the client arouses the therapists' own generic conflicts. Also, some therapists will consistently be too removed from their clients because of their own enduring conflicts and coping styles. This distance renders treatment ineffective. When therapists are not emotionally available to the client, the working alliance is lost; therapy loses its intensity and often becomes intellectualized. Therapists lose their intuition and creativity and, when this occurs, the client does not have the supportive relationship or holding environment necessary to explore affect-laden conflicts. Counseling is more than an intellectual discourse, and little enduring change will occur in an emotionally disengaged relationship.

Without the impetus provided by genuine personal involvement, a corrective emotional experience will not occur.

A therapeutic relationship in which the therapist is too distant or uninvolved will also recapitulate many clients' generic conflicts. For example, for clients whose aloof, self-centered, or authoritarian parents could not meet their childhood needs for emotional support, a distant therapist might be frustrating but ultimately safe. Their anxieties over their insecure attachments—and their conflicts associated with being close to others—will not be activated in this therapeutic relationship. Unfortunately, just as the clients' conflicts are not aroused in a disengaged relationship, the opportunity to resolve these conflicts is lost as well. Thus, when therapists defend themselves against their own anxieties that are aroused by an authentic relationship and genuine emotional involvement with clients, many clients will not be able to change. *Novice therapists who are consistently unwilling to risk genuine involvement with clients are much less likely to become effective therapists than are those who are willing to risk a real relationship but tend to become overinvolved at times.*

Effective Involvement

In order to be most effective, the therapist is trying to find an optimum middle ground of involvement. For two reasons, maintaining this middle ground of involvement throughout the course of treatment is one of the most challenging aspects of therapy. First, therapists are continually confronted with anxiety-arousing issues in their own lives by the affect-laden material that clients present. All therapists can become more aware of how they tend to respond when clients arouse their own anxieties. For example, some therapists will withdraw from the client; some will distance themselves by intellectualizing—for example, by interpreting the client's affect before the client fully experiences it; others become controlling and overly prescriptive, telling clients what they should do in a particular situation. Therapists can identify their own countertransference propensities and learn to hold them in check.

A useful exercise for beginning therapists is to write down or role-play their initial reaction propensities in different anxiety-arousing situations—for instance, situations in which a demeaning client is critical of the therapist's abilities or a competitive client questions whether the therapist is really capable enough to be of help. In small discussion groups of three or four, trainees can provide each other with feedback and share their observations of the other therapists' responses while role-playing these challenging situations.

Second, the balance between overinvolvement and underinvolvement is difficult because clients often try to move the therapist along this continuum of involvement as part of their interpersonal coping strategy. When the therapist has been responding effectively, some clients may become threatened by the anxiety-arousing material that begins to emerge for them. To defend against the difficult emotions that are being aroused by the therapist's effectiveness, clients may try to create either distance or enmeshment in the therapeutic relationship. For example, some clients may try to bore the therapist by talking about issues they are not really concerned about; others may try to elicit too much involvement from the therapist by exaggerating their distress and escalating their demands for help. Thus, forces from within the therapist and from the client's eliciting behaviors may press the therapist away from an effective middle ground of involvement.

If they can maintain this middle ground of involvement, therapists can provide a more even therapy with no stormy ups and downs or lengthy impasses. It is also safer for the client to be vulnerable and trust when the therapist is dependably available without swinging to either side and becoming overreactive or unresponsive. When therapists maintain an optimum interpersonal balance, the client does not have to be concerned about disappointing an overinvolved therapist or eliciting the emotional accessibility of a technician. The attachment researchers have best-appreciated the importance of the caregiver's *consistent* response in developing security and trust.

The essential balance is for therapists to maintain their own personal boundaries while still being responsive to the client. They are striving to fully enter the client's subjective worldview—to be fully present and connected—but still maintain appropriate self-boundaries and self/other differentiation. Regardless of the therapist's theoretical orientation, treatment is usually successful when the therapist can maintain this optimum interpersonal involvement. Conversely, treatment usually reaches an impasse—and ends before the client's work is complete—when the therapist and client become enmeshed or do not emotionally connect sufficiently to establish a strong working alliance.

To help therapists recognize the process they are enacting with their clients and assess their balance of involvement with clients, therapists can consider the following questions after every session:

1. What are my feelings and reactions toward this client?
2. How do my reactions parallel those of significant others in the client's life?
3. What is the client doing to arouse such reactions in me, and how do these behaviors tie into my conceptualization of the client's generic conflict and interpersonal coping strategies?
4. Do my feelings typify my reactions to significant others in my life and/or to other clients, or are they more confined to this particular client?

By writing answers to these questions after each session, therapists ensure that they are attending to the process dimension. *Asking themselves such questions will also help therapists to distinguish their own countertransference reactions toward the client—therapist-induced countertransference reactions—from the client's eliciting behaviors and interpersonal coping style—client-induced countertransference reactions.* Building upon these four questions, beginning therapists are encouraged to utilize the guidelines for writing process notes in Appendix A after each session.

As we have already noted, all therapists will sometimes become overinvolved with a client. Indeed many theorists have described the therapeutic process as therapists' ability to become immersed in clients' modes of relating and then to work their way out (Gill & Muslin, 1976; Levenson, 1982; Mitchell,

1988). However, beginning therapists are especially prone to become overidentified with their clients and to perceive their clients' problems as fundamentally the same as their own. This is the most common way in which clients' maladaptive relational patterns come to be reenacted in the therapeutic process. Beginning therapists often feel discouraged when their supervisors accurately point out that, here again, they are metaphorically reenacting aspects of the client's conflict in their interpersonal process. Beginning therapists must be patient with themselves and accept that it is difficult to maintain the effective middle ground of involvement that safeguards against recapitulating the client's conflict. By gaining more clinical experience, examining their own experiential range, and tracking the separateness-relatedness dialectic in the therapeutic relationship, beginning therapists will become increasingly capable of maintaining an optimum interpersonal balance.

When therapists realize that they have become overinvolved or underinvolved with a client and find that they cannot realign the relationship on their own, it is best to consult with a colleague or supervisor. In fact, the most productive use of supervision is to help therapists regain control of their own emotional reactions so that they can reestablish an effective degree of involvement. Maintaining this optimum interpersonal balance is the best way to keep from recapitulating the client's conflict in the therapeutic relationship.

The Ambivalent Nature of Conflict

The Two Sides of Clients' Conflicts

Therapists can better conceptualize their clients' problems when they recognize the ambivalent nature of conflict. There are two sides to the client's generic conflict, and therapists must respond to both sides before problems can be resolved. Too often, therapists fail to recognize both sides of the conflict and only respond to one side of the issue. Therapists are more helpful to their clients when they can identify the two opposing feelings or competing beliefs that comprise the double-binding conflict.

Colloquial expressions convey the two-sided nature of conflict. We often say that someone is "stuck between a rock and a hard place," that someone has grabbed "both horns of the dilemma," or that a difficult relationship is "like a double-edged sword." These expressions capture the push-pull nature of conflict—that you are "damned if you do and damned if you don't." For example, therapists often fail to appreciate that many clients have concerns both about being hurt by the therapist and others, and about hurting others. The field has well-developed concepts regarding clients' concerns about being hurt, misunderstood, and so forth. However, far less attention has been paid to clients' concerns about hurting others by succeeding, finding their own voice, advocating on their own behalf, and so forth. When clients improve in therapy or successfully pursue their own interests in life (for example, by graduating from college, receiving a promotion, or having a satisfying love relationship), many will retreat from these success experiences by sabotaging themselves, be unable to enjoy their success, or be unable to sustain their progress. Often, they will become depressed as a result of separation guilt and/or survivor guilt, or become anxious as success threatens their attachment ties or challenges their familial role.

To highlight the two-sided or double-binding nature of many clients' enduring conflicts, let's examine these elements in two previous case studies. As we will see, once the therapist is able to recognize the two-sided structure of the conflict and reflect the ambivalent feelings that result from the client's bind, the client's seemingly irrational or self-defeating behavior makes sense and becomes more resolvable.

Recall Anna, the young adult client who was discussing her dream with her therapist at the end of Chapter 4. Anna wanted to end her chronic depression, become more involved with friends, and find a career path that she could pursue. Although she had expressed these wishes for many years, she could not follow through and realize them. She rarely initiated activities with peers and often declined the invitations she received from others. She repeatedly enrolled in college but would eventually "lose interest" and drop out after one or two semesters. Without recognizing the two-sided conflict that plagued Anna, her behavior seems irrational.

On the one hand, Anna wanted to grow up and have her own independent life with her own friends and career. At the same time, however, she felt a responsibility to remain close to her mother and provide continuing support. Anna's repetitive relational scenario is that whenever she starts to pursue her own goals and succeed with friends, in school, or with life in general, her mother will withdraw emotionally, look hurt, and induce guilt. Unable to break this tie that binds, Anna continues this maladaptive relational pattern by repeatedly failing and having to become dependent again and return home. Only upon her return home does her mother stop acting like a martyr and again become available to her. Having to sacrifice their own independence to maintain their attachment ties, clients like Anna become depressed over their inability to have their own lives and experience internalized anger over such binding guilt.

After a few months, however, the same cycle will be reenacted again. Anna will start to act on her own need for autonomy, seek peer contacts, and reenroll in school. As before, however, her mother will withdraw and act hurt, and Anna will not be able to individuate without her mother's support. In a few months, she will again return home feeling hopeless and a failure. Completing this scenario, her mother will become comforting and supportive of Anna in her defeat. This type of maladaptive relational pattern can be reenacted for decades, unless both sides of the client's conflict are made overt and worked through in therapy.

Without understanding these family dynamics (see Haley, 1980), the therapist and others might conclude that Anna is lazy, unmotivated, or receiving too many secondary gains from being home and depressed. Once we understand the maladaptive relational pattern that immobilizes her, however, we see that such pejorative labels are unwarranted. Clients' contradictory or self-defeating behavior does make sense, but not until we understand the two-sided, push-pull, or double-binding nature of their conflict.

Now recall Jean, the client in Chapter 5 who began to change only when she could risk experiencing her painful feelings with the therapist. Until we recognize her underlying conflict, Jean's behavior also seems irrational and self-defeating. Jean's primary symptom was her continuous involvement with selfish and exploitative men. Jean recognized that she felt bad about herself in these demeaning relationships, but she was unable to disengage from them. Following her relational templates, she remained in the relationship no matter how bad things got for her. Paradoxically, when the man decided to leave her, she would feel profoundly insecure, rather than relieved, and would quickly become reinvolved in a similar relationship. This maladaptive relational pattern had continued throughout her 20s and 30s, and finally led her to seek help for her dysthymia and budding agoraphobia.

Jean entered therapy describing herself as "stupid." Her friends described her as "hopeless" for continually reinvolving herself in the same type of bad relationships. As with Anna, however, Jean's seemingly irrational and self-defeating behavior makes sense once the two-sided nature of her conflict is recognized.

For years, Jean had said she wanted to be able to meet her own needs, set limits and say "no" sometimes, and leave relationships that were not supportive. But, year after year, she continued to meet a man's needs at the expense of her own well-being. She could not say "no" and set self-preservative limits, even though she had learned the skills to do so in an assertion-training class. Why? A countervailing force covertly opposed all of Jean's attempts to change. As soon as Jean started to act on her own needs, set limits, or try to leave a relationship, she felt cold, small, and alone. In other words, she was gripped by painful separation anxieties whenever she made any movement toward her own individuation. Her only means of abating these separation anxieties and feeling connected to others again was to retreat to the familiar dependent/victim role. Thus, her symptoms preserved attachment ties, and improving in therapy evoked the distress of her painfully insecure attachments.

This paralyzing conflict developed in Jean's family of origin. Growing up, Jean was a parentified child. She was expected to respond only to her parents' needs, and she was harshly rebuffed or overtly ridiculed for expressing her own "selfishness." Throughout her childhood, she received so much criticism from both of her parents, and so little emotional support, that she needed whatever thin threads of approval she could earn by submerging herself to her parents' demands. Afraid of suffering further rejection, she automatically complied with whatever they asked of her. By meeting the caregivers' needs, Jean and other parentified children take responsibility for the parent-child relationship. This coping strategy, thrust upon so many female clients, is aimed at fashioning a more secure attachment. Such offspring, laboring under the myth of being able to control the attachment tie by meeting their caregiver's needs, may struggle mightily with control conflicts such as bulimia.

Further, to preserve some minimal sense of relatedness, Jean identified with her parents' critical rejection of and excessive control over her. That is, she internalized her parents' contempt for her and began loathing herself for being bad and selfish. (See component 3 in Figure 7.1.) As an adult, she recapitulated the feelings and experiences of her childhood by selecting self-centered and demanding men who confirmed her feelings of worthlessness.

Understanding this developmental context, we can appreciate how Jean's seemingly irrational behavior does in fact make sense. Her only sense of herself as a worthwhile person—and her only means of being responded to in a relationship—is to meet another person's needs. When she eventually becomes depleted by this role and begins to respond to her own needs (set limits, ask for what she wants, leave a hurtful relationship), anxiety and guilt result. Lacking any better working models, she tries to cope with these difficult feelings by retreating to frustrating relationships and reenacting the same maladaptive relational patterns. Once again, she complies with her boyfriend's demands and suffocates her own needs, feelings, and authentic voice. However, the same cycle will soon start again, as her frustration over having her own life denied begins to reemerge. She will start to respond to herself and her own needs, only to be immobilized again by the same anxiety and guilt.

A cyclic dynamic such as this is challenging to change. As Jean illustrates, however, change is possible if two things occur. First, Jean experienced a relationship with the therapist in which her own needs were listened to and valued for the first time. *This was difficult to provide, however, because Jean was adept at rejecting the concern that she simultaneously sought from the therapist. In this way, she tended to elicit the same feelings of helplessness in the therapist that her parents had originally engendered in her.* Second, the therapist was also able to help Jean understand the two-sided structure of the conflict that had bewildered her throughout her life. Understanding this maddening conflict helped Jean change by allowing her to feel more compassionate toward herself. In turn, feeling more empathic toward her own predicament, rather than contemptuous of herself, Jean was empowered to feel more accepting of the choices and compromises that she had been compelled to make throughout her life. This self-acceptance was a precursor of change and helped her to disengage from her own self-hatred and value herself in a way that she had not been able to do before. Indeed, these more accepting and understanding responses from the therapist led to important internal changes for Jean and, soon thereafter, she became involved with a supportive and responsive man for the first time.

Ambivalent Feelings

As we have seen, the client's conflict includes two opposing forces that the client has not been able to reconcile. Oftentimes, therapists cannot understand the client's problem until they recognize its two-sided structure. In addition to conceptualizing the client's problem as a two-sided conflict, therapists need to respond affirmingly to the opposing feelings that accompany both sides of the client's conflict. Therapists will not be effective when they only respond to one half of the client's ambivalent feelings, as the following case vignette illustrates.

Marie, a depressed, 25-year-old graduate student, entered time-limited (12-session) treatment at the student counseling center. During the first few sessions, Marie explored how angry she was at her mother. Her mother expected

Marie to be perfect, insisted that Marie never had any problems, and always needed Marie to be happy. In therapy, Marie began to realize how much she resented her mother for denying so many of her true feelings and for having so many expectations of how she should be. Marie's therapist resonated with her protest and actively supported her anger.

At first, it was liberating for Marie to have her long-suppressed anger affirmed. She was both excited and relieved to be realizing that she no longer needed to fulfill her mother's unrealistic expectations. This good feeling was short-lived, however. In the next few sessions, Marie's new-found freedom gave way to a growing despondency. The therapist thought that Marie must be feeling guilty about defying her mother's expectations and depressed because she was defensively turning the anger she felt toward her mother back upon herself.

While repeatedly offering this interpretation, the therapist continued to draw out and encourage Marie's anger. Moving in a different direction than the therapist, however, Marie began to share fond memories about her mother and recalled with longing special times they had spent together making cookies after school. The therapist was not very responsive to Marie's positive feelings toward her mother, however. The therapist thought that Marie's fond recollections were part of denying her anger toward her mother and avoiding the conflicts in their relationship, and was concerned that Marie would only continue to meet her mother's problematic expectations.

In contrast to Peter's therapist in Chapter 7, Marie's therapist only recognized one side of her ambivalence and failed to respond to the feelings on both sides of her conflict. Yes, Marie was angry at her mother, and these feelings needed to be explored and affirmed. At the same time, Marie knew that her mother had offered her many fine things as a parent and she did not want to forgo the good things they had shared. Marie's conflict was that she wanted to keep the good things she had experienced with her mother without having to internalize the problematic aspects of their relationship as well. Thus, on the one hand, Marie wanted to reject the unrealistic demands to be perfect and happy that her mother placed on her. But, on the other, Marie became depressed when it seemed that the only way to do this was to give up her identification with her mother altogether. This had been the most important relationship in her life, and Marie did not want to risk losing the good parts of this relationship as well.

Realizing that time was short and therapy was stagnating, her therapist sought consultation from a colleague. The colleague quickly recognized both sides of Marie's conflict and encouraged the therapist to allow Marie to simultaneously have contradictory feelings of appreciation and anger toward her mother. As a result of this helpful consultation, the therapist began to invite and affirm the other side of Marie's feelings as well.

Therapist:
You feel two different ways toward your mother at the same time. You're angry at her for expecting you to be perfect, but you also treasure many of the fond times you've had together. Tell me about both your angry and your loving feelings toward her.

As the therapist encouraged Marie to explore both sides of her feelings, her depression improved. During their final sessions, Marie began to clarify what aspects of her mother she wanted to keep and make a part of herself and what aspects of her mother and her mother's expectations she wanted to discard. This process of *differentiation* also allowed Marie to clarify which issues she wanted to confront and try to change in her current relationship with her mother and which issues she preferred to let go of and simply accommodate herself to. Marie progressed in therapy as soon as the therapist allowed her to come to terms with both sides of her ambivalent feelings. Therapy was successfully terminated at the 12th session as Marie better integrated aspects of her own identity as a woman and initiated a rapprochement with her mother.

To summarize, therapists help clients change when they recognize the two-sided structure of their clients' conflicts, and help clients explore and integrate both sides of their ambivalent feelings. Trained to look for pathology, therapists tend to miss positive aspects. However, clients will make far more progress when the therapist can actively encourage clients to keep the good parts of the parental or spousal relationship while they are realistically examining the problematic parts. *The therapist can tell clients explicitly that they are not being disloyal if they talk about problems with a parent or spouse, and that the therapist knows that their complaints do not reflect all of their feelings about this person.*

Just as therapists in individual practice must affirm both sides of the client's conflict, couple therapists must affirm both sides in an interpersonal dispute. Their primary task is to resist the couple's eliciting maneuvers and keep from taking sides by participating in the good guy/bad guy, overadequate/inadequate, healthy/neurotic, or other polarizations that the couple presents. When couple counseling fails, one member of the couple usually stops treatment because she or he accurately perceives that the therapist covertly has taken sides in the marital conflict.

Closing

In this chapter, we examined three dimensions of clients' current interpersonal functioning: how clients' problems with others are brought into the therapeutic relationship through eliciting maneuvers, testing behaviors, and transference reactions; how therapists can establish an optimum degree of interpersonal relatedness and enact a corrective balance of separateness-relatedness; and the two-sided structure of clients' conflicts. This discussion may help therapists better understand clients' self-defeating or contradictory behavior and their seemingly irrational feelings. In turn, this greater understanding may enable therapists to be more compassionate toward the conflicts with which clients struggle. These interpersonal factors, together with the developmental/familial perspective and interpersonal schema presented in the previous two chapters, will help therapists to better understand and intervene with the problems clients present.

With these conceptual guidelines in mind, we now return to intervention strategies. The next chapter examines how therapists can use the interpersonal process they enact with clients to begin resolving clients' problems in the therapeutic relationship. If they are to utilize process-oriented interventions and the therapeutic relationship to effect change, therapists must be able to talk forthrightly with clients about their current interaction. Working with the process dimension or "talking about you and me" is essential to all forms of relationally oriented therapy. Initially, however, it may feel uncomfortable for beginning therapists to metacommunicate in this way. Thus, in Chapter 9, we explore therapists' potential concerns about working with the process dimension, and examine guidelines to help counselors intervene more effectively within the therapeutic relationship.

Suggestions for Further Reading

1. In a managed care milieu, writing effective treatment plans becomes an essential skill for clinical trainees. Illustrating the guidelines for writing case conceptualizations in Appendix B, Chapter 8 of the Student Workbook provides another case example.

2. Readers are encouraged to learn more about clients' testing behavior by reading Chapters 4–6 of *The Psychoanalytic Process* by J. Weiss and H. Sampson (New York: Guilford Press, 1986). Two popular press versions of this important work, which also emphasize the role of unconscious guilt in many clients' symptoms and problems, may be found in *Hidden Guilt* by L. Engel and T. Ferguson (New York: Pocket Books, 1990) and in *How Psychotherapy Works* by J. Weiss (New York: Guilford, 1993).

3. Chapter 2 of Jay Haley's *Leaving Home* (New York: McGraw-Hill, 1980) elucidates the family dynamics and structural family relationships that impede individuation and emancipation for many young adult clients.

4. Aspects of Margaret Mahler's developmental theory—in particular, her model of parent-toddler interaction—can help therapists establish an optimum degree of interpersonal relatedness with their clients. Readers are encouraged to read Chapter 6, "Rapprochement," in *The Psychological Birth of the Human Infant: Symbiosis and Individuation* by M. Mahler, F. Pine, and A. Bergman (New York: Basic Books, 1975). Family systems concepts of cohesion and adaptability can also be used to describe an optimal degree of interpersonal relatedness. Interested readers may examine *Families: What Makes Them Work* by D. H. Olson et al. (Newbury Park, CA: Sage, 1983).

PART FOUR

Resolution and Change

CHAPTER NINE

An Interpersonal Solution

Conceptual Overview

Once the therapist has conceptualized the client's dynamics, how does change occur? Are the client's problems resolved through an emotional catharsis of childhood traumas, from having unmet developmental needs fulfilled, or from delving into illuminating insights about the past? No. Therapists do obtain essential information from understanding familial relationships and developmental experiences, which is useful in creating case formulations and treatment plans. Also, clients need to develop their own narrative that allows them to make sense of the influences and events that shaped their development and who they have become. Unless they do so, it will be harder for them to guide where they are going in the future and to maintain changes after treatment has stopped. A consistent focus on the past will not produce change, however, and often serves to defend clients against the anxiety of addressing current problems. Certainly, information on the interpersonal scenarios that clients experienced in the past does help both the therapist and clients to understand how faulty relational templates have developed, and why current relationships are being shaped in a particular way. However, resolution must be found in the current relationship with the therapist. The primary reason for exploring developmental and familial relationships is to throw light on clients' coping strategies and characteristic ways of responding in current relationships. In other words, understanding formative relationships is a necessary component of change but is, in itself, insufficient to produce enduring change. Enduring change comes about when clients enact a resolution of their conflicts in their current interaction with the therapist, and the therapist then helps clients generalize this experience of change to others.

If the primary focus of therapy and fulcrum of change is the current interaction between the therapist and the client, what must the therapist do to help clients change? Does the therapist need to provide clients with the love they never received? Is the therapist's role to re-parent the client? No. The therapist cannot meet clients' unmet developmental needs, although many clients may wish that were possible. Instead, the therapist has more modest but attainable goals. The therapist can help clients enormously by responding empathically to their unmet developmental needs. This allows these unacceptable or anxiety-arousing parts of the client to be acknowledged and integrated rather than to remain split-off and disavowed. The therapist must also provide clients with a different and better response to their maladaptive relational patterns than they have received in the past and come to expect from others. Enduring change results as the new and more satisfying relationship with the therapist expands clients' internal working models, alters their beliefs about themselves and expectations of others, and increases their interpersonal range. This corrective emotional experience creates interpersonal safety. This newly felt safety frees clients to become more authentically connected to themselves, and fosters greater choice and flexibility in their construction of current relationships.

The corrective relationship with the therapist evokes the conflicted emotions that accompany clients' maladaptive relational patterns. However, it also allows clients—to remember and experience more fully than before—those feelings and experiences that were too threatening or painful for clients to contain on their own. In turn, the holding environment provided by the therapist's understanding allows clients to integrate these conflicted feelings that they have not been able to resolve on their own. In the safety of the holding environment, clients resolve their conflicts by facing honestly and grieving openly for the emotional support they missed or for the hurt they experienced. Clients also develop the capacity to feel compassion for themselves, which was not possible before. As a result, clients feel empowered to take more responsibility for their own choices, to act on behalf of their own wishes, and to relate more authentically to themselves and others. Therefore, the issue is not just reexperiencing what happened in the past. Instead, as clients experience a corrective relationship with the therapist, this interpersonal safety allows clients to come to terms with their pathogenic beliefs and unwanted feelings.

We have already seen in Chapter 8 that the client's problems are not just talked about abstractly in therapy; rather, they are brought into the therapeutic relationship and reenacted in the interpersonal process. In order for the client to change, the therapist and client must not recapitulate the client's maladaptive relational patterns in their interpersonal process. Instead, they must mutually work out a resolution to them in their real-life relationship. As the attachment researchers put it, you do not get an empathic child by teaching or admonishing the child to be empathic; you get an empathic child by being empathic with the child (Karen, 1997).

As we will see in Chapter 10, once this corrective emotional experience becomes an ongoing pattern of the therapeutic interaction, it will be relatively easy for the therapist to help the client *generalize* this emotional relearning to other arenas in the client's life where the same themes and patterns are being enacted. Conversely, if the therapist and client talk about important issues and dynamics, but their interpersonal process does not enact a resolution of the conflicts they are discussing, enduring change will not occur. Thus, the purpose of this chapter is to further clarify the process dimension and illustrate how the therapeutic relationship is the best avenue available to resolve clients' problems.

Chapter Organization

This chapter is divided into two sections. In the first section, we examine how the interpersonal process that therapists enact with clients can be used to resolve clients' problems. First, we will review the course of therapy up to this point in treatment and summarize how clients' conflicts are brought into the treatment setting and reenacted with the therapist. Next, we look at five case vignettes that illustrate more concretely how therapists can use the process dimension to help clients resolve these conflicts that have been activated within the therapeutic relationship.

In the second section, we explore why some beginning therapists may be reluctant to use the process dimension and address how the clients' conflicts may be occurring in the therapeutic relationship. We examine six reasons why therapists-in-training may have difficulty utilizing process comments that focus on the current interaction between the therapist and the client, along with guidelines to help therapists begin working in this way.

Enacting a Resolution of Clients' Conflicts in the Interpersonal Process

Bringing Clients' Conflicts into the Therapeutic Relationship

Before we examine how therapists can use the process dimension to resolve problems, let's briefly review the course of therapy. This review serves to introduce the next stage of therapy and to place it within the overall course of therapy.

Maladaptive relational patterns with caregivers give rise to developmental conflicts that clients are unable to resolve on their own. Symptoms and problems emerge as the client tries to gain control of these difficult feelings and beliefs by adopting fixed interpersonal coping strategies. Although these coping strategies once were necessary and adaptive, they are no longer necessary and are maladaptive in most current relationships. The therapist's first task is to establish a collaborative or working alliance with the client by making accurately empathic connections. They then work together to identify and change the maladaptive relational patterns that are occurring with others and in the therapeutic relationship, as well as the conflicted emotions and pathogenic beliefs that accompany them. As this work proceeds, the therapist must not comply with the client's strong pull to reenact these relational themes in the therapeutic relationship. Instead, the therapist's goal is to provide a different type of relationship that resolves, rather than reenacts, the client's recurrent relational themes. By responding in more empathic and affirming ways, the therapist gives clients the opportunity to experience new ways of relating and to evaluate the costs and gains of their habitual coping styles. In this process, the client becomes empowered to choose more flexibly how to cope with current relationships, and to develop a more realistic and affirming self-concept.

The therapist's primary role is not to give advice, explain, reassure, self-disclose, or focus on the behavior or motives of others, although these responses will all be effective at times. Instead, the therapist encourages the client to set the direction of therapy. What issues does the client feel are most important? While following the client's lead, the therapist participates in shaping the course of treatment and providing a *focus* by listening for (1) the relational patterns, (2) pathogenic beliefs about self and faulty expectations of others, and (3) affective themes that recur throughout the material the client presents. The therapist continues to focus the client inward, reflects the core affective messages, links the client's maladaptive relational patterns with others to their own current interaction, and maintains a collaborative alliance by ensuring that the client is an active participant who feels ownership of the change process. Continually refining working hypotheses, the therapist will be able to conceptualize the generic conflicts and interpersonal coping styles that each client presents.

This conceptual formulation provides a treatment focus and direction for the ongoing course of therapy. For example, the therapist will be able to do the following:

- Point out when and how clients employ their interpersonal coping strategy to rise above their conflicts
- Help clients become aware of how they block their own needs and feelings by responding to themselves in the same way that others originally responded to them
- Help clients explore why they become anxious at a particular time, what types of responses they tend to elicit from the therapist and others to manage this anxiety, and how they systematically avoid certain interpersonal modes or feelings

In these and other ways, the therapist helps clients become aware of their generic conflicts and the interpersonal strategies they have adopted to cope with them. Although this conceptual awareness will be an important part of helping clients change, something more is needed.

At the same time as the therapist and client explore the content of the client's concerns, they also tend to play out these same relational themes in their interpersonal process. As we have seen, the client and therapist do not just talk about issues in therapy; they actually relive them in the therapeutic relationship. This occurs in three ways.

Transference reactions   First, transference reactions bring the client's generic conflict into the current relationship with the therapist. As the client's conflicts begin to emerge in therapy, the client will become increasingly concerned that the therapist has been responding—or is going to respond—in the same unwanted ways that significant others have in the past. These fears and misperceptions follow from the client's relational templates, and are most likely to occur when strong feelings have been evoked and the client feels distressed or vulnerable. In some cases, therapy will reach an impasse until the therapist can help the client differentiate the therapist's actual intentions from the problematic responses that the client expects and has in fact received in most other relationships.

Eliciting behaviors   Second, clients will systematically elicit responses from the therapist that metaphorically reenact their relational expectations. When therapy stalls or ends prematurely, clients have usually recapitulated their generic conflict in the therapeutic relationship. This unwanted reenactment occurs when clients elicit responses from the therapist that are thematically similar to those they have received from others in the past—responses perceived as rejecting, idealizing, competitive, critical, and so forth. Clients elicit these familiar but problematic responses from the therapist for different reasons. Sometimes defensively—to avoid the internal aspects of their conflicts—and sometimes more adaptively—by testing whether they can obtain a better response to their familiar relational scenario than they have come to expect from others.

Testing behaviors   Third, the manner in which the therapist responds to clients—the therapeutic process—may unwittingly reenact their conflicts. Clients' relationship with the therapist must provide a more helpful response to their conflict than they have received in the past. It is not always easy to provide this corrective emotional experience, however. Clients will expect, and often successfully elicit, the same problematic responses from the therapist that they have received in the past. The task for the therapist, then, is *to identify how the interpersonal process with clients may be reenacting their conflict and, when this is occurring, to use a process comment to make this interaction overt as a topic for discussion.* The therapist then works together with clients (and with the help of a supervisor when the clients' issues have evoked the therapist's own matching or reciprocal conflicts) to establish a different pattern of interaction that does not recapitulate the old relational scenario. As we have seen, however, *this idea is far easier to understand than to put into practice.*

Thus, if it is to lead to change, the therapeutic process must enact a resolution of clients' conflicts rather than a repetition of them. When this occurs, clients have received far more than just an explanation for their problems. They have experienced a meaningful relationship in which their old conflicts have been aroused but, this time, they have found a better outcome than the maladaptive relational pattern they have come to expect. They have been able to reveal themselves, to act more boldly, to express a need, and so forth, without receiving the unwanted response they have come to expect (for example, being ignored, exploited, or diminished). When it occurs in the context of an important relationship, such in-vivo relearning is powerful.

Experiencing an interaction with the therapist that is incompatible with their maladaptive relational templates does not make up for clients' childhood deprivations or disappointments, of course. However, it does behaviorally demonstrate that change can occur—that at least some relationships can play out in a different way, at least sometimes. This corrective emotional experience broadens clients' internal working models and interpersonal modes of responding with others. At this critical juncture in treatment, when clients are having a meaningful experience of change with the therapist and the therapeutic process is providing a resolution of the clients' conflicts, *all other intervention techniques become more effective.* Cognitive, interpretive, self-monitoring, educational, skill-development, and other interventions all can be utilized more productively by clients because the useful new content in these interventions is congruent with the corrective process they are enacting. In addition, as clients have repeated experiences of change with the therapist, the therapist can help clients begin to generalize this experience of change to other arenas in their lives where similar conflicts are being enacted. This next period of therapy—generalizing this emotional relearning that has occurred with the therapist to other relationships—is called the *working-through* phase of treatment and will be discussed in the next chapter.

With this overview of the change process in mind, let's now examine more closely how therapists can use the interpersonal process they enact with their clients to effect change.

Using the Process Dimension to Resolve Conflicts

In order to change, clients must not reenact their maladaptive relational patterns with the therapist. Although such reenactments are inevitable at times, they cannot come to characterize the ongoing interaction with the therapist. To alter clients' relational templates and expand their interpersonal range, clients must have the real-life experience of a relationship that does not go down the same old problematic lines they have known before and come to expect. To provide the in-vivo or experiential relearning necessary to propel this reparative experience, it is not enough for the therapist to listen empathically and provide core/facilitative conditions, or to suggest apt interpretations and facilitate client insights about the roots of their problems. Nor is it sufficient to train new skills, offer cognitive re-framing, or prescribe new, more adaptive behavioral responses for clients to try. Although these and other interventions all will be useful, they are usually insufficient to produce enduring change.

Something else is needed. In addition, clients must live-out with the therapist a relationship in which their core conflicts are activated but the therapist does not respond in the problematic way that clients expect. This is difficult to do. Therapists often fail to recognize how they are metaphorically reenacting aspects of their clients' conflicts in their interpersonal process. Clients do not just talk about their problems in the abstract; they reenact the same relational themes with the therapist, in the three ways described earlier. When treatment fails, the working alliance has been ruptured because the therapeutic process has recapitulated some aspect of the same interpersonal problem that the client has been struggling with in other relationships, and that originally brought the client to therapy. Further, this usually occurs because the client's eliciting maneuvers, testing behavior, or transference reactions have tapped into the therapist's own personal dynamics and created a yet-to-be resolved mutual conflict. For example, perhaps the client's need to control others taps into the therapist's strong concerns about being controlled. As we will examine more closely, it can be personally challenging for the therapist to utilize the process dimension and enact a solution to the client's old scenario rather than a repetition of it. We now look at five different case examples to illustrate how the process dimension can be utilized to effect change.

Example 1: An inability to recognize the process   Therapists usually find it easy to understand the process dimension conceptually and relatively easy to observe when a client's maladaptive relational patterns are being reenacted with another therapist. However, it is far more difficult for therapists to recognize how the interpersonal process they enact with their own clients may be recapitulating the conflict rather than resolving it.

An alcoholic client began therapy by expressing how much pain he was in as a result of his drinking problem. His wife was threatening to leave him and he was on probation at work. His 10-year-old son had recently seen him in an embarrassing situation while intoxicated and had asked him not to drink anymore. The client was distraught that, even though the structure of his life was collapsing, he could not stop drinking. During their first session, the client sobbed, "Nothing ever works out; relationships all go bad. Everything I try to do falls apart in the end."

The therapist was moved by the client's plight but became overinvested in helping him change. The therapist found it difficult to listen to how hopeless he sounded and she began reassuring him that things would get better. The therapist disclosed that she had had a drinking problem of her own years ago and that she knew what he needed to do in order to stop drinking. The therapist and client quickly established a friendly relationship in which the client successfully elicited a great deal of nurturance from the responsive therapist. The client enjoyed this support immensely and sincerely tried to follow the therapist's advice.

Things went better for about a month but then the client "fell off the wagon" and began drinking again. When he started arriving late for his therapy sessions, and then missed two of them, the therapist began to feel let down. It was the final straw for the therapist when the client arrived drunk for therapy. The therapist felt angry and betrayed because she had extended herself to him more than she usually did with other clients. She felt that he was letting them both down and told him so. Unwittingly, the therapist became punitive toward the client in this way, and she induced guilt over the impact of his "irresponsible" drinking on his family. The client was contrite and tried, unsuccessfully, to elicit her sympathy and support again. The client did not show up for his next appointment, however, and did not return to therapy.

This interaction recapitulated the client's conflict in two ways. First, the client initially elicited sympathy from the therapist, but the therapist eventually turned to criticism and control. This confirmed the client's disillusionment that "nothing ever works out; relationships all go bad"—the specific relational pattern about which he had forewarned the therapist during their initial session. Second, the therapist benevolently tried to rescue this client, who was behaving as a victim. The caretaking, advice-giving response from the therapist inadvertently defined the client as incapable of managing his own life. This response paralleled a similarly diminishing attitude that the client had received from his indulgent but belittling parent. The rescuer-victim relationship that was enacted in treatment prevented the client from addressing his real concerns over adequacy and achievement, which he defended against, in part, through drinking.

In reviewing this case several weeks later, the therapist reported that "the client just wasn't ready to stop drinking yet. He's going to have to get worse and really hit bottom before he'll be able to stop denying there's a problem and do something about it." The therapist did not recognize that their interpersonal process enacted in therapy unwittingly reenacted the client's relational patterns and prevented change from occurring. As is often the case when treatment fails, the therapist attended only to the content of what they talked about and not to how their process might be recapitulating the client's interpersonal problems.

Beginning and experienced therapists alike will reenact the client's conflict in their interpersonal process at times without realizing that this is occurring. From an emotionally neutral vantage point—that of a therapist's supervisor or colleague looking at the session on videotape—it is relatively easy to see how the old pattern is being reenacted in the interpersonal process. In humbling contrast, it can be exasperatingly difficult for therapists to see their own interpersonal process while engaged in an intense, affect-laden relationship. This is complex work, and it is challenging to perceive and assess at both the content level and the process level simultaneously. Early in clinical training, it is especially difficult for beginning therapists to attend both to the content of what is being discussed and to the process that is being enacted as the discussion proceeds. With more experience, trainees will become more effective at observing the process level. The next challenge, however, is *using* the new understanding they now possess. That is, trainees may have difficulty making process comments and intervening with the interpersonal process they now see occurring with their clients. Clearly, therapists-in-training must be patient with themselves. The ability first to recognize and then to respond to the process level usually is acquired gradually over a period of two or three years. In their attempts to adopt a process-oriented approach, therapists will benefit from further reading on this topic (see the suggestions at the end of the chapter) and the assistance of a supportive supervisor.

Example 2: Process comments keep therapists from reenacting conflicts   In the previous example, the therapeutic process recapitulated the client's conflict. In the next example, a similar dynamic begins, but the therapist recognizes their interpersonal process and effectively realigns this problematic interaction by means of a process comment.

Many clients enter therapy because of depression. They feel sad, believe they are bad, and see themselves as helpless to change their circumstances. In many cases, these clients communicate their very real suffering to the therapist with emotional pleas for help. However, whatever the therapist tries does not work. Often, the therapist's efforts to meet the client's request for help are met with "Yes, but . . ." A problematic cycle begins: The client feels increasingly distressed and intensifies his plea for help. For example, the client might even exclaim that they "can't go on living" if things don't change. In response, the concerned therapist becomes more active and tries even harder to respond in some helpful way. However, nothing the therapist does has any impact or provides any relief for the client.

As the client rejects the help he elicits, the therapist's own need to help is frustrated. The therapist's own sense of adequacy as a therapist may be threatened when the client's eliciting behavior has tapped into the therapist's own personal issues. In response to this, some therapists will become punitive and critical toward the client; others will withdraw and emotionally disengage. When this occurs, the therapeutic process thematically repeats the client's mal-adaptive relational patterns.

Treatment will stall at an impasse or terminate prematurely when the therapist *remains* critical toward or disengaged from the client. The depressed client's conflict is recapitulated, as the client again feels dependent on a relationship with someone who is critical, is inconsistently available, or threatens to terminate the relationship. What is the alternative? In successful therapy, the therapist is able to make a process comment (or *metacommunicate*) and begin a dialogue with the client about what is going on in their present interaction. Rather than focus solely on the content of what they are talking about (for example, depression), the therapist can wonder aloud or inquire about how they seem to be responding to each other. This is one effective way to change the reenactment.

Therapist:
>Let's talk about what is going on between us right now and see if we can understand what's happening in our relationship. It seems to me that you keep asking me for help, but keep saying "Yes, but . . . " I think that I get frustrated with you then, feel that you don't ever allow me to help you, and ultimately become more demanding of you. What do you see going on between us?

With a process comment of this type, the therapist is inviting the client to join in trying to understand their mutual interaction. The therapist is responding to the client's need by helping the client look at the depression and how it is expressed to and experienced by others. As we will see, the process comment also allows the therapist to break the cycle, at least for the moment, of the client's escalating need and the therapist's increasing frustration.

How does the process comment help the client to change? In this example, the client is experiencing a relationship with the therapist that provides a new and different solution to his old relational pattern. The client is not rejected or abandoned, as has happened in the past, but remains engaged in a relationship with someone who finds other ways to keep responding to the client's concerns. The client has not experienced this consistent availability, or the holding environment it provides, in past relationships. If this corrective emotional experience is repeated in many other ways—large and small—throughout therapy, the client is living out a resolution of his generic conflict in the relationship with the therapist. When clients have this real-life experience of change, rather than passively or intellectually hearing advice, reassurance, or interpretations, they are able to make enduring changes.

A key concept in the interpersonal process approach is that the capacity for change increases as the therapeutic relationship becomes more meaningful to the client. When an interpersonal solution is enacted in a valued relationship with the therapist, the therapeutic alliance is strengthened and change is facilitated in two ways. First, this new type of relationship provides interpersonal safety for the client. This safer context permits clients to come to terms with the core conflicted feelings and pathogenic beliefs that accompany their maladaptive relational patterns. In other words, clients have the supportive relationship necessary to experience and integrate those feelings and situations that were too threatening to be dealt with before, rather than continuing to split off or disavow them. Second, the experience of change with the therapist *demonstrates* to clients, rather than merely tells them, that current relationships can be different than they have learned to expect. It does this by expanding their internal working models of relationships—by altering their basic beliefs about the possibilities that relationships hold for them (for example, "I do matter, and can be cared about"; "I can say what I want and not be left"; "I do have the right to set limits and stop letting my children run over me"). Next, with the therapist's assistance, the client can begin to establish better relationships with others along these more flexible and self-affirming lines.

As an example, let's consider the depressed client's response to the therapist's process comment:

Client:
Yeah, I do keep saying "Yes, but . . . " to you—like my wife always says I say to her. This is ridiculous. If I'm doing this with you, too, then there's no way I'm going to get better. Maybe we should just forget it and stop now?

Therapist:
Oh, no—that's not what I'm suggesting at all! It makes sense that you are having the same problems in here with me that you have with other people. In a way, it is a problem, but in another way, it gives us the opportunity to resolve your problem right here in our relationship.

Client:
How?

Therapist:
If you and I can work out a better relationship than you have had in the past—one that doesn't repeat this same old pattern—I think it will go a long way toward helping you change with others.

Client:
Do you think we can do that now if we haven't been able to up to this point?

Therapist:
Yes, I think it's very possible. In fact, I think we are breaking that pattern right now, just by talking about the way we interact together. Tell me, how is it to be talking with me about our relationship and the way we respond to each other?

Client:
It's different, but I like it.

Therapist:
I like it, too. I feel like I'm working with you, rather than being pushed away as I was feeling before.

In this example, the therapist has used a process comment to effectively alter the current interaction with the client. For the moment, the process comment has kept their interaction from recapitulating the client's maladaptive relational patterns. Some variation on the old "Yes, but . . . " pattern will probably soon reappear, and the therapist will have to make another process comment and work through a similar cycle again. With the process comment, however, the therapist is temporarily providing the client with a different response that does not recapitulate the client's old scenario. If the therapist continues to find ways to provide this response throughout treatment, the client will experience a corrective relationship and change will begin to occur in other relationships as well. The next example illustrates a macroperspective that tracks the process dimension over the course of therapy.

Example 3: How the process dimension bridges differing theoretical orientations   There are many ways to help clients change. Therapists with different personal styles and therapists working from different theoretical orientations can all help clients change. Every theoretical approach clarifies some aspects of clients' problems, and every theoretical approach has limitations. Within each theoretical orientation, the primary component of change is not the theory per se, but how effectively individual therapists apply it in their therapeutic relationships (Bergin & Garfield, 1994; Teyber & McClure, 2000; Walborn, 1996). Regardless of theoretical orientation or treatment approach, change is likely to occur when therapists can maintain a strong working alliance and the therapeutic relationship provides a resolution of the client's conflicts rather than a repetition of them.

After two years in analytically oriented therapy, Rachel still could not take charge of her life. She was always complying with her husband's demands and it was impossible for her to get her children to do what she asked. She also complained of her stultifying daily routine as a homemaker, yet she was never able to do anything to improve it. At her husband's insistence, she would periodically enroll in a class or interview for an office job, but she never followed through on any of his suggestions.

In therapy, Rachel had spent many hours exploring her childhood. Her therapist had an uncanny ability to see unifying themes in the recollections she shared with him. For example, her therapist astutely observed how she was subtly discouraged from initiating activities on her own as a child. Subtly, she was not allowed to feel good about her success experiences. Anytime she expressed an enthusiastic interest, tried out a new venture, or achieved something she was proud of as a child, her parents did not notice it—or, if they did notice, they somehow didn't seem very happy about it.

Rachel's therapist once explained to her, in a sensitive and nonjudgmental way, that she had a passive-dependent personality style. Rather than feeling labeled or put down, Rachel was impressed that her therapist seemed to understand her so well. He knew so many things about her without her even having to tell him. Although her problems had not changed much yet, she still believed her therapist would cure her. He was so bright and insightful, and he seemed genuinely to care about her. He didn't like to tell her what to do but, when things became too much, he could usually help by explaining what the problem really meant. It was comforting to be with him and Rachel did not know what she would do without him.

It was Rachel's husband, Frank, who finally ran out of patience with the slow course of treatment. After two years, he was fed up with the unending therapy bills and his wife's unremitting discontent. His wife's helpless dissatisfaction felt like an unspoken but unending demand for him to love her more, give her more, or somehow fill up her life. He was tired of these subtle, nagging demands, and he wanted a change.

A man of action, Frank obtained the name of a behaviorally oriented therapist on the faculty of a nearby college. A friend told Frank that this therapist was a "problem-solving realist" who could make things happen quickly. This sounded like just the right approach to him. Frank insisted that Rachel stop treatment with her present therapist and begin immediately with the new therapist. Rachel was sickened at the thought of leaving her therapist, but she sensed that Frank was truly at the end of his rope. Although she still believed in her therapist and felt very loyal to him, she was afraid that Frank might actually leave her if she did not comply with his demand.

After her first two sessions with the new therapist, Rachel was surprised to think that perhaps Frank might have been right after all. The new therapist wasted no time in taking charge of the situation. It was actually encouraging to have the therapist outline a treatment plan with steps for her to follow. The therapist discussed a list of specific treatment goals with her, and they planned a set of graduated assignments for meeting these goals on a scheduled timetable.

In their first hour together, the therapist had Rachel role-play how she responded to her children when they disobeyed her. Then, with the therapist serving as a model and coach, they rehearsed more assertive responses that Rachel could try with her children. The therapist also had Rachel enroll in an assertiveness-training class that the therapist was running for some other female clients. Each week Rachel was also to complete a homework assignment. For the second week, she was to call one new person she might like to get to know and ask her to lunch. She was supposed to report back to the therapist on this assignment at the beginning of their next appointment.

Frank was encouraged by this practical, problem-solving approach to his wife's problems. He began to think that something might change after all. Rachel was surprised to find that she actually felt hopeful, too. She felt reassured by her new therapist's goal-oriented, problem-solving approach. In fact, Rachel became determined to make this therapy work, even though she didn't want anything to do with it initially. She promised herself that she was going to try to do everything the therapist asked of her.

Although therapy progressed well for the first few weeks, things soon started to slow again. Although she did not really know why, Rachel began to find it hard to muster the energy to attend the assertiveness class. She knew her therapist would be disappointed in her, but she just couldn't seem to help it. Although she felt guilty and confused about it, Rachel began to come late to her therapy sessions. Over the next month, she began to miss them altogether and faded out of treatment.

Both of these therapists failed to have a significant impact on Rachel's problems, even though the treatment approaches they used were seemingly very different. Theoretically, the analytically oriented therapist might attribute the unsuccessful outcome to the great difficulty in restructuring the passive dependency needs of a basic oral character. The behaviorist might note that Rachel was not sufficiently motivated to change because she was receiving too many secondary gains from her help-seeking behavior. If we look at the interpersonal process that transpired between both therapists and the client, however, a different picture emerges. These two therapists actually responded to Rachel in a very similar—and problematic—way.

Let's look carefully at the interpersonal process. With both therapists, Rachel reenacted the same maladaptive relational pattern that she had with her children and with her husband. Her presenting problem was that her children ran over her and she could not make them listen to her. Her disturbing degree of compliance with her take-charge husband was a profound example of the same problem. Therapy failed because her pattern of being passive and compliant recurred with both therapists. Unwittingly, both therapists provided a hierarchical relationship in which she remained the passive helpee led by an all-knowing helper. This interpersonal process reinforced her pathogenic belief that the source of strength and the ability to solve problems did not reside in her but in the therapist or others.

In order for Rachel to change, she needs to experience a therapeutic relationship that behaviorally affirms her own efficacy. This would be a relationship in which she is able to initiate what she wants to explore in treatment, set personal limits with and share control with the therapist, and is more actively supported by the therapist in making her own decisions and choices. The client's symptoms will visibly improve in the context of a new interpersonal process in which the therapist (1) encourages the client to act more independently *in their relationship* and (2) joins with her in collaboratively exploring the anxiety and guilt that arises each time she tries to act more assertively or independently with the therapist or others. This experience of change with the therapist behaviorally shows her that her own strengths are valuable and can be utilized, in conjunction with the therapist's skill and understanding, in a more productive relationship.

Unfortunately, both of her therapists were comfortable with the helper-helpee relational process and did not bring it up as a focus for treatment. Neither therapist offered the metacommunication that, in certain ways, their current interaction was reenacting the same maladaptive relational pattern that was problematic for Rachel in other relationships. Such a process comment, tactfully delivered, would have allowed Rachel to begin looking at this maladaptive relational pattern in a supportive environment. More importantly, it would have given her permission to change her dependent coping style and to begin expressing her own feelings, interests, and authentic voice. The sequence is important: this would occur first in her relationship with the therapist, and then in her relationships with others.

What do these changes in the relationship with the therapist have to do with solving the problems that brought her to treatment? As long as the therapists were telling her what to do, Rachel could not set limits with her children, take a more assertive stance with her husband, or follow through and do what she wanted to do for herself. That is, the highly relevant content of what she discussed with both therapists (issues about autonomy, assertiveness, and her own identity) was not matched by the interpersonal process they enacted. *Unless Rachel has the actual experience of behaving as an active, equal participant in her relationship with the therapist, she will not be able to adopt this stronger stance in other areas of her life.* In other words, the process must be congruent with the content.

To enact a more egalitarian or collaborative relationship with someone like Rachel is not a simple task for the therapist, however. It requires therapeutic skill and thoughtful monitoring of the process dimension. In response to her life experiences, Rachel has become accomplished at getting her therapists, her husband, her children, and others to lead, direct, and control her. Both the behaviorally and analytically oriented therapists could have successfully used their differing theories and techniques to help Rachel if they had enacted a different process in their relationship. If they had encouraged her to initiate more in the session, and then focused on her reluctance to lead, Rachel's generic conflict would have emerged quite overtly in the therapeutic relationship. That is, as soon as either of the therapists invited her to follow her own agenda and bring up whatever she felt was most important to talk about, encouraged her to disagree with or express any dissatisfaction she may have with the therapist, or celebrated whenever she was acting stronger or behaving competently in the therapeutic relationship, Rachel would have become anxious. At that moment, either therapist could have focused her inward on this anxiety so that they could begin exploring together what the threat or danger was for her to step out of her compliant and help-seeking mode.

Simultaneously, either therapist could have given her permission to act more assertively within the therapeutic relationship and responded to her in ways that facilitated this new behavior in their interaction together. The therapist could do this in two ways. First, the therapist could watch for and affirm this effective new stance whenever it emerged in their relationship or with others. Second, the therapist could respond to instances in which Rachel "undid" herself by retreating to the safer, nonassertive mode just after she had risked acting in a stronger way with the therapist. Given Rachel's life experiences and interpersonal coping style, she was likely to become apologetic or confused, or to act dependently and ask for direction, soon after disagreeing with the therapist, making an insightful connection on her own, or redirecting the session more toward her own interests or concerns at that moment. If therapists have formulated working hypotheses about her maladaptive relational patterns, they can track this potential reenactment in their interpersonal process. If so, they will be prepared to help her identify this problematic pattern as it is occurring, and then explore the threat or danger she feels when she has just retreated from her new, stronger stance with the therapist.

Therapist:
What do you think I might be feeling, and how are you afraid I might respond, after you have just acted more assertively by disagreeing with me like that?

With both of her therapists, however, Rachel only continued to defend against her anxiety over being more assertive and independent by retreating to her dependent, help-seeking role and, in their interpersonal process, merely reenacting this maladaptive relational pattern.

In most therapeutic relationships, the therapist and client will temporarily reenact the client's generic conflict in their interpersonal process. In successful therapy of every theoretical orientation, however, the therapist and client do not *continue* to reenact the maladaptive relational pattern. Instead, they are able to recognize this problematic pattern (and perhaps find a name to help them talk about it) and work out a different type of relationship that changes the course of the old familiar scenario. Once clients find that their conflicts can be activated with the therapist, but do not have to result in the same hurtful or frustrating outcomes they have come to expect, they are able to begin generalizing this experiential relearning to other relationships in their lives.

Example 4: Resolving conflicts by working with the process dimension   Excessive performance demands often leave trainees feeling pressure to do something to make their clients change. Unfortunately, these internal pressures on the therapist are usually translated into interpersonal pressures on the client to change. This is often evidenced by a premature emphasis on intervention techniques. Unless the therapist has conceptualized what has gone awry for the client in other relationships, and considered how these maladaptive relational patterns could be reenacted in the therapeutic process, intervention techniques will often fail. In contrast, once the therapist has several good working hypotheses about the interaction between the therapist and client—and the specific relational experiences that this client needs in order to change—it is usually relatively easy to find effective ways to intervene. Therapists are encouraged to ask themselves repeatedly "What does this mean?" rather than "What should I do?" As we will see, the second question is usually answered by the first.

Therapists can employ a wide range of techniques from different theoretical modalities (Wachtel, 1982, 1993). They can then observe the client's response to what they have just said or done. That is, the therapist evaluates the effectiveness of each successive intervention in terms of the client's ability to utilize this type of response to make progress in treatment. Based on this behavioral feedback from the client, *therapists need the personal flexibility to modify their interventions* to provide the responses that work best for this particular client. Sadly, researchers find that many therapists do not flexibly alter their interventions to match the client's needs, but dogmatically stick to the same approach whether the client finds it helpful or not (Strupp & Navitts, 1994).

To illustrate the need to use our understanding to guide our interventions, we examine two critical incidents in the treatment of an incest survivor. In these incidents, the therapist uses intervention techniques that have a highly significant impact on the client: validating the resistance and role-playing. It is not the intervention techniques in themselves that facilitate change, however.

Both interventions are effective because they follow from the therapist's understanding of what is transpiring in the current interaction, and because the therapist then is able to work with these dynamics in the immediacy of the client-
therapist relationship.

Sandy had confided to her male therapist that she was an incest victim, and
he anticipated that trust would be a central issue in their relationship. Therapy
had gotten off to a good start but, several months into treatment, progress
began to slow down as the material that Sandy presented became repetitious.
About this time, Sandy recounted two different narratives in which the relational theme was feeling unsafe with men. The therapist hypothesized that,
without being aware of it, Sandy was using these vignettes to address the topic
of trust in their relationship. As their relationship was becoming more important to Sandy, it seemed that her profound concerns about safety and betrayal
were now being activated with him.

The therapist began to talk with Sandy about trust in their relationship
and asked about the different thoughts and feelings she was having toward
him. Sandy genuinely liked the therapist and was finding him helpful. However, when the therapist pursued the trust issue and asked specifically whether
she felt safe with him, her affirmative response seemed distant and unconvinc-
ing. It soon became clear to both of them that Sandy was emotionally remov-
ing herself from the therapist as they talked more directly about safety in their
relationship.

The therapist responded to Sandy's concern by accepting both sides of her
ambivalent feelings and tying her concerns directly to their relationship.

Therapist:
> I know that one part of you likes and trusts me, but it makes sense to me
> that another part of you doesn't feel safe. I think that both sides of your
> feelings toward me are valid and are important for us to work with.

Sandy sheepishly nodded and gestured vaguely to indicate that this was true.

Rather than trying to talk her out of these concerns or convince her of his
trustworthiness, however, the therapist validated her distrust and went on to
characterize it more fully:

Therapist:
> If I violated your trust in some way after you had taken the risk to ask me
> to help you, it would be very bad for you. If I tried to approach you sexu-
> ally, or foster any other kind of relationship between us, it would hurt you
> very much. In fact, I think it would hurt you so much that you might not
> be able to risk trusting or asking for help again.

Sandy began tearing, looked at the therapist, and slowly nodded in agreement. The therapist continued to elaborate Sandy's concern in terms of their
relationship:

Therapist:
> If I took advantage of our relationship in some way, I think you would feel
> hopelessly betrayed. I think you would become very depressed again, enter

into other relationships that would be destructive for you, and may even start thinking again that you do not want to be alive.

As the therapist spoke and made Sandy's concern overt in terms of their relationship, her whole demeanor changed. She remained tearful but became alert and present, agreeing with what he was saying. As they continued to talk about this, the therapist went on to describe his own attitude:

Therapist:
I can see how much it would hurt you if I betrayed your trust, and no part of me wants you to have that terrible experience again. In fact, I like the part of you that doesn't trust me. That distrustful, removed part of you is your ally. We need her—because she is committed to not letting you get hurt again. That's why it's so important that we go at your speed in here, that you have control over what we talk about, and that you can say "no" to me and know that I will honor your limits.

After the therapist responded so affirmingly to Sandy's resistance and worked with this dreaded expectation of betrayal in terms of their relationship, important changes occurred. Whereas in the past Sandy had only alluded vaguely to her childhood abuse, she now began to share it more explicitly. Over the next weeks, she recalled in vivid detail, and with strong emotions, how she had been repeatedly sexually abused over a period of years by her stepbrother, who was 12 years older. When she first went to her mother for help, her mother was not supportive, denied that the abuse was occurring, and told Sandy never to talk about anything like that again. Sandy did not even consider seeking protection from her stepfather, who had always been distant and unresponsive. After failing to receive help, Sandy recalled sitting alone on the floor of her closet for long periods with the door closed, feeling afraid, ashamed, and deeply alone. In response to this deep sharing, the therapist was comforting, validated her experience, and began to help her with the many significant connections she began to make between this familial tragedy and her current life.

During one of these sessions, the therapist used a role-playing technique to provide a different response to the trauma she had suffered.

Therapist:
I wish someone could have been there to stop him and protect you. No one was there for you but, if I had been there, I would have walked into your bedroom when he was there, turned on the light, and in a loud voice commanded: "Stop it! Get away from her and leave her alone right now! I see what you're doing and it's not fair. You're hurting her and I won't allow it."

After speaking this loudly and forcefully, as if he were actually saying it to the perpetrator, the therapist stood up and rolled Sandy's coat into a ball. Using it as a little Sandy doll, he held it close and spoke reassuringly:

Therapist:
You're safe now, and he's gone for good. I'm going to call the police now and keep protecting you so that you'll never have to worry about him hurting you again.

As the therapist symbolically gave Sandy the affirmative and protective response that she desperately longed for—but did not get—as a girl, the full intensity of her pain and rage was evoked. When these powerful feelings ran their course, Sandy became more composed and said, "I'm going to be all right." The therapist, still holding the rolled-up jacket tenderly, carefully tucked it in Sandy's arms.

Therapist:
> This is the little girl you were. She needs you to open up your heart to her and give her a home. You need to take care of her—and not push her away anymore like they did and you have done. You need to hold her, talk to her, and listen to what she tells you. I want you to join me in taking care of this part of you, so this little girl is no longer alone behind the closet door.

Sandy readily accepted this responsibility and later bought a doll that she used to represent the part of her that needed to be cared for but had not been protected.

Sandy entered therapy almost 25 years after the abuse, anxious, dependent, and almost incapable of leaving home alone. However, two weeks after this session, Sandy got her driver's license renewed and began driving for the first time in several years. Soon afterward, she got a job as a waitress—her first paid employment in six years. Although she had always foiled any type of success for herself, she enrolled in a local college and began receiving As in many of her classes. Whereas Sandy had felt helpless, had characteristically acted as a victim, and had been repeatedly taken advantage of by others, she gradually became more assertive and expansive. All of Sandy's problems did not go away, of course, but long-standing symptoms were resolved and did not return.

What allowed Sandy to get stronger in these significant ways? Several factors made this role-play intervention effective. By using this technique, the therapist brought Sandy's conflict into the therapeutic relationship and provided her with the reparative experiences she needed. In sharp contrast to what occurred in her family, this time Sandy experienced protection, validation, appropriate boundaries, and a supportive holding environment. The therapist's compassionate response also helped Sandy shift from an identification with her parent's rejection of her—and her own resulting shame and self-hatred—to an identification with the therapist and his compassion for her. Eventually, this healthy new identification led her to be able to care better for herself and believe that she did matter for the first time in her life. These far-reaching consequences were set in motion by the therapist's affirming response to Sandy's vulnerability. Even though it was enacted in fantasy, the therapist provided Sandy with a corrective emotional experience that played an important part in allowing her to become stronger and capable of protecting herself.

Example 5: Recognizing the process dimension in all interpersonal relations
As we have seen, the most likely reason for therapy to fail is that the therapist and client unwittingly reenact aspects of the client's generic conflict in their interpersonal process. Therapists-in-training often find it challenging at first to

track the process dimension and to distinguish the process that is enacted from the content that is discussed. For our final example, let's look at how the process dimension operates outside the psychotherapy setting—at a sporting event.

In a championship basketball game, the defending champions confidently face an underdog team made up of talented but younger players who are far less experienced. The defending champions are heavily favored to win.

Late in the fourth quarter, the game is tied. The fans, on their feet, are cheering at the prospect of an upset. The coach of the young team can feel that the game is about to slip away, however. His players are becoming rattled by the screaming fans, the pressure of time, and the unrelenting play of the opponents. To help compose his less-experienced players, the coach calls a 2-minute time-out. His team has played well up to this point, but the seasoned coach knows that, in these pressure-packed closing minutes, the other team's experience is likely to make the difference.

Crouched on the sideline, the coach positions his five players in a semicircle in front of him. With the roaring crowd only a few feet away, the coach quietly makes eye contact with the first player on his left. Without speaking, he calmly holds the first player's gaze for 12–15 seconds (a very long time amid such pandemonium). Having compelled the first player's attention, the coach turns to the second player and maintains eye contact with him for another 12–15 seconds, and so on with each of the five players. Each time, as he turns to engage the next player, the previous player's attention remains riveted on the coach. Finally, after making contact with each player in this way, the coach holds up a chalkboard with one word written on it: POISE.

The players and coach sit together silently for the remaining seconds of the time-out. The referee blows his whistle, the players return to the court, and the underdogs defeat the defending champions. The players did not make mental errors in the closing minutes of the game. They did not turn the ball over to the other team; they made clutch-free throws in the final seconds, and they smoothly executed the plays they had practiced all year. In a word, they played with poise.

This coach astutely identified the central issue that could defeat his team in the final minutes: being intimidated by a more confident and experienced team, losing their composure, and making mental mistakes that would cost them the game. Having recognized the central issue, however, the coach did not just *tell* his team that they had to remain poised. Instead, he gave them the *experience* of composure during the time-out. The players were able to finish the game with such composure because the coach used the relationship he had developed with his players to give them the experience of composure during the final tense minutes. During the time-out, the players were able to generalize this experience to their performance on the court and play with poise.

Similarly, it is far more effective if the therapist-client interaction provides an experience of change than if the therapist merely tells the client what to do or explains what something means. It can be challenging to provide a relationship that enacts a resolution of clients' conflicts, however, and there are no simple formulas that tell the therapist how to do so. Every client is unique. The therapist must be able to take the personal risk of entering into a significant emotional relationship with each client, and have the flexibility to provide the new relational experiences that this particular client needs in order to change.

Working with Clients' Conflicts in the Therapeutic Relationship

As we saw in Chapter 8, an essential element of interpersonal psychotherapy is to recognize how clients' conflicts are played out in the therapist-client interaction. Working together, therapist and client then find new ways to relate that resolve, rather than reenact, the client's maladaptive relational patterns. For example, cultural, ethnic, or racial issues that have caused problems for the client in relationships can be made overt in the therapeutic process and resolved differently with the therapist.

Therapist:
> I know that we all have different cultural experiences and they affect how we cope, what we are sensitive to, and many other things. I'm wondering how your cultural experiences and our differences may be influencing this issue in our relationship?

Thus, the best way to begin this work is by asking clients in an open-ended way how the problem that they are talking about in relation to others may be occurring in the relationship with the therapist:

Therapist:
> How does this battle for control you have been describing with your partner go on between us? Where do you see the control issue in our relationship?

By bringing the client's conflicts directly into the therapeutic relationship in this way, the therapist creates the opportunity for a corrective emotional experience. In order to change, this client must find that the therapist is neither trying to control, nor willing to be controlled by, the client, as occurs for the client in other relationships. Only if the client has this real-life experience of change can the client resolve this conflict, find that control can be shared in close relationships, and learn that some relationships can be different from those in the past. Although this immediate, real-life experience of change is a powerful way to intervene, some beginning therapists find it uncomfortable to make this observation or process comment. Thus, the purpose of this section is to help therapists take the risk of judiciously using their real-life relationships with clients to facilitate change.

In the first part of this section, we review how therapists can use process comments to link clients' problems with others to the current interaction with the therapist and work with them in the immediacy of the therapeutic relationship. In the second section, we examine why these metacommunications that produce such immediacy can be anxiety-arousing for beginning therapists, and consider helpful guidelines.

Intervening within the Therapeutic Relationship

To some extent, most of the interventions discussed so far have asked the therapist to work on the clients' issues within the immediacy of the therapist-client relationship. At the beginning of treatment, for example, therapists can speak directly to clients about their *current interaction* when they are trying to establish a collaborative relationship (Chapter 2):

Therapist:
Am I understanding what is most important to you here?

Talking about what is going on between the therapist and client is even more important in working with *resistances* to entering treatment (Chapter 3):

Therapist:
What's it been like for you to talk with me today?

Working directly with the current interaction is also helpful when establishing an *internal focus* (Chapter 4):

Therapist:
You keep talking about others and I keep asking about you. What do you see going on between us?

Therapists need to work with the *client's affect* in a way that brings the full intensity of whatever feelings the client is experiencing into the immediacy of the therapist-client relationship (Chapter 5):

Therapist:
I can see how much this has hurt you and how sad you are feeling right now.

Finally, this focus on the interaction between the therapist and the client is central to working with *transference reactions* (Chapter 8):

Therapist:
How do you think I am going to react if you do that?

The unifying theme in all of these interventions is to link the client's issues and concerns with others to what is currently going on between the therapist and the client. Thus, the client's conflicts are not just talked about in abstract discussions about others; they are reexperienced and resolved in the real-life relationship with the therapist. Let's return to the client whose basic relational theme is a control battle. The best way to resolve the client's conflict is to make it overt and address it directly in terms of the client-therapist relationship.

Therapist:
Does that control battle ever get going between us, too? Does it ever feel like you and I are fighting over control in our relationship?

Making such a comment may seem a dreadful prospect to the beginning therapist. However, by taking this risk, the therapist can bring real-life immediacy to the therapist-client interaction. Problems with others are not just being talked about intellectually; instead, for the first time perhaps, they are being addressed directly and constructively with someone *as they occur*. If the therapist waits for good timing and responds respectfully the potential for change is at hand.

Let's consider a range of three progressively more challenging responses to the therapist's query.

Response 1: Least challenging to the therapist is a disconfirming response.

Client:
No, I don't feel like we are in a control battle.

The therapist can use this response to begin exploring what is different about their relationship.

Therapist:
Good, I'm glad that's not a problem for us. Any ideas about what makes our relationship different?

The client's answer may provide useful information:

Client:
You treat me with respect; that's what's different.

The therapist can follow up on this comment by exploring the client's concerns about not being respected in other relationships, and can build on it to establish that other kinds of relationships can exist.

Response 2: Clients may avoid the immediacy of their conflicts by discounting the relevance of the therapist.

Client:
We're not in a control battle because this isn't a real relationship. You're just my therapist.

Clients make far more significant gains in therapy when this distancing defense is resolved. At this point, it is essential to clarify that the therapist and the client really are two people who have been having a relationship for some time, even though there are constraints on their relationship. The therapist can then begin to explore the threats aroused for the client by more meaningful involvement—such as being controlled or not being respected. The therapist and the client can then agree to watch for these concerns and address them if they arise in their relationship. It is usually most effective if therapists do not bring up these relational patterns in the abstract, however, but wait until they think the client is experiencing them in their current interaction before addressing them.

Response 3: Clients may respond to the therapist's question affirmatively.

Client:

Of course you're in control. You insist that we stop at 10 to the hour, whether I'm finished or not. And you're subtle about it, but you try to take control by making things go where you want.

Most beginning therapists fear such a response. They may believe that they have done something wrong if the client disapproves of them in this way. This possibility must always be considered but, in most cases, the client's reactions have as much to do with the client's own internal working models as they do with the reality of how the therapist has responded. Thus, rather than posing a problem, this affirmative response actually provides an important window of opportunity. Openly acknowledging that the same conflict the client has with others also exists with the therapist creates the opportunity to explore this issue directly and work together to try to resolve it in their real-life relationship. The therapist can do this in the following ways:

1. By remaining nondefensive and accepting the validity of the client's concerns whenever possible

Therapist:

Yes, we do have to stop at 10 to the hour and that is an unnatural ending for you. I can see how you would feel controlled by those time constraints.

2. By exploring the client's perceptions further and working with the client to understand them better

Therapist:

What do you think is going on for me when I try to take things somewhere? What does it seem I am trying to do at those times?

3. By differentiating the therapist from other figures in the client's life

Therapist:

Yes, I have had ideas about where we should go, but I have also been interested in following your lead, too. In fact, I may be different from some other people in your life because I genuinely liked it when you disagreed with me last week and told me what you thought. I like it when neither of us feels controlled and we can both say what we want. It makes our relationship feel more alive to me.

4. By offering to be sensitive to this concern in their relationship, expressing a willingness to handle the issue differently in the future, and inviting the client to tell the therapist whenever this conflict occurs between them

Therapist:

I don't want you to be controlled in our relationship as you have been in others. That's no good for you or anybody else. Let's try to do something about it. From now on, I will watch the clock and let you know when it's 5 minutes before we have to stop. And any time you feel like I am directing you away from where you want to go, or being controlling in any way, tell me, and we'll stop right then and change it. What else could we do?

Adopting these approaches will work easily with some clients, whereas other clients will insist on reestablishing the same constricted relational patterns. However, if the therapist is willing to remain nondefensive and able to tolerate personal discomfort, most clients will become more engaged in the therapeutic relationship and motivated to explore this and other issues more deeply. As clients repeatedly find that the expected but unwanted old scenario does not recur with the therapist, they experience greater interpersonal safety than they have known before. As a result, important new issues, pathogenic beliefs, and previously threatening feelings can emerge—providing new material for exploration and a clearer treatment focus. As we will see in the next chapter, clients also begin to relate to others in new and more effective ways.

The real-life experience of change within the therapeutic relationship is powerful for clients, but working so directly and forthrightly can be uncomfortable for therapists. Next, we explore several reasons for such discomfort, and some practical guidelines to help therapists work with the process dimension more effectively.

Therapists' Initial Reluctance to Address the Process

Although the best vehicle for effecting change is the therapeutic relationship, it can be anxiety-arousing for beginning therapists to take the plunge and say, "Does that problem ever go on between us here in therapy?" Talking together so directly about issues between two people breaks the cultural norms and family rules that most therapists grew up with. Because of this, and the immediacy that such forthright communication creates, some beginning therapists are reluctant initially to make process comments and explore how clients may be experiencing similar problems with the therapist that they are having with others.

A few beginning therapists find it easy to work with clients using a process approach. They have always been able to talk forthrightly with others about problems in their relationships, and it makes sense to them that what goes on for clients in other relationships will also come into play with the therapist at times. For these therapists, it is enlivening to make potential problems or misunderstandings overt and to talk about them openly with the client. The process approach simply gives them the permission they may need to work with clients in this engaging way.

For other trainees, however, the process approach is more difficult at first. Although bringing clients' conflicts into the therapeutic relationship makes sense to them intellectually, they find it harder to actually put into practice. Because they have had so little direct clinical experience, most first- and second-year graduate students have little confidence. Understandably, they are not eager to have what little self-confidence they may possess shaken by stepping outside of familiar bounds. Further, even if they think that this type of intervention might be useful with a particular client, all beginning therapists struggle to discern whether they are really observing the client's conflict or whether their perceptions simply reflect their own countertransference. Finally, even if they feel confident that it is indeed a maladaptive relational pattern, they may be unsure of how best to address this reenactment and find a helpful way to make it overt.

A few therapists feel concerned that to be forthright with clients could be hurtful because it would expose clients' pain, or they have seen directness used hurtfully to induce shame or guilt. Therapists never want to intrude on their clients, of course, and diminishing clients' sense of personal safety will impair their ability to make progress in treatment. Therapists can address the process dimension in sensitive and respectful ways that actually reassure clients, rather than make them feel uncomfortable. We now examine six of the most common concerns of beginning therapists' about using the therapeutic relationship to work with clients' conflicts, and suggest some possible solutions.

1. Uncertainty about when to intervene   Because they have not had much experience conceptualizing clients or tracking the process dimension, beginning therapists are often unsure when or if the client's conflicts are being reenacted in the therapeutic relationship. Trainees will become more confident of this as they begin to integrate the complex new information they are learning, and more comfortable intervening with the process dimension. It is something of a personal risk for therapists to suggest, "I'm noticing something that may be going on between us right now, and wondering if. . . . " Although therapists will be accurate with some of their observations, the client will not resonate with others. Therapists have not failed when they are inaccurate; they are simply trying to understand and clients usually appreciate these good intentions. Nevertheless, novice therapists should not be in a hurry to make these interventions until they feel ready to do so. Forcing such responses will not feel good to the therapist or be effective for the client. When therapists are disempowered because they do not feel in control of choosing how they intervene with the client, they will not be capable of empowering the client.

How can therapists discern when the client's conflicts are being reenacted between them? Let's follow a five-step sequence. First, it is helpful if the therapist has already formulated working hypotheses that alert the therapist to how the client's maladaptive relational patterns are likely to be replayed in the therapeutic relationship. Second, this conceptualization will help therapists make sense of their own experience and recognize when they are starting to feel or respond as others do in the client's old scenario. Third, this is a good time for the therapist to consult with a supervisor and check out together what may be occurring in the therapeutic interaction. If it does seem as if a reenactment is underway, supervisees can explore any concerns they may have about trying to make this interaction overt and discussing it with the client. Looking for effective ways to explore this possibility with the client, therapists can also role-play or rehearse alternative responses with the supervisor. Fourth, with this preparation, therapists can wait to broach this possibility with the client until the next time they feel this reenactment may be occurring. It is always more effective to intervene at the moment when the interaction is occurring rather than bringing it up in the abstract. Fifth, beginning therapists are encouraged to wait until they feel that this intervention is likely to work. That is, it is important that therapists *choose* to make this or any other intervention—compliance is just as problematic for therapists as it is for clients.

It will often take two or three years before these responses feel natural and come effortlessly. These are new ways of responding for most beginning therapists that often break cultural norms and familial rules. Again, good advice for the novice therapist is to be patient.

2. Fear of offending the client   Some beginning therapists are concerned that the client will feel that therapists are being too blunt or personal if they directly address what may be going on in the relationship. Therapists might worry that the client might feel hurt, not like them, or—worse yet—just stare blankly back at them! Of course, process comments, like any other interventions, can be made in blunt, insensitive, or otherwise ineffective ways. Warmth, tact, and a good sense of humor can all go a long way toward making every intervention more effective.

Therapists may also be worried about trespassing on social norms and expectations. Speaking directly about what may be going on between two people often breaks unspoken social rules, and most trainees are concerned that clients may be surprised or taken back by this. To help ease the transition toward this more genuine dialogue, therapists can offer contextual remarks that facilitate the bid for more open communication. For example, the first few times the therapist addresses the process dimension with a client, the therapist can create safety for the client by offering an introductory remark that acknowledges the shift to another level of discourse:

Can we be forthright with each other and speak directly about something?
Let me break the social rules for a minute and ask about something that might be going on between us.
I know that people don't usually talk together this way, but I think it would help if we could talk about . . .

Therapists will not be too blunt or direct for clients if they respond respectfully and provide transition comments such as these to help clients shift from the socially polite to this more direct approach. Rather than being threatened by this invitation for more forthright communication, the vast majority of clients are hungry for a more authentic exchange. Welcoming it from the beginning, they are reassured by the therapist's willingness to get down to business and move closer to their real experience and problems.

3. Therapists' insecurity and countertransference issues   To work with clients in the manner described here, therapists are personally challenged to risk engaging in a genuine relationship with the client. In this approach, therapy is a two-way street where each person is willing to be effected by the other. Process comments fail when therapists are not willing to be emotionally present and engaged in the real relationship with the client. For example, the client receives a mixed message from the therapist when the therapist makes a process comment and invites a more authentic dialogue, but then retreats to the therapist's role and is unwilling to consider the potential validity of the client's criticism or honestly explore the client's dissatisfaction with something the therapist has done. In contrast, when therapists are willing to talk with the client about their participation in problems and what is occurring between them, sessions become more intense and productive. Therapists' willingness to be personally affected by the client, and to look at their own contribution to their interpersonal process, will be anxiety-arousing at times. However, a corrective emotional experience cannot occur unless the relationship is significant to *both* the client and the therapist.

By looking honestly at their own contribution to problems in the relationship, therapists facilitate an egalitarian relationship that holds genuine meaning. Of course, this increasing mutuality will activate therapists' own conflicts at times (Jordan et al., 1991). For example, in order to respond more authentically, therapists must relinquish hierarchical control over the relationship, which will be anxiety-arousing for some therapists. Therapists may also be concerned that this genuine emotional contact with the client will lead to a loss of appropriate therapeutic boundaries and result in overinvolvement or acting-out on the part of the client or therapist. Clearly, every therapist will have moments of overinvolvement. However, by applying the guidelines on optimum interpersonal relatedness in Chapter 6, therapists will be able to recognize when their own countertransference issues are prompting them to become too close to—or too distant from—the client.

Further, in making process comments, how do therapists know whether what is going on is due to their own personal issues or to the client's dynamics—to client-induced or therapist-induced countertransference? Therapists have two useful methods of distinguishing whether the client's dynamics—eliciting maneuvers, testing behavior, transference distortions, and so forth—or the therapist's own dynamics are operating. In order to help sort out their own issues from the client's dynamics, supervisees need a supportive supervisor who helps track the therapeutic process. In the beginning, it is unrealistic to expect trainees to be a participant-observer who can maintain a balance of emotional relatedness and objectivity without on-going assistance.

Another way beginning therapists can safeguard against confusing their own and clients' issues is by not jumping in and making observations about the therapist-client relationship as soon as they see something significant happening. Following the five-step sequence suggested above, it is more effective for the therapist to generate working hypotheses first, and then wait to see whether the observation also applies to subsequent exchanges between therapist and client. If the issue or concern is relevant for the client, it will systematically recur as a theme or pattern. If it does not keep coming up, then it is likely that the therapist's own issues are involved, or that it is not an important concern for the client and this hypothesis should be discarded. Thus, therapists are not cavalier about making process comments. When in doubt about whether an issue is the therapist's or the client's, it is best for the therapist to wait, gather additional information about the therapeutic process, and/or consult with a supervisor before raising the issue with the client. Remember, too, that all process comments are simply observations. They are offered to the client tentatively, as possibilities for joint exploration and mutual clarification, not as truth or fact.

Across all treatment approaches, the therapist's relationship with the client is the most significant means of effecting change (Strupp, 1991; Teyber & McClure, 2000). At the same time, the most likely reason for therapy to terminate prematurely is that the client's interpersonal coping strategies have activated the therapists' own conflicts. Therapists need to be open to the possibility of their countertransference issues, and should generate working hypotheses about how each client's conflicts could be recapitulated in treatment (see Appendix B, Part 6). The therapeutic enterprise requires that beginning and experienced therapists alike must:

- Make a lifelong commitment to being open to their own countertransference propensities
- Consult with supervisors and colleagues as an ongoing career activity
- Seek their own personal therapy when countertransference issues persist

Countertransference issues are most likely to create problems when therapists disregard them. Counselors who are aware that they are susceptible to countertransference—both client-induced and therapist-induced—and who discuss possible instances of countertransference with supervisors as they arise should feel unfettered about working with clients in a process-oriented manner.

4. Concern about appearing confrontational   Some therapists are concerned that, if they address issues directly, the client will misconstrue their behavior as an intimidating confrontation. Such a confrontational stance is not our intent. Process interventions can be presented in an aggressive manner, just like any ineffective intervention, but there is nothing insensitive or disrespectful about simply communicating forthrightly. If therapists think these (or any other) interventions may be perceived by the client as demanding, intrusive, or unwanted in any way, they should not use them. Therapists can either wait for what feels like a good time to try this particular intervention or simply respond in other ways.

Similarly, some therapists may worry that, by responding more directly in therapy, they may "hurt" the client in some way. These expectations of client vulnerability are usually inaccurate. Virtually all clients will welcome the invitation for a more straightforward dialogue. As clients find it safe to speak with the therapist about what goes on between them, novice therapists will feel more secure that it is possible to be sensitive and direct at the same time.

Fear about hurting the client takes on more personal significance for a few therapists who grew up with highly authoritarian parents (Chapter 6). Such therapists may have been exposed to double-binding family communications in which there was a great discrepancy between what was being done (for example, the child may have routinely been threatened, humiliated, or hit) and what was being said (for example, "We are a close and loving family. There are no problems here, and everyone is happy"). The essential element in the double-bind is the clearly understood but unspoken family rule that the child cannot *acknowledge* the incongruency in any way. For example, the child in such a double-bind cannot make these contradictory messages overt by metacommunicating and saying, "You're telling me to clean my plate, but later you'll make fun of me for being fat. Stop it, you're driving me crazy!" Children who have grown up with these double-binds are frightened to make them overt because further rejection or debasement has been threatened or implied.

Such double-binding communications occur in many dysfunctional families and are routine in physically and sexually abusive families. Therapists who have suffered such mistreatment themselves may find it threatening (or at other times liberating) to use process comments and make overt what is occurring. An enduring legacy of such painful developmental experiences is that these therapists may fear that, if they speak directly about what is going on, the client will be damaged by them. That is, their pathogenic belief in their own inherent destructiveness or badness, instilled by the abuse they suffered, will be confirmed. Thus, old threats can be evoked—but also resolved—by breaking double-binds and violating unfair family rules about how people have to communicate. Such dynamics are most appropriately resolved in the therapist's own therapy, rather than in the supervisory relationship.

5. Therapists' fear of revealing their own inadequacies   Therapists may fear that, if they bring out how the client's conflict is being replayed in the therapeutic relationship, their own mistakes or inadequacies will be revealed. For example, the therapist may think: "Yes, there it is, the same conflict that he has with his wife is going on between us right now. It sure is getting in our way, but I don't have any idea what to do about it." In that case, making the conflict overt and trying to talk together about it may seem like the last thing in the world the therapist would want to do.

Whether the therapist chooses to address them or not, such reenactments will occur at times in most therapeutic relationships—for novice and experienced therapists alike. By making them overt, however, therapists do not assume responsibility for *causing* the client's conflicts or pain because their effective responses have revealed them (a common pathogenic belief for many who enter the helping professions) or assume sole responsibility for changing them. Rather, the therapist is simply clarifying what may be happening in the relationship and suggesting that the therapist and client work together to rectify the maladaptive relational pattern.

Effectively addressing potential reenactments, therapists work together with the client to *clarify the repetitive sequence of interactions* that typically unfolds in problematic scenarios with others. Continuing this collaborative effort, therapist and client can explore if and when these problematic patterns have been occurring in the therapeutic relationship. If they have, therapists can acknowledge their own participation in this shared conflict, express their willingness to change their part in the old scenario, and try to make this relationship come out better for the client. Still working collaboratively, the therapist and the client search for new ways of relating that do not continue the old relational pattern. By doing this, the client no longer feels invisible, controlled, responsible, and so forth, at least in this relationship, and is empowered to begin constructing more rewarding relationships with others along these new lines.

When the therapist makes these reenactments overt, clients may feel discouraged initially—which can evoke feelings of inadequacy in the therapist. Initially, the client may feel hopeless because this reenactment confirms the clients' expectations that relationships cannot be different than they have been in the past. Intellectually, the client may recognize that other ways of relating are possible. However, this possibility holds little meaning experientially when the current enactment with the therapist confirms the client's relational templates and follows the problematic interpersonal scenario that he has experienced repeatedly. Therapists' feelings of inadequacy will be resolved when therapists:

Realize they are already responding adequately to the client by recognizing that this is how the client's relationships have gone in the past

Understand that this has been an unsolvable and painful problem that has characterized the client's life

Affirm the historical validity of the client's hopeless feelings, based on the reality of what the client has actually experienced with others in the past

By doing this, the therapist is *behaviorally demonstrating* that the relationship between the therapist and client is no longer following the same well-worn, problematic patterns and is giving clients the corrective relationship they need in order to begin changing with others.

By thoughtful preparation—mentally formulating and revising working hypotheses, keeping written process notes, and writing case conceptualizations—beginning therapists will be able to resolve legitimate concerns about their own adequacy and performance. In the beginning, therapists cannot successfully track the process dimension just by thinking on their feet. Therapists must be prepared by considering in advance various hypotheses about what may be occurring along the process dimension if they are to effectively bring the client's conflicts into the therapeutic relationship. As we have already noted, therapists should not make a process comment until they have seen an issue occur several times and until they have some tentative understanding of what it may mean. When the client's coping strategy elicits anger, discouragement, or some other strong emotion, therapists are encouraged to wait to respond until they understand their reaction and can reestablish their neutrality. Finally, therapists who have prepared themselves with working hypotheses will feel more confident and be less defensive, better able to hear their clients' concerns, and adjust more flexibly to their clients' needs.

As developing therapists become more confident, it will be easier for them to explore more open-endedly what may be occurring in the client-therapist relationship. When feeling confused, for example, an experienced therapist may be able to inquire simply, "How did we get into this anyway?" In the beginning, however, therapists may want to wait until they understand better what is transpiring before addressing it with the client.

6. Concerns about owning personal power A pervasive but generally unacknowledged issue in clinical training is therapists' conflicts about owning their own personal power. Many new therapists feel uncomfortable about allowing themselves to become someone who is important to clients and having a significant impact on their lives. Especially in the beginning, it is often anxiety-arousing to make strong interventions that can influence clients profoundly and to accept the responsibility that comes with exercising such personal power. As a result, many beginning therapists overqualify their comments and dilute the emotional impact of their interventions.

There are many reasons why the therapist's own effectiveness can be anxiety-arousing. As discussed in Chapter 6, some therapists were parentified or aggrandized as children. For such therapists, legitimate competencies can be readily exaggerated into unrealistic all-powerful or all-responsible grandiosities. Initially, it may have been exciting to be specially powerful in this way, but it soon becomes lonely, intimidating, or burdensome. As a result, these therapists may undo, or quickly retreat from, strong interventions they make or may avoid them altogether.

A more common reason for beginning therapists' reluctance to have a strong impact on clients is therapists' own separation anxiety or separation guilt. These therapists' experience, in their family of origin, was that competent or independent functioning threatened their emotional ties to parental caregivers. As the young child began to explore and to move toward greater autonomy, certain caregivers may have looked sad, acted indifferent, demanded more, ridiculed attempts at mastery, or otherwise paired anxiety with competence strivings. Therapists with these developmental experiences also tend to retreat from their own effective interventions, and may be heard saying, "I don't know what to do," "I'm afraid of hurting the client," or "I'm so screwed up myself that I have no right to try to help somebody else." All therapists certainly have their own personal problems and limitations. When such comments persist, however, it may indicate that the therapist wants to avoid the anxiety evoked by clearly and effectively addressing the client's problems.

What can help trainees with these concerns? An affirming supervisory relationship is crucial if therapists are to embrace their own personal power and be as effective as they can be. The supervisor-supervisee relationship must be such that supervisees:

- Feel personally supported by the supervisor
- Receive conceptual information and practical guidelines when needed
- Obtain nonjudgmental assistance in sorting out their own conflicted reactions toward the client that have been triggered by the client's eliciting maneuvers or by the therapist's own personal issues
- Are able to address and resolve the conflicts that are likely to arise in the supervisor-supervisee relationship

If the supervisee feels that the supervisor is solely directing the case, the interpersonal process between the supervisor and supervisee is problematic. Even though the supervisor is ultimately responsible for the client, supervisor and supervisee should be consulting together collaboratively in a way that allows the supervisee to feel ownership of the treatment process. Supervisees must have primary responsibility for what occurs in sessions and must be free to act on their own ideas, while taking into account the supervisor's. Otherwise, supervisees will not be able to experience their successes and failures as their own and be able to learn from them.

Ideally, the supervisor will help the supervisee evaluate the effectiveness of interventions and consider alternatives without taking away the supervisee's own initiative. When the supervisor and supervisee become stuck on a conflict in their relationship, the same issues often carry over to the supervisee-client relationship and will be reenacted there. In other words, the interpersonal process between the supervisor and supervisee is often paralleled in the supervisee-client relationship. Often, the therapeutic relationship will not progress until the conflicts in the supervisory relationship are resolved. In contrast, when the supervisor and supervisee can maintain a collaborative relationship, the supervisee is better able to enact a working alliance with the client as well.

Finally, same-gender supervisor-supervisee relationships may be preferable at times, although they are not essential for supervisees to develop their own professional identities and sense of efficacy. Similarly, people of color may feel more personal safety and fewer performance demands with a supervisor of the same race, especially as they begin their training or in the event that their experiences with individuals of other races have been disempowering. However, positive interpersonal processes are possible with supervisors of any gender, race, or theoretical orientation. Issues of similarity or difference relate only to supervisors' initial *ascribed credibility*. Supervisors of all cultural backgrounds can obtain *achieved credibility* by their sensitivity to the familial and cultural backgrounds of both the supervisee and the supervisee's clients and to the ways in which these may impact the therapeutic process. As we have seen, it may take several years before therapists can allow themselves to have as significant an impact on clients as possible and to fully possess their own personal power; this is a gradual developmental process. If beginning therapists do not progress along this feeling-of-adequacy dimension as they move through their training, they should discuss this issue with their supervisors and/or seek treatment for themselves.

Closing

Therapists will find that, as in individual therapy, working with the process dimension is also the central focus in marital counseling, group therapy, and family therapy. The interpersonal patterns and relational conflicts between couples, family members, and group members similarly begin to involve the therapist. As in individual therapy, process comments that describe the current interaction with the therapist, or what is transpiring between group or family members, can be used to identify and change these maladaptive relational patterns. In couple counseling, for example, the therapist may repeatedly make explicit observations about the way in which the couple is interacting or talking together.

Therapist:
> You two are arguing about who takes out the garbage, but it seems like the issue is really about who has the power to make decisions or the right to tell the other what to do. I'm wondering if it would be helpful for us to talk about that issue more directly?

Finally, the underlying assumption in all interpersonal/dynamic therapies is that the therapeutic relationship will come to resemble other prototypic relationships in the client's life. Clients' conflicts will emerge in the therapeutic relationship, especially if the therapist tracks the process dimension and attends to the reactions that clients elicit from the therapist and others. Therapists do not need to artificially construct situations to recreate clients' conflicts and make them accessible for treatment. In other words, therapists can respond to clients in genuine ways, and do not need to contrive or manipulate events in order to strategically recreate clients' conflicts and thereby create "therapeutic experiences" for clients. This undermines the authenticity and integrity of the therapeutic relationship and diminishes the client's efficacy by taking control from the client; such treatment strategies arise from the therapist's agenda rather than from a collaborative working alliance. Commonly, this strategic process merely recapitulates the client's developmental history and the client is again forced to comply with others' covert control or attempt to decipher incongruent metacommunications. In contrast, if therapists are willing to work with the natural expression of the client's conflicts in the therapeutic relationship, they can offer the client a meaningful, real-life experience of trust, empowerment, and change.

Suggestions for Further Reading

1. Drawn from the film *Ordinary People* (directed by Robert Redford, 1980), a case conceptualization is provided in Chapter 9 of the Student Workbook.

2. Practical guidelines to help therapists learn how to metacommunicate and when to intervene with the process dimension are found in Chapter 15 of *Handbook of Interpersonal Psychotherapy,* edited by J. Anchin and D. Kiesler (New York: Pergamon, 1982); see especially pages 285–294.

3. S. Cashdan's *Object Relations Therapy* (New York: Norton, 1988) provides superb clinical illustrations of the interpersonal process, helps therapists recognize and respond to metacommunicative messages, and demonstrates how the therapeutic relationship can be used to effect change; see especially Chapters 3 and 5.

4. Useful illustrations of how therapists can integrate other theoretical modalities with the interpersonal process approach are provided by P. Wachtel

in Chapter 3 of *Handbook of Interpersonal Psychotherapy,* edited by J. Anchin and D. Kiesler (New York: Pergamon, 1982). Readers are also encouraged to examine two other texts by Wachtel that provide a practical integration of psychodynamic and cognitive-behavioral approaches, with effective case illustrations: *Psychoanalysis and Behavior Therapy: Toward an Integration* (New York: Basic Books, 1982) and *Therapeutic Communication: Principles and Effective Practice.* (New York: Guilford, 1993).

CHAPTER TEN

Working Through and Termination

Conceptual Overview

Working through and *termination* are two distinct phases of therapy. As clients continue to have a corrective emotional experience with the therapist, the task is to generalize this experience of change beyond the therapy setting. The conflicts that are being resolved in the therapeutic relationship must now be worked through in other relationships as well. The working-through phase of treatment is an exciting period of growth and change as clients try out with others the emotional relearning that has occurred with the therapist. As the therapist helps clients resolve with others the conflicts that they have already been able to resolve with the therapist, therapy evolves to a natural close. The termination phase provides one last opportunity to reexperience and resolve old conflicts, to internalize the helping relationship with the therapist, and to successfully terminate treatment.

Chapter Organization

This chapter has two sections. In the first section, we consider the working-through process. Following an overview of client change that places the working-through phase within the context of the overall course of therapy, we examine the details of the working-through process. Sample therapist-client dialogues illustrate how the emotional relearning that has occurred with the therapist can be generalized to the conflicts that the client is experiencing with others. We conclude this section by highlighting the client's transition from concerns about problems and conflicts to future plans that reflect life-enhancing aspirations and goals. This transition from conflict- to growth-oriented concerns is often marked by the emergence of the client's Dream.

In the second section, we discuss the critically important phase of termination. We examine specific guidelines for effecting successful terminations and common pitfalls that occur around termination.

Working Through

The Course of Client Change: An Overview

Beginning therapists usually lack a conceptual overview of how client change comes about. Because they have not worked with many clients or may not have experienced their own successful treatment, most beginning therapists do not have a sense of the ordered sequence in which change often occurs. In the previous chapter, we reviewed the sequence of therapist activities over the course of therapy. We begin this section by extending this framework for conceptualizing the course of therapy. To get an overview of the change process, we look at when and how clients tend to resolve their conflicts and adopt new, more adaptive responses with others. Clients often change in a predictable sequence of steps, and this unfolding course of change is reviewed from the beginning of therapy to the final working-through and termination phases.

For some clients, change begins as soon as they decide to enter treatment. That decision involves acknowledging that there really is a problem and that they are no longer going to deny or avoid the problem. It is also an acknowledgment that clients cannot resolve the problem alone and need help. Some clients recognize that their resolution to seek help is a healthy step forward and they feel good about making this decision. Sadly, others may experience it as a failure experience, evidence of their inadequacy, or as disloyalty to their family or a violation of cultural beliefs. When clients can allow themselves to feel good about the decision to seek help, or the therapist can help clients re-frame their need with more self-empathy, some initial relief of symptoms—such as anxiety or depression—may result. This internal commitment to seek help is a crucial first step.

Some changes in feeling and behavior may also occur during clients' initial sessions with the therapist. Clients are reassured when the therapist:

- Invites them to express their concerns directly and fully
- Listens intently with respect and concern for their distress
- Is able to enter their subjective worldview and grasp the core meaning that these concerns hold for them

As clients find that the therapist can understand their experience and respond compassionately to them, a working alliance is established that also reduces anxiety and depression. Such validation and support does not resolve most clients' conflicts, but it does engender hope and helps to relieve their distress. Taken together, these initial responses—which in fact are complex interventions that require skillful application—comprise the essence of crisis intervention treatment. They also form the bedrock of therapeutic effectiveness for counselors working within every theoretical orientation and treatment modality.

More substantial changes begin when the therapist succeeds in focusing clients inward on their own thoughts and feelings and away from their preoccupation with the problematic behavior of others. When clients examine their own internal and interpersonal reactions, they often recognize how their behavior contributes to an interpersonal conflict and may be able to change their participation in it. Adopting an internal focus for change is also an important way to re-frame how clients think about their problems. Redefining the conflict with the other person as, in part, an internal problem usually reveals a wider array of new and more adaptive alternatives that clients can adopt.

Next, as the therapist focuses clients inward on their own experience, the conflicted emotions that are central to most clients' problems will emerge. The pace of change accelerates as clients begin to express each sequential feeling in their affective constellations. Clients have the opportunity to resolve their problems when they stop defending against their core conflicted emotions, feel the safety to begin expressing them, and receive a more affirming response from the therapist than they have come to expect from others. Far-reaching changes are set in motion as clients begin to integrate their affective constellations through the therapist's understanding.

As clients' conflicted emotions emerge, the therapist is also better able to conceptualize clients' relational templates and the pathogenic beliefs that accompany them. The therapist then focuses treatment along these conceptual lines and clarifies how clients are reexperiencing the same relational patterns, faulty assumptions, and affective themes throughout the various issues and concerns they present. Some further change may occur as the therapist helps clients to identify when each of these three patterns occur in their current interaction, and to begin to recognize when and how they are occurring with others. This conceptual awareness may or may not lead to significant behavioral change, but it does facilitate the next and most significant point of change.

The same conflicts that originally led clients to seek therapy, and that they have been discussing with the therapist, will usually be reenacted in the therapeutic relationship, especially along the process dimension—that is, the way in which the therapist and client interact. This reenactment in the therapeutic relationship provides both an opportunity for therapy to fail—by recapitulating the client's conflict—and an opportunity for therapy to succeed—by resolving the conflict that has emerged in the therapeutic relationship. Treatment reaches a critical juncture when clients feel the therapist is responding in the same problematic manner as significant others have done in the past. This usually occurs in one of three ways.

1. *Transference:* At a point when clients are distressed or feeling vulnerable, they are likely to systematically misperceive what the therapist has said or done, in line with their relational templates.
2. *Client-induced countertransference:* The client's eliciting maneuvers, interpersonal coping style, or testing behavior elicits the same problematic response from the therapist that the client has received in the past.
3. *Therapist-induced countertransference:* The client activates the therapist's own issues or personal conflicts that recapitulate the client's conflict.

These types of reenactments predictably occur in most therapeutic relationships. However, most clients cannot make significant progress in resolving their conflicts with others until this reenactment with the therapist is resolved. The therapist and client do this by altering the old relational scenario that usually occurs with others and working out a better solution to the client's conflict in their real-life relationship (for example, the client is not controlled, dominated, and so forth, as she has come to expect from others). Only when the therapist and client can resolve this familiar but problematic scenario in their relationship does the client have *a real-life experience of change.* This experience of change, as opposed to insight, cognitive re-framing, or challenging faulty assumptions, is the pivotal step that determines whether enduring change can occur. One episode of such emotional relearning will not usually be sufficient, however, as clients repeatedly need to find that their maladaptive relational patterns continue to unfold differently with the therapist than they have with others.

At this point, the therapist begins in earnest to start generalizing the client's emotional relearning to other relationships. Let's now consider the working-through phase of therapy, in which client change with the therapist is generalized to others.

The Working-Through Process

When clients change, it begins in the therapeutic relationship. With just the therapist's encouragement and advice, some clients who function well can successfully adopt new behavior. When the new behavior is integrally linked to their generic conflict, however, most clients will need to practice this new response in the therapeutic relationship. Thus, therapists must encourage clients to try out new ways of responding in the therapy setting and reassure clients that they intend to greet clients' new responses in affirming ways, rather than in the problematic ways that clients have grown to expect. If therapists give this permission verbally *and* enact it behaviorally, clients will soon test the therapist and try out anxiety-arousing new responses (for example, by being more assertive with the therapist or asking the therapist to respond to a need or personal request). If the therapist passes the clients' test—responds affirmingly to the new behavior, rather than metaphorically recapitulating unwanted but familiar responses—clients can resolve their conflicts and change.

This corrective emotional experience is a powerful agent for change, but, as we have noted, one trial learning is not sufficient to effect change. Most clients will need to reenact this and other relational themes over and over again in the therapeutic relationship. As a rule of thumb, the more they have been hurt, the more often they will need to reexperience a new and safer response that doesn't fit the old template. Once clients have seen that relationships can be different, however, the therapist must help clients generalize this new experience of change to other relationships beyond the therapy setting. Typically—but by no means always—clients try out new ways of responding with others in the following progression.

First, change occurs in the client's relationship with the therapist. Clients then change with acquaintances they do not know well or with others who are not especially important to them. This is often followed by change with supportive others who are important to the client, such as caring friends, teachers, or mentors. Next, clients often change old response patterns with the historical figures with whom the conflicts originally arose, such as caregivers and important family members. Finally, clients change their behavior with primary others with whom the conflict is currently being lived out, such as spouses or children. This final arena is the most challenging because the consequences have so much greater import.

During the working-through stage of therapy, some clients will rapidly assimilate the new ways of responding and readily apply them throughout their lives. Other clients, who have more pervasive conflicts or who have been traumatized or deprived more severely, will work through their conflicts more slowly. These clients must confront the same fears and expectations over and over again. Each time, the therapist must patiently clarify how, here again, the same relational patterns, affective themes, or faulty beliefs are being played out in this situation and help clients resolve this particular manifestation of their generic conflict. In this regard, the generic conflict will be repeatedly expressed in four areas of client functioning:

1. In the current interaction with the therapist
2. When reviewing crisis events in the client's life or the crisis events that originally prompted the client to seek treatment
3. In current relationships with friends and significant others in which the client's emotional problems are being aroused
4. In historical relationships with family members

For many therapists, working through is the most rewarding phase of treatment. The therapist often serves as a cheering squad who takes pleasure in clients' newly obtained mastery and encourages them as they successfully adopt new responses in progressively more challenging situations. At this late stage of therapy, the therapist can also become more actively involved in providing information and suggestions to help clients enact new behavior. Some therapists find that cognitive interventions and behavioral procedures for rehearsing alternative responses are especially helpful during this phase. Once the interpersonal process is enacting a corrective experience, the client becomes more responsive to other types of intervention, such as assertiveness training and parenting education; self-monitoring and self-instructional training; role-playing and other modeling techniques; and educational inputs or readings. The nature of the therapeutic relationship also changes during this period. As clients improve through the working-through period, they will increasingly perceive the therapist in more realistic terms. As this occurs—that is, once the transference projections and eliciting maneuvers have been jointly identified and are being resolved—the therapist can disclose more personal information to the client and enjoy further mutuality with the client. However, the therapist must still be prepared to respond to the reenactments, relational patterns, or resistance the client may present.

At this point in treatment, clients are living out a new and different relationship with the therapist—a relationship that is resolving rather than reenacting their maladaptive relational patterns. As clients find that new ways of relating with others are possible, they will begin—on their own—to try out these more adaptive responses with others in their lives. This is a critical juncture in treatment, and *therapists must actively help clients to anticipate and negotiate the successes and failures that are likely to follow.* Clients need to be informed that, although some individuals will respond positively to the clients' new changes, others will not. It will be exciting when others respond affirmingly to clients' changes, but therapists must also help clients anticipate that it may be deeply discouraging when they receive the same unwanted but familiar responses. Most clients will have experiences with others that both affirm the new ways of being and other experiences that painfully reenact the old relational patterns or expectations. Thus, we must examine what therapists can do to help clients anticipate these disappointments, which are an inevitable part of the change process, and how therapists can prepare clients to respond more effectively to such disappointments than they have in the past.

As clients find that they can safely respond in new ways with the therapist, and then make successful changes with certain individuals in their current lives, their expectations of change in some other relationships may become unrealistically high. For example, suppose an adult survivor of childhood abuse has been deeply understood and validated by an effective therapist. Following this success with the therapist, the client risks disclosing this shameladen secret to her husband, and then to her best friend, and they both respond affirmingly as well. Next, the client's lifelong wish that the abusing parent can now acknowledge the reality of what occurred long ago is evoked, or that the nonabusing parent, who could not be protective at the time, will now hear or believe the client. With expectations buoyed by successes with others, this client may address parents or other family members, only to encounter the same invalidation, scapegoating, and threats of ostracization that she received decades ago. Or, some clients may be profoundly discouraged to find that, unlike the therapist and some others in their lives, their marital partner cannot change or respond positively to this new behavior (for example, being more assertive or limit setting). Therapists must help clients examine their expectations of others *before* they try out new responses. Also, therapists must prepare clients by anticipating (1) how others are likely to respond and (2) how the clients are likely to respond (for example, what they have said and done in the past) if they receive the same unwanted responses they have received before. Thus, in order to manage these disappointments that are an inevitable part of the change process, therapists must prepare clients in the following ways:

1. By helping clients *realistically anticipate* how each new person is likely to respond to their changes. For example, if the client is a married woman, what will her husband probably say and do when she acts more assertively with him for the first time?

Client:

> He's likely to smile at me like I'm a cute child or something, and then just change the topic as if I didn't even say anything—or didn't even matter.

2. By helping clients spell out in detail how they are likely to feel and what they are likely to say and do if they receive the old unwanted response to their healthier new behavior.

Client:

> I think I'd feel ashamed, feel stupid for trying to stand up for myself, and just give up and withdraw inside—like I've always done before.

3. By role-playing and providing new, more adaptive responses that, after rehearsed in therapy, clients can use at this discouraging moment, instead of repeating what they have always done in such circumstances.

Client:

> I'm telling you something important right now and you are not taking me seriously. I don't like your smile, and I don't like feeling dismissed like this. Are you willing to listen seriously or should I stop now?

Most clients will continue to have experiences with others that reenact maladaptive relational patterns and confirm their pathogenic beliefs. *By helping clients walk through these three steps, however, the therapist can help them find ways of changing their own internal and interpersonal responses—even when others are unable to change.* For example, clients can learn that they are not to blame for, or deserving of, the way they were treated—even though family members still cannot affirm them. The married woman in our example learns that she can still advocate for herself. She does not have to comply with her husband's dismissal, and she can establish new relationships in which her limits or opinions can be respected, even though her spouse cannot do this. In this process, clients learn that they can change their own internal and interpersonal responses, even when significant others in their lives are not supportive.

Thus, the familiar but unwanted reactions that clients sometimes receive as they try out new ways of responding with others can be deeply disappointing, but they also provide an opportunity to further work through and resolve the conflict. Through each setback with a significant other, the therapist and client gain further awareness of the client's maladaptive relational patterns and deeper access to the painful feelings and faulty assumptions that accompany them. Working together, *therapists and clients can recognize how the client's old response patterns are being evoked again, and formulate new responses that clients might try out the next time they are in that situation.* More specifically, the therapist might explore each successive step with the client:

> What were you thinking and feeling inside when he dismissed you like that?
> Tell me what you did when he said that to you?
> What would you like to be able to say to him instead the next time this comes up?"

In this way, therapists can help clients recognize what they say and do to contribute to these prototypic problematic interactions, practice new responses that would change their role in the maladaptive relational pattern, and feel empowered by learning that they can alter the unwanted scenario by changing their own responses, even when the others cannot change. In addition, clients can begin to establish new, more affirming relationships with others that encompass these important changes. In other words, *clients' resolutions do not rest on changes in the parent, spouse, or others, as most clients believe when they enter treatment, but on changing how they respond to themselves and others in current interactions.*

It is not always easy for therapists to provide a corrective emotional experience, however. Although clients are improving during the working-through period, they will still be struggling with their generic conflicts. Eliciting behavior, testing behavior, and transference distortions that slot the therapist into the old relational templates will still have to be negotiated. Especially when clients are distressed or feeling vulnerable, therapists need to work through the distortions that stem from the old schemas, such as accusations that the therapist "doesn't really care about me," "isn't strong enough to handle me," "feels burdened by my needs," "will take advantage of me," and so forth.

Furthermore, most clients will become discouraged at times by the repetitious working-through process. Seeing that the same old conflict is confronting them again, clients may feel that nothing has changed, that it is futile to keep trying, and that they should consider dropping out of therapy. Therapists can make no guarantees of change, of course, and they cannot assume responsibility for the clients' motivation to continue. At these critical junctures, however, therapists must actively reach out and extend themselves to their clients. Only when therapists have grasped the profound influence of schemas or templates in shaping our experience of "reality," can therapists be empathic with the clients' discouragement. Although clients may realize it intellectually, experientially it really does seem to them as if this is the only way that relationships can be.

Therapists must hold steadfast that there can be another way—some relationships can be different some of the time. Therapists can point out possible instances where such other ways have occurred and convey that they remain committed to working with the client, even though they appreciate how frustrated or discouraged the client is feeling right now. It is therapists' belief in the potential for change, and their personal commitment to helping this particular client change, that pulls the client through these expectable crisis points in the working-through phase.

As an illustration of the working-through phase, we now consider a case example in which the client changes first in her relationship with the therapist and then, with the therapist's assistance, is able to extend that change to other conflict areas in her life. The episode in which clients' change first emerges—the *critical incident*—may be a compelling, transference-laden conflict. More likely, however, as in this example, it is a subtle aspect of the interaction with the therapist that could easily go unnoticed unless the therapist, informed by working hypotheses, has been prepared to look for it.

Our case example concerns Tracy, who habitually responded to others in a compliant, pleasing manner. Tracy usually went along with others—acting as if her wishes and feelings didn't matter. Tracy said that she didn't believe that others would be very interested in listening to her or doing what she wanted to do. Her therapist linked Tracy's compliant behavior to her presenting symptom of crying spells and went on to explore Tracy's low self-esteem and feelings of worthlessness.

In addition, the therapist encouraged Tracy to respond differently in their relationship. The therapist behaviorally demonstrated that she cared about what Tracy had to say and invited Tracy to express her own opinions in therapy—even if that meant disagreeing with the therapist. Two weeks later, Tracy risked this significant new behavior in the therapeutic relationship:

Tracy:
You're right. I guess I never have felt very good about myself.
Therapist:
From other things you've said, it seems to me that your mother just wasn't there for you as a child. You didn't get the support you needed from her. Maybe we should explore that further.
Tracy:
No, I don't really want to talk about her. I remember her more fondly, but my father was pretty harsh with us.
Therapist:
You just said no, you don't want to talk about your mother, and that I was wrong about her.
Tracy:
I'm sorry; sure we can talk about her. What do you want to know?
Therapist:
Forget your mother; you just disagreed with me! You just told me you didn't want to do what I wanted to do and suggested what you thought would be better. That's great!
Tracy:
What?
Therapist:
You just did what we have been talking about. You expressed your own opinion, said what you thought, and disagreed with me. I am so happy for you!
Tracy:
Aren't you mad at me? Didn't that hurt your feelings?
Therapist:
Oh no, we can disagree and still be close. That was a brave thing you just did.
Tracy:
So . . . I guess what you're saying is that maybe it is OK for me to say what I think sometimes?
Therapist:
Sure it is. I care about what you think. I want to know what you feel, and I want to do things your way, too.

Tracy:
   (tearing) But you're safe. It's easy to do that with you. You're not like other people.
Therapist:
   Yes, I am "safe," and it is easier to do things like that with me than with other people. But you have changed to be able to do that with me; you've grown. You couldn't do that before, you know.
Tracy:
   Yeah, that's right.
Therapist:
   And if you can value yourself enough to express what you think in here with me, then you can begin to do that out there with other people as well.
Tracy:
   Do you really think so?
Therapist:
   Yes, I'm sure you can. If you can do it with me, you can do it with them.
Tracy:
   Oh, I would love to be able to do this with my boyfriend.
Therapist:
   Tell me about how this usually plays out with your boyfriend.
Tracy:
   (describes the usual scenario)
Therapist:
   OK, what do you want to be able to say to him instead?
Tracy:
   (describes an alternative scenario)
Therapist:
   Would it be helpful to you if we role-play that together? We might be able to work out together where the problems might develop if you try that with him.

Why is the therapist so excited about such a seemingly small change? Is this single manifestation of a new behavior all that is necessary for Tracy to resolve her problems? No, but this highly significant event brings her to the working-through phase of therapy. Within the safe orbit of the therapeutic relationship, Tracy was able to try out a new behavior that taps into her generic conflict and arouses intense anxiety. For Tracy, adopting such a seemingly insignificant self-assertive behavior arouses lifelong feelings of worthlessness. At the same time, she is discarding the primary means she has developed to protect herself from these painful feelings—that is, to "move toward" and go along with others. In order to resolve her conflict, Tracy will have to repeatedly experience the same sequence:

1. Her old relational pattern—having to go along—is aroused.
2. She tries out a new and more adaptive response with the therapist, which is to assert her own preference.
3. She receives a different and more satisfying response from the therapist: The therapist affirms Tracy's initiative.

4. She generalizes the new behavior to others beyond the therapy setting by considering how she typically behaves with significant others in her life and, specifically, what she might want to say and do instead the next time she is in this situation with her boyfriend.

This repetitious reworking of the new behavior and the old conflict is called *working through*.

When Tracy returned for her next session, she made no mention of this incident with the therapist and said nothing about what went on with her boyfriend. Recognizing this as resistance, the therapist waited for the opportunity to acknowledge what had occurred between them the week before and to reassure Tracy that she was supportive of her new assertive response.

Therapist:
We haven't talked yet today about the important new way that you responded to me last week. Did you have any thoughts about that during the week?
Tracy:
No, not really.
Therapist:
Fine, but I just wanted to say again how happy I was for you. It felt very good to me to see you express your own opinion and be able to disagree with me.

Evidently, this reassurance must have given Tracy the permission she needed. The following week, Tracy was able to adopt the same type of assertive behavior with her boyfriend for the first time. This successful experience, in turn, encouraged her to confront the same issue in another more threatening relationship—with her father.

Tracy:
(beaming) Guess what! I did it! I've been dying to tell you all week!
Therapist:
Ha! What did you do?
Tracy:
I did what we talked about with my boyfriend. He was talking about something he thought we should do, and I disagreed with him and told him what I thought we should do instead. It felt great, and I don't think he minded either.
Therapist:
Good for you! You're on your way.
Tracy:
I can't believe how easy it was. I wish it could be that easy with my father. He's always putting me down when I say something.
Therapist:
Maybe you're ready to start changing your relationship with your father as well. How do you respond to him when he puts you down?
Tracy:
I get real quiet when he does that. I feel like crying, but I don't. I wish I could just tell him that I don't like it and I want him to stop.

Therapist:
> Yeah, confronting him at the time he does that and setting limits with him would certainly change your relationship. It would be saying to him—and to yourself—that you value who you are and what you have to say and you're not going to let him hurt you like that anymore.

Tracy:
> But I could never do that. I'd like to, but I just couldn't.

Therapist:
> You haven't been able to do that in the past, but you have been changing in here with me, and now with your boyfriend, too. So maybe you can do something different with your father as well. What holds you back from setting limits with him? What are you most afraid of?

Tracy:
> He wouldn't take me seriously; he'd just laugh at me.

Therapist:
> How would that make you feel?

Tracy:
> Worthless. *(begins crying)*

Therapist:
> So that's where that awful feeling comes from.

Tracy:
> Yeah. *(long pause)* It makes me mad, too.

Therapist:
> Of course it does! It hurts you very much when he diminishes you like that, and you have every right to be angry!

Tracy:
> It's not fair! I don't want to let him do that anymore. It's not good for me.

Therapist:
> That's right, it's not good for you, and you don't deserve it. Now that you can see that clearly, maybe you don't have to go along with it the way you used to.

Tracy:
> What can I do?

Therapist:
> What would you like to do the next time he does that?

Tracy:
> Well, maybe I can tell him to stop it—tell him that I just don't want to be treated like that anymore.

Therapist:
> I think that would be a strong and appropriate response on your part. But before you try it, let's explore further how that conversation might play out. Maybe we can identify the trouble spots and role-play some responses to help you through them.

In ever-widening circles such as these, clients confront and work through the same relational patterns in the different spheres of their lives.

Notice that the critical incident in this case was very subtle: When the therapist suggested that Tracy talk about her mother, Tracy said she didn't want to.

Therapists could easily miss such slight shifts in behavior unless they have conceptualized the client's dynamics and hypothesized how the client's conflict might be reenacted in the therapeutic relationship. If Tracy's therapist hadn't *anticipated* that Tracy would tend to comply with her, and if the therapist hadn't been alert for any budding sign of self-direction or assertiveness from Tracy, the opportunity for change would most likely have been lost. That is, the therapist might have responded by focusing on the content (by saying, for example, "But it's important to look at your relationship with your mother because . . . ") rather than on the process (Tracy's self-assertion). In that case, Tracy would have complied with the therapist and dutifully talked about her relationship with her mother. The interpersonal process between Tracy and her therapist would then recapitulate her maladaptive relational pattern, and no change would result.

When working with a client who seems unable to change, the therapist may eventually grow discouraged and become critical toward, or disengaged from, the client. In those circumstances, the therapist should consider whether the therapeutic interaction is subtly reenacting the client's maladaptive relational patterns. That would explain why the client is unable to change, even though the therapist and client continue to talk about adopting more assertive behaviors with others, the therapist sincerely supports and encourages this, and the client learns useful skills that should facilitate the new behavior. Such subtle reenactments along the process dimension are predictable. If therapists utilize process notes (Appendix A) and generate working hypotheses (Appendix B), they can get an idea of how each particular client's conflicts may be reenacted in the therapeutic relationship. *By anticipating what to listen and watch for, therapists will be able to recognize such reenactments more quickly and respond to them more effectively.*

As emphasized above, clients will experience nonaffirming responses at times when they are generalizing change to relationships outside the therapy setting. As we will see later, for example, Tracy's father did not respond as well as her boyfriend did. Therapists can use this experience as an opportunity to help clients see that some relationships can be different and some can't—and that nonvalidating responses from certain people do not constitute evidence of the clients' lack of inherent worth. In this way, the working-through phase focuses on both interpersonal and intrapersonal change.

From Present Conflicts, through Family-of-Origin Work, and on to Future Plans

In the working-through phase, the therapeutic action is primarily in the present; the significant events are current interactions with the therapist that enact resolutions of maladaptive relational patterns. Simultaneously, these reparative events propel internal reworkings of clients' templates, which will be critical in assisting the clients to assimilate this new and more authentic way of relating. These corrective experiences are then generalized, as clients try out with others the new ways of relating they have found with the therapist, which are now also congruent with the new ways in which they perceive themselves and how relationships can be. As we have seen, clients will have both successes and disappointments; some individuals will welcome the clients' changes, whereas others continue along well-worn, problematic lines. As clients explore the potential for change in current relationships and come to terms more realistically with the possibilities and limits, two new subphases in the working-through process may emerge. First, clients, on their own, often start to look back in order to better understand the formative experiences that shaped these problems they are resolving. In the first part of this section, we examine therapeutic guidelines to help with this family-of-origin work. Second, as clients develop a more realistic narrative that helps them better understand their life experiences and how they have become who they are, they also begin to look ahead: to think more about what they want their life to be in the future; and to reformulate life plans to better fit the person they are becoming. This subphase, often ushered in by the emergence of the Dream, will be explored in the second part of this section.

Family-of-origin work   It can be liberating for clients to explore the familial interactions and developmental experiences that shaped their current conflicts. As they gain understanding of their childhood dilemmas, they become more accepting of the choices, compromises, and adaptations they have had to fashion in their lives. They also begin to feel empathy—both for themselves and for the personal limitations or difficult life circumstances that led their caregivers to respond in problematic ways. Clearly, this developmental perspective can be profoundly enriching. In most cases, however, it is not productive for therapists to lead clients back in time, make historical interpretations, or attempt links between current relational patterns and historical family relationships. Although the following comment may be accurate, few clients will find it meaningful.

Therapist:
> I keep hearing the same theme: that what you are complaining about with your spouse seems in some ways to be similar to what used to occur with your father.

In contrast, a mutual exploration of familial and developmental experiences can teach therapists a great deal about the corrective experiences that clients need in treatment, and help therapists to begin generating working hypotheses. Towards this different purpose, for example, the therapist might ask:
> What was it like to be a child in your family?
> Tell me what your parent would say and do when you were successful or acted stronger. What was the look on your mother's face when you showed her that spelling award?
> Tell me about your parents' marriage. Can you bring it to life for me?

Early in treatment, this information will inform therapists about the formative experiences that shaped the client's relational templates. This provides therapists with invaluable guidelines to begin formulating the kinds of responses that are likely to reenact problematic relational patterns for this client and the types of therapeutic responses that are more likely to be reparative for this client. Early in treatment, however, most clients will not be able to make meaningful bridges between current and past relationships from this developmental exploration. When the therapist highlights seemingly obvious connections, the relational patterns do not usually come alive for clients until they have experienced a corrective response to these old relational patterns with the therapist. Later in treatment, many clients will spontaneously lead the therapist back to developmental experiences. These client-initiated explorations, *which usually occur immediately after clients have a corrective emotional experience or have successfully adopted a new response with others,* continue throughout the working-through period. They are enlivening for clients and routinely lead to productive behavior change with others.

An important sequence of change is being emphasized here. In contrast to other counseling theories, the interpersonal process approach does not suggest that behavior change leads to insight or, conversely, that insight leads to behavior change. Although both of these change processes will occur for certain clients, a different mechanism of change is proposed here. *Both meaningful insight and sustainable behavior change follow or result from clients' new or reparative experience with the therapist.*

As clients make their own meaningful links between formative and current relationships, they often ask the therapist what they should do about their current relationships with family members. We now examine two broad guidelines to help clients with this family-of-origin work. First, attending to the internal focus for change described in Chapter 4, we see how clients must change how *they* respond to maladaptive relational patterns in their current interactions with family members. Second, we look at the concept of *grief work*, in which clients mourn and come to terms with what they have missed developmentally. Working in both of these areas results in an interpersonal and internal resolution of family dynamics.

Clients who struggle with more pervasive or long-standing problems usually have family-of-origin work to do. Some clients do not wish to discuss problems from the past with caregivers or family members, and therapists should not press clients to do so. However, some clients will want to acknowledge past conflicts with parents and other family members and try to resolve them. It is enormously gratifying—and a powerful impetus for change—when such rapprochements succeed. As we have been emphasizing, however, such current resolutions of historical problems often do not occur. Occasionally, caregivers have grown and changed but, in many cases—especially with more serious problems—the client will often receive the same invalidating, blaming, or hurtful responses as in childhood. This is especially painful when the client's unacknowledged motivation for talking with the parent is a wish to get the parental affirmation or protection that the client has always longed to receive. If the therapist has not prepared the client for this potential disappointment, the client may come to feel hopeless about change occurring in any relationship. Oftentimes, the client also will be breaking family rules by addressing conflicts with parents directly, and in many cases punitive homeostatic mechanisms will be set in motion. For example, when Tracy confronted her father about always putting her down, he again disparaged her—just as he had for the past 20 years. In addition, her mother threatened to disown her for being "disrespectful" and "ungrateful," and her depressed and obese sister telephoned and tried to make her feel guilty for "hurting Daddy" and "stirring up trouble"! Thus, the problem is not simply the parent's mistreating the child. Instead, more serious consequences—such as the self-hatred, bulimia, and dysthymia that Tracy suffered—result because the parent and extended family system typically make the adult offspring again feel responsible for or deserving of the parent's original and current mistreatment.

Clients usually feel hopeless and blame themselves when parental caregivers and the broader family system cannot change. Although these clients may go on to act helpless in other current relationships, in actuality they are not. Their interpersonal resolution does not rest on the parent changing, but on how they change and respond in current interactions with living parents or, equally important, in their ongoing internal relationship with deceased caregivers. In order for clients to make enduring changes in deep-seated relational conflicts, they must change their own responses in these prototypic interactions. For example, Tracy's father was too limited to be able to hear her concern and to talk with her about the problem. However, Tracy stopped going along with her father's disparagement of her. Instead, Tracy began setting limits with her father and bravely metacommunicated with him, by saying: "I keep asking you to 'Stop it' when you put me down, and you laugh at me and keep doing it. I can't make you stop, but I don't have to pretend it's OK anymore. It hurts and I don't like it." Although her father stayed in the same disparaging mode and again made fun of her, things changed profoundly for Tracy when she changed how she responded to him. As she was able to sustain this stronger stance toward him and the family system over the next few months, she felt less insecure and intimidated with them and others in her life than she had ever been. Her longstanding feelings of worthlessness and depression significantly improved, and her intermittent battles with bulimia largely disappeared. These far-reaching changes occurred for Tracy even though her father—and the broader family system—could not change, because her attitude toward herself had changed.

In this way, clients make progress by changing how they respond in current interactions with significant others in their lives. Clients also change by doing grief work—coming to terms with the feelings left in them by hurtful parent-child interactions. Clients do not need to discuss with caregivers problems in historical relationships. However, they do need to stop disavowing and acknowledge to themselves what was legitimately wrong, stop their own participation in any ongoing mistreatment, and grieve for what they have missed. Regardless of whether formative attachment figures can change—and whether they are living or deceased—clients can resolve these conflicts through their own internal work.

Testing the reality of new possibilities with family members is always important, but it may or may not lead to improved relations. Often, aging caregivers cannot change, so therapists need to prepare clients that when they introduce healthier new behavior into their family systems it will often be met with great resistance. Even if the family system has improved, however, grown offspring will still function better in current relationships once they can realistically acknowledge what was wrong in the past and mourn what they missed developmentally. The same model applies again: The therapist provides a supportive holding environment that allows clients to come to terms with the sadness, anger, shame, and other feelings that were too threatening or unacceptable to contain before. To succeed in this work, however, therapists are encouraged to remember that clients will be safe to discuss the "bad" part of the caregiver (or spouse) only after knowing that therapists also appreciate and affirm the "good" parts of these relationships as well.

In another way, the therapist must also help clients come to terms with both the good news and the bad news about their current situations. The good news is that, through grief work, clients can disconfirm the pathogenic belief that they were in some way to blame for their rejection, parentification, and so forth. The bad news is that clients' developmental needs—such as the need for secure attachment ties—were not adequately met then, and these same developmental needs cannot now be met in current adult relationships (for example, with a spouse, friend, or therapist). Only after such losses or deprivation can be acknowledged as real, that is, mourned rather than disavowed, can clients go on for the first time to get adult versions of their needs met in current relationships. As long as developmental needs are operating unacknowledged, clients are unable to get legitimate adult needs met in current relationships. Finally, once clients have accepted both the "good news and the bad news" associated with their caregivers, they are then free to accept those same characteristics in themselves. That is, by giving themselves permission to acknowledge both their "good" and "bad" self perceptions, their own perfectionism will subside, leaving a greater sense of self-worth.

Two countertransference propensities can prevent therapists from helping clients achieve resolution of such family-of-origin work. First, because of their own splitting defenses, therapists sometimes do not have the breadth to help clients accept both what was good and what was problematic in their development. Therapists' countertransference propensities lead them to make one of two errors. Some therapists characteristically want to downplay the extent and continuing influence of painful experiences.

Therapist:
That was then. You need to forget about what happened in the past. Let's try to do something about the problems you're having now.

In contrast to this denial, other therapists only want to blame the caregivers and ignore the strengths and contributions that were present in the relationship.

Therapist:
Your father was stupid and sick to do that to you and your sister!

Instead, therapists need to affirm the full reality of what went wrong in formative relations, while appreciating the (internalized) child's need to maintain whatever ties can be preserved. Except in extreme circumstances, therapists should discourage the client from ending all contact with caregivers. When a client breaks off all communication with family members, splitting defenses are usually operating and lasting change is unlikely. In such cases, the therapist will be idealized and clients will be compelled to recreate the split-off, bad side of their conflicts in other relationships or to make themselves "bad." Therapists will be far more effective—and clients will be better able to work with this guilt-inducing material—when therapists can acknowledge the bad news without taking the parental figure away.

Therapist:
I can see how much it hurt you when he did that.

VERSUS

Therapist:
I can't believe he would do that to you. He's a Nazi!

The second countertransference propensity that interferes with family-of-origin work is that therapists often want to bypass the work of mourning. When clients acknowledge the ways in which they were hurt or missed what they needed, painful feelings are aroused. Therapists may find it difficult to help clients come to terms with those feelings. Most clients do not want to end the hope—or fantasy—of getting what they missed; some may even become angry when therapists acknowledge these limitations. Therapists' reluctance to address these issues will be reinforced if, as routinely occurs, such material arouses therapists' own unfinished family-of-origin and grief work. However, if the therapist tries to bypass the work of mourning and move too quickly to a problem-solving approach—for example, by looking for substitute supportive relationships— clients will not be able to internalize much from these practical interventions. Only after original losses have been acknowledged and integrated (that is, grieved for in a supportive holding environment with the therapist) can clients open up their adult emotional needs and feel sustained by new relationships in a way they have not been able to risk before. As clients improve in these very significant ways, they become able to leave the past behind, to turn the page, and to begin looking ahead to a different future.

Next, we examine the emergence of the Dream and the second subphase of the working-through period.

The Dream   Clients make behavioral changes and feel better as they work through their conflicts. As they improve, the focus of therapy moves away from developmental conflicts and current interpersonal problems. The therapeutic dialogue moves toward plans for the future and existential issues regarding personal choices: How can clients create more meaning in their lives, live more authentically, and make the most of the time and relationships they have? This exciting transition from psychopathology to health is often signaled by the emergence of the Dream.

In *The Seasons of a Man's Life,* Levinson (1978) describes the profound influence of the Dream in shaping the structure of adult life and the course of personality development. As Levinson points out, the Dream does not refer to casual waking or sleeping dreams but, in the largest sense, to the kind of life that one wants to lead. At first, the Dream may be poorly articulated and tenuously connected to reality, but it holds imagined possibilities of self in the adult world that generate vitality. It is the central issue in Martin Luther King's historic "I Have a Dream" speech or in Delmore Schwartz's story, "In Dreams Begin Responsibilities."

The Dream has roots in the grandiose and unrealistic hero fantasies of adolescence, but it is more than that. In early adulthood, the Dream still has the quality of a vision—of who and how we would like to be in the world. In this way, the Dream is inspiring and sustaining for the individual, even though it may be mundane to others. For example, individuals have Dreams to be a good husband or wife, a respected community leader, an ethical attorney, a skilled craftsperson, a successful but honest businessperson, an artist, or a spiritual leader. *If the individual is to have purpose and a sense of being alive, occupational and marital choices must incorporate some aspects of the Dream. When the Dream has been abandoned or set aside, life will not be infused with vitality and meaning, even though the person may be successful.*

Up until the working-through phase of therapy, therapists are usually responding to the despair of broken dreams, the disillusionment of unfulfilling dreams, and the cynicism of abandoned dreams. In many cases, the life-infusing Dream that Levinson has articulated has not yet been addressed in therapy. In the working-through phase, however, clients are resolving their conflicts and emerging as healthier individuals. At this point, therapists must help clients to articulate and renew their Dream and to find ways of incorporating aspects of their Dream in their everyday lives. To achieve this, therapists must first be able to differentiate the Dream from the broken dreams that underlie clients' presenting symptoms. These broken dreams often reflect failed attempts to rise above their conflicts and become special, as discussed in Chapter 7.

In the process of working through, therapists repeatedly help clients recognize their compensatory strivings to be special and to rise above their conflicts by pleasing others (moving toward), achieving success and power (moving against), or becoming safely aloof and cynically superior (moving away). Therapists help clients progressively relinquish these inflexible coping styles and adopt a wider interpersonal range. In doing so, clients stop blocking the expression of their generic conflict. This presents the opportunity for clients' conflicts to be expressed and resolved within the therapeutic relationship. However, therapists must anticipate that when clients relinquish these attempts to rise above their conflicts, they may experience a sense of failure or loss of self-esteem. Thus, one important component of helping clients resolve their conflicts is to replace these defensive, grandiose strivings with clients' own attainable, yet sustaining, Dream.

Arthur Miller's (1949) Pulitzer Prize–winning play, *Death of a Salesman*, poignantly illustrates how the neurotic striving to rise above conflicts inevitably fails, how it can lock clients in their generic conflicts for a lifetime, and how it stifles the Dream. The protagonist, Willy Loman, has adopted a moving-toward interpersonal style to cope with his profound feelings of inadequacy and shame. With his smile and his freshly shined shoes, Willy pathetically strives not just to be liked but to be well-liked. In a brilliant exposition of multigenerational family roles and relationships, Miller shows how Willy's grandiose strivings to rise above his shame-based conflicts by being well-liked are also acted out through his oldest son, Biff. Willy exaggerates Biff's accomplishments as magnificent, excuses or ignores his shortcomings, and never responds to the reality of his son's actual feelings, needs, or wishes.

Near the end of the play, both father's and son's defensive strivings to rise above fail. Willy is fired from his job—a crushing humiliation that shatters his myth of being well-liked and his lifelong coping strategy of winning approval. That same day, Biff fails to secure an important business opportunity that his father has encouraged. With this setback, Biff realizes that he has been living out his father's myth that he would become a fabulously successful entrepreneur. In the closing scene, father and son respond very differently to their respective crises.

Biff is able to relinquish the unrealistic script of superachievement and success that his father has demanded of him. In the last scene, Biff begs his father to release him from this grandiose role, to "speak the truth" in the family for the first time, to acknowledge that he is just another regular guy—nothing more, nothing less—and to let that be good enough. However, Willy cannot relinquish his own rising above defense and, therefore, cannot grant Biff's plea to "let me off the hook." Instead, Willy alternately rejects Biff and idealizes him as magnificent.

Biff finally recognizes the maladaptive relational pattern that he and his father have been enacting all of their lives. By addressing it with his father, Biff is able to change his part of their conflict. Although this honest communication does not lead to an interpersonal resolution with Willy, it does lead to an internal resolution for Biff. Soon afterwards, Biff decides to leave home. He says that he now knows who he is and what he wants to do with his life. He is going to pursue his own Dream: to leave the city, move out west, and work outdoors where he can feel the sun on his back and smell the grass. Whereas Biff could relinquish the neurotic strivings to be special and allow his own attainable Dream to emerge, Willy could not.

In one last attempt to change their relationship and have a more authentic connection, Biff reaches out and puts his arms around his father. Horrified, Willy recoils in fear and disgust. Although his lifelong obsession has been to win the approval and respect of others, he rigidly rejects exactly what he has always wanted from his son. To accept the love that Biff was offering—to get what he wants—would reveal Willy's own generic conflict—his unmet needs to be respected and wanted, his profound sense of being unlovable, and his intense feelings of inadequacy and shame. With no understanding of what these disavowed feelings are about, and no supportive relationship to help him contain this flooding, his unacceptable feelings would be overwhelming. Thus, Willy is now trapped in his conflict: He is no longer able to maintain his own (or Biff's) defensive strivings to rise above, and he is unable to let his true feelings and concerns emerge. Desperately bound in that unresolvable conflict, Willy's only solution is to end his own life.

As clients' neurotic strivings to be special and rise above their conflicts fail, their generic conflicts are revealed. Without a holding environment to help them contain their feelings, some individuals—like Willy—feel hopeless and see suicide as the only way out. In contrast, if the therapist can provide a different and more satisfying response to the conflict, clients can begin to come to terms with these feelings in a new way and change. As clients begin to improve during the working-through phase, the Dream will often emerge.

In order to help clients change, therapists must support and encourage the Dream. They can do this in many ways. First, therapists must listen for the issues that genuinely enliven clients, engage their intrinsic interest, and bring them pleasure. By attending closely to what clients really want to do and inquiring about what feels right for them, therapists can help clients discover and articulate their Dream. Some clients' Dream is completely undeveloped because their caregivers never accepted their feelings or responded to their interests. These clients, with a less developed sense of self, are unaware not only of what they want to do in life but, more basically, even of what they like or dislike. Therapists can help these clients discover and successively articulate their feelings, interests, values, and ultimately their Dream by repeatedly:

- Asking clients to attend to what they are experiencing at that moment
- Orienting clients to listen to themselves and simply be aware of what they want to do in different situations—even if they cannot act on their wishes (for example, to realize "I'm not enjoying this right now.")
- Being interested in and entering into clients' subjective experience
- Calling attention to, and trying to participate in, what seems to interest or hold meaning for clients
- Acknowledging and taking overt pleasure in what clients do well
- Encouraging clients to act on their own feelings or preferences when possible (for example, even something as simple as saying, "I'm sleepy. I'm going to take a nap.")

Although many clients have a more developed self than this, some will need to begin attending to their own internal experience in these very basic ways. It is rewarding for the therapist to help clients differentiate a self by responding—perhaps for the first time in their lives—to their own feelings and interests.

As clients begin to experience their inner life more fully, and the therapist responds affirmingly, their Dream will start to emerge. At this point, the therapist can do several things to help clients bring aspects of their Dream into their current lives. For example, the therapist must provide a practicing sphere in which clients can discuss and explore various possibilities of their Dream without having to become committed prematurely to any action. The therapist can also help clients tailor their Dream to reality and find ways to express aspects of their Dream in their everyday lives. The therapist can also help clients identify training or education routes that will facilitate realization of their Dream, and must give clients permission to develop close relationships with others who can act as mentors in their chosen fields and pursuits.

Finally, clients will confront the same generic conflicts in exploring their Dream that they have experienced in other aspects of their lives. Thus, the therapist will also have to help clients work through their old conflicts that are aroused by pursuing their Dream. For example, the therapist will have to help clients differentiate between their Dream and their interpersonal coping style and defensive strivings to rise above (for example, strivings to feel secure by being loved by everyone; to gain a sense of personal adequacy by becoming wealthy and powerful; or to escape criticism or rejection by being safely aloof and superior). *Progress will not be simple or unimpeded for many clients, however, because pursuing what they really want to do or getting what they want may threaten object ties and arouse separation guilt and separation anxiety.* However, as clients successively work through these conflicts and find that they do have a right to their own life, they can follow through on realistic plans for attaining aspects of their Dream. With this success, treatment begins evolving to a natural conclusion.

Termination

Termination is an important and distinct phase of therapy that needs to be negotiated thoughtfully. Ending the therapeutic relationship will almost always be of great significance to clients—for some, this may be the first positive ending of a relationship they have ever experienced. The way in which this separation experience is resolved is so important that it often influences how well clients will be able to resolve future losses and endings in their lives. It also helps determine whether clients leave therapy with a greater sense of self-efficacy and the ability to successfully manage their own lives more independently. Most beginning therapists greatly underestimate how powerful an experience termination is for clients. It holds the potential either to undo or to confirm and extend the changes that have come about in therapy. Therefore, therapists must be prepared to utilize the further potential for change that becomes available as therapists and clients prepare to end their relationship.

Early in their training, therapists often ask, "How will I know when it's time to end therapy?" It is time to end when clients have achieved significant relief from their symptoms, can respond more flexibly in current situations rather than slotting all experience into predisposed categories, and have begun to take steps toward promising new directions in their lives. Therapists know that clients are ready to terminate when they have converging reports of client change from three different sources: (1) when clients report that they consistently feel better and can respond in more adaptive ways to old conflict situations; (2) when clients can consistently respond to the therapist in new ways that expand their old coping styles and do not reenact their maladaptive relational patterns; and (3) when significant others in clients' lives give them feedback that they are different or make comments such as, "You never used to do that before." With this convergence of perceptions that important changes have occurred, it is time for therapy to end. Before going on to discuss successful termination, we must distinguish between two types of endings: natural and unnatural.

The termination sequence just described is a natural ending because clients' work is finished. One of the therapist's primary goals in these natural endings is to affirm both sides of clients' feelings about ending: to take pleasure in clients' independence and actively support their movement out on their own; but also to let clients know that the therapist will accept their need for help or contact in the future. In contrast to clients' old templates, therapists can clarify that they will not be disappointed or burdened if clients want to send a letter to stay in touch, call to share important life events such as marriages and births, or recontact the therapist for help if they have problems in the future. Many clients did not have permission to be both related and individuated at the same time. Leaving (for example, pursuing their own interests, goals, and relationships) often signified a loss of emotional support and connectedness, while staying meant not having their own voice, views, or feelings distinct from those of the people they were with. The termination phase provides these clients with an important opportunity to further resolve the separateness-relatedness dialectic and can be a potent corrective emotional experience. Assured of the therapist's support for both sides of their feelings, clients can further internalize the therapist and keep all that she has offered, solidify their own sense of self and personal efficacy, and successfully end the relationship.

Unfortunately, most graduate-student therapists do not have the satisfaction of seeing many cases evolve to a natural close. Instead, therapy ends before the client is finished because the graduate student must move on to another placement, the school year ends, or the training clinic provides only time-limited services. Sometimes the therapist and client have made substantial gains at the point when external constraints demand termination; at other times, the treatment process may be only in midstream. In either case, significant therapeutic gains can still be made. To accomplish this more difficult task, however, it is essential that the therapist and client address and work through the complex issues that an unnatural or externally imposed ending arouses. Because graduate-student therapists often have to initiate terminations before clients are ready, we will focus on unnatural endings in the discussion that follows. *It is necessary to examine unnatural or imposed endings so closely because they routinely evoke blaming and distancing reactions in the client, which, in turn, activate guilt and other countertransference reactions in the therapist that can undo the significant gains that have been made.*

Accepting that the Relationship Must End

It is often difficult for both the therapist and the client to end their relationship. In many cases, bringing up the topic of termination will arouse emotional responses that are far more intense and difficult than either the therapist or the client has anticipated. As a result, therapists and clients regularly collude to deny the reality of the impending ending and to avoid the difficult feelings it arouses for both of them. When therapists and clients stumble in this way and do not address the ending squarely, their interpersonal process will often recapitulate clients' generic conflicts and prevent clients from resolving their conflicts as fully as they could have. Thus, *the single most important guideline for negotiating a successful termination is to unambiguously acknowledge the reality of the ending.*

Therapist:
> We have three more sessions left before the school year ends and we have to stop. Let's talk about what this ending is going to mean for both of us.

The therapist and client can then discuss all of the client's positive and negative reactions about the ending, especially the client's emotional reactions toward the therapist. Although the therapist and client may both want to avoid this topic, the therapist cannot let that happen. Once the therapist and client mutually decide that the client is ready to terminate, or that the therapist must terminate because of outside constraints, they need to establish the specific date for the final session (for example, Therapist: "Our last session will be in two weeks, on Thursday, December 9th, at 4:00"). Because the therapist and client need time to work through their ending, the final session should be scheduled at least two weeks hence.

In order to have a successful ending, the therapist and client mutually establish a final date and then explore together the client's reactions to ending their relationship. In some cases, it will be difficult for the therapist to set a specific termination date and discuss the ending because this arouses the therapist's own conflicts over this ending—or more commonly, endings in general. In addition to the therapist's own countertransference propensities, which are especially likely to be evoked around terminations, setting a clear termination date will often bring up the client's presenting symptoms and generic conflicts again. In order to further resolve their conflicts, some clients may need to work through again, in the termination phase, aspects of the same issues that have been dealt with before. In other words, when termination becomes a reality, some clients will temporarily retreat from or undo the changes they have made. For all of these reasons, therapists need to acknowledge the termination forthrightly, count down the final sessions, and repeatedly invite clients to discuss their reactions to the termination:

> After today, we will have only three sessions left. How is it for you to hear me say that?
> We have only two sessions left now, and I think it's important that we talk together about our ending. What comes up for you when I acknowledge that soon we are going to have to stop working together?
> Next week will be our last session. We've accomplished some things together, but problems remain and our work is unfinished. What do you find yourself thinking about our relationship and the time we have spent together?
> This is the last time we are going to be able to meet. I'm feeling sad about that, and wondering what kind of feelings you're having.

This type of countdown precludes any ambiguity that clients may have about the ending. For example, it will prevent the client from asking, at the end of the last session, "Will I be seeing you next week?" This approach also keeps the therapist from acting out and avoiding the separation as well—for example, by saying at the end of the last session, "Oh, I see our time is up. Well, I guess that's it." When therapists do not address the termination directly they are often acting out their own separation anxieties or guilt over letting someone else down. In particular, their own unresolved feelings from past problematic endings in their own lives—parental divorces, deaths, and other unanticipated or unwanted leavings—are being reevoked. Such countertransference reactions become even more problematic when they often dovetail with clients' issues. For example, many clients are also struggling with their own unresolved feelings over past painful endings, and others are avoiding the topic because they sense it is difficult for the therapist. Unresolved conflicts over ending relationships may be the most common type of countertransference. When this occurs, therapists can reassure themselves that this is a common and understandable problem to have, but they also need to do something about it by seeking consultation from a supervisor or colleague.

Because both clients' transference distortions and therapists' countertransference reactions are especially likely to occur around unnatural terminations and at the end of time-limited treatments, we must explore these issues further. Both therapists and clients often have experienced painful endings with significant others as "just happening to them." In many cases, they

- Were not prepared in advance for the separation
- Did not understand when or why this particular ending was occurring
- Were not able to participate in the leave-taking by discussing it with the departing person or by saying good-bye

Such problematic endings have left many therapists and clients feeling powerless in regard to some of the most important experiences in their lives. In contrast, the approach suggested here enhances clients' efficacy by allowing them to be active, informed participants in the ending. Clients know when the ending is occurring; have the safety to express their sadness, anger, or disappointment about it; and have the opportunity to say good-bye. Thus, in order to achieve a successful leave-taking the therapist must: (1) acknowledge whatever therapeutic work remains unfinished; (2) remain affirming or nondefensive in the face of the client's disappointment, angry feelings, or accusations of betrayal and other guilt-inducing responses that may occur; and (3) explore how this termination may be evoking other unwanted or unexpected endings in the client's past. Why is this so important? Especially with unnatural endings, it often will activate old templates and seem to the client that this relationship has just failed in the same way that others have in the past. Let's elaborate this expectable problem more fully.

As the old saying goes, children are supposed to grow up and put their feet under their own table, but it's hard to get up from the table when you're still hungry. The emotional deprivation that many clients bring to therapy becomes activated when treatment ends prematurely or before they feel ready to stop. Clients' relational templates may be such that their anger and disappointment—or even their sense of abandonment and betrayal over unmet attachment needs—will be directed toward the therapist, no matter how clearly the treatment parameters were explained at the beginning of treatment. Clients' guilt-inducing accusations can be difficult to hear:

You're just like my mother. You're abandoning me, too.
Our relationship never meant anything to me, anyway.
Nothing changed for me in therapy. This was just a waste of time.

The success of treatment in these cases depends in part on therapists' ability to remain nondefensive and tolerate clients' protests—rather than feel guilty, become defensive, and make ineffective attempts to talk clients out of their feelings:

Therapist:
> But you *have* changed. You're not as depressed as you used to be, and now you can do _____, and before you couldn't do that.

Instead of these futile attempts, therapists first need to acknowledge clients' feelings:

Therapist:
> You're really angry at me as our relationship is ending, and I respect those important feelings. Right now, it just seems like nothing meaningful has occurred and that nothing has gotten better for you.

By accepting clients' angry protests in these ways—without agreeing that their factual content is correct—therapists give clients a response that doesn't match their relational templates. In turn, this affirmation of their current frustration or disappointment often allows clients to connect their strong emotional reactions to the developmental figures who originally disappointed or left them. Thus, *what therapists often find hardest about unnatural endings—and about time-limited therapy—is managing their own guilt or defensiveness in the face of clients' anger over the ending.* By drawing on the guidelines outlined here, however, therapists can help clients see how this ending is in fact different in some important ways from other unwanted endings in the clients' past, even though it feels to them as if the same maladaptive relational pattern is occurring. Therapists can help make the termination of therapy different from past problematic endings—and, in doing so, help resolve this unfinished grief work from the past in the following ways:

- Talking with clients about the ending in advance
- Inviting clients to share their angry, disappointed, or sad feelings and being able to accept those feelings without becoming defensive
- Talking with clients about the meaning the termination holds for them
- Talking about the meaning this relationship has held for the therapists and sharing their feelings about ending this relationship
- Validating clients' experience by acknowledging the ways that this ending may evoke other painful or unwanted endings with others
- Ensuring that therapists and clients say good-bye to each other

In other past endings that have been problematic, clients were not able to have such experiences with the departing person. Although most clients cannot recognize these differences initially, the therapist offers clients an opportunity to resolve long-standing problems by helping them differentiate this type of mutually acknowledged ending from the incomplete or unsatisfying endings with which they have had to cope in the past. In this way, successfully managing the termination becomes the prototype for coping effectively with future losses.

In the face of clients' withdrawal or blaming accusations that their termination is just like prior losses and disappointments, therapists, too, may lose sight of the real differences that exist between this termination and clients' problematic past endings. In that case, therapists may accept the clients' blame, feel guilty, and avoid the ending—or invalidate the clients' sad, disappointed, or angry feelings by trying to talk clients out of them. In that case, paradoxically, therapists make the clients' accusations come true: They metaphorically reenact the clients' past conflicted endings. Therapists do better by acknowledging the similarity of the clients' feelings at termination and at past separations, and then exploring with clients how they can make this termination different from such problematic partings.

During the termination period, clients often have a range of sad, hopeful, grateful, and other feelings toward the therapist. As therapists track clients' reactions to the ending, they will often see how the clients' generic conflicts are simultaneously being aroused in two contexts. First, the clients' primary feelings about termination (anxiety, helplessness, betrayal) may match the clients' feelings about the formative relationship in which the generic conflict originally arose. Second, clients' reaction toward termination may also match the feelings that were aroused in the crisis situation that originally brought them to treatment. By responding affirmingly to the clients' feelings, and clarifying their connection to these two sources when appropriate, the therapist is helping them further resolve their generic conflicts.

Finally, another useful intervention for the termination phase is to go through a Review-Predict-Practice sequence with all clients. Therapists *review* progress, accomplishments, successful changes and transitions, and unfinished issues with clients. Especially important, therapists help clients *predict* the developmental tasks or anxiety-arousing relationship issues that are likely to evoke their dysfunctional patterns in the future. It is essential to anticipate these potentially problematic situations or relational scenarios as specifically as possible. As illustrated with Tracy, earlier in this chapter, the therapist might ask, "What is he likely to say, how is that going to make you feel, and what are you most likely to say and do at that moment?" *Practice* in dealing with these issues or patterns also prepares clients to succeed on their own. Therapists can provide this by role-playing effective and ineffective responses to the threatening or unwanted interactions that clients are likely to face in the future.

To sum up, the therapist must address the client's feelings about ending the relationship and help the client work them through. This is difficult to do, however, when both the therapist and the client may wish to avoid the ending. The therapist and client have shared much in their work together and the relationship holds real meaning for both of them. In order to avoid the conflicts that are activated by the termination, clients—and therapists, too—may try to diminish the importance of the ending. Some clients may try to devalue the significance of their relationship with the therapist or to diminish the importance of the work that has been accomplished. Other clients will become symptomatic again and anxiously communicate that they are still too troubled to terminate and make it on their own. A few clients will ask the therapist if they can become friends. These and other common responses during the termination phase keep clients from facing current and past losses, preclude clients from receiving a new and more reparative response from the therapist, and keep clients from being able to internalize the therapist and carry forth with others the gains they have made in treatment.

Ending the Relationship

Although some clients will reexperience their presenting problems at the time of termination, many will not. Both clients and therapists still need to talk about the ending, however. For example, therapists can talk with the client about the different ways in which they have seen the client change over the course of treatment. Therapists can also acknowledge the limitations of their work together and the unfinished issues that clients will need to continue to work with on their own. Recollections of close moments, awkward misunderstandings, risks ventured, and humorous incidents can be shared. This is also a time when therapists can share their own feelings about the client: what they have learned about themselves and life from the client; how the client has enriched or influenced them; ways in which therapists have cared about or have enjoyed the client; and how therapists will remember the client.

Finally, in natural and unnatural endings, termination brings the therapist and client back to the separateness-relatedness dialectic. In the case of natural endings, the therapist must give clients permission to leave. Clients must know unambiguously that the therapist enjoys their success, is pleased by their independence, and takes pleasure in seeing them become committed to new relationships. A few clients may need to be reassured that they can have future contact with the therapist if necessary. Clients will be reassured to know that, whenever possible, the therapist is available to help them with future crises, should the need arise. Knowing that the therapist is available if needed, but also supports their decision to leave, clients are able to keep and apply the help the therapist has provided. This type of natural termination supports clients' own strivings for individuation in a supportive relational context, and helps clients claim their own self-efficacy and authentic voice.

Graduate-student therapists often cannot be available in this way, however, because endings with their clients are of necessity unnatural. In these cases, the finality of the ending must be honestly acknowledged. Therapists, often with the help of a sensitive supervisor, want to be prepared to manage their countertransference reactions and respond acceptingly, rather than defensively overreacting, to the client's disappointed, dismissing, and other reactions. The unfinished business that remains must also be realistically acknowledged, and the therapist may need to help the client find a new therapist to finish the work that was just begun. If the therapist can do this, most clients will be able to accept the limitations of the relationship without having to reject the therapist or undo the work that was accomplished. Although often incomplete, when clients can live-out with others what they have experienced with their therapists, and therapists can appreciate what they have learned from and given to their clients, both have been enriched by this successful relationship.

Closing

This book has followed the course of therapy from beginning to end. Just as clients still have unresolved conflicts when treatment ends, many questions about therapy still remain unanswered for therapists-in-training. Complex issues and interpersonal processes have been introduced here, but therapists will need further reading, supervision, and experience to help with the myriad exceptions that occur in every therapeutic relationship. Despite these limitations, the reader has learned essential guidelines for working with the process dimension and the relational aspects of therapy. The conceptual framework presented here should enable beginning therapists to function more effectively as professionals and, hopefully, to better incorporate training experiences from different theoretical perspectives.

In closing, the central theme of this text is that growth and change occur within the context of significant emotional relationships. Accordingly, therapists are encouraged to utilize themselves and the relationships they provide their clients as their most effective means of helping clients change.

Suggestions for Further Reading

1. At the end of Chapter 4, readers were encouraged to complete the Self-Assessment Scale for Interpersonal Skills. Having completed this text, students may wish to assess themselves again on these essential clinical skills (see Chapter 10 of the Student Workbook).

2. Student therapists who have found the ideas presented here to be helpful are likely to enjoy *Between Therapist and Client* by M. Kahn (New York: Freeman, 1997) and *How Psychotherapy Works* by J. Weiss (New York: Guilford, 1993). For more experienced therapists, *Relational Concepts in Psychoanalysis* by Stephen Mitchell (Cambridge: Harvard University Press, 1988) and *Coping with Conflict: Supervising Counselors and Psychotherapists* by W. Mueller & B. Kell (New York: Appleton-Century-Crofts, 1972) both provide an informative presentation of interpersonally oriented psychotherapy.

3. Useful information about identifying clients' generic conflicts and how they can be reactivated during the termination phase of treatment can be found in J. Mann's *Time-Limited Psychotherapy* (Cambridge: Harvard University Press, 1973) and in Chapter 9 of H. Strupp and J. Binder's *Psychotherapy in A New Key* (New York: Basic Books, Inc., 1984).

4. The Dream is an important psychological dimension in people's lives that has received too little attention. The reader is encouraged to read Daniel Levinson's groundbreaking book, *The Seasons of a Man's Life* (New York: Ballantine, 1978). See especially pages 91–109. *The Seasons of A Woman's Life* (New York: Knopf, 1996), published soon after Levinson's death, tracks developmental stages and transitions in women's lives. The Dream, as it relates to women, is given special attention on pages 238–239.

Process Notes

Name: _____ Date: _____

1. Client Behavior: What is the client doing in this session?

What were the principal concerns expressed or new issues raised? Identify the emotions tapped and relational themes expressed.

_____

_____

_____

_____

_____

_____

_____

_____

_____

_____

2. Therapeutic Relationship: What can I learn from our current interaction?

How did I feel about my effectiveness in this session? What were my personal reactions toward the client? Do my emotional reactions typify the reactions of others toward the client (e.g., boredom, frustration, protectiveness), or do they suggest my own countertransference? How is our current interaction similar to (reenacting) or different from (resolving) what transpires with others?

_____

_____

_____

_____

_____

_____

_____

_____

_____

3. Treatment Plans: What are my goals?

What issues do I need to explore to clarify the focus of treatment? How are the client's maladaptive relational patterns (and the dysfunctional beliefs and conflicted feelings that accompany them) linked to his or her presenting problems? What are the specific relational experiences I need to provide for this client in order to enact a reparative experience? Have there been any significant changes in the client's life situation or therapy to alter my treatment plans? Have I missed any factors that have an impact on my treatment plan?

_____

_____

_____

_____

_____

_____

_____

_____

4. Intervention Strategies: Which interventions were effective in this session and which were ineffective?

What did I do to try and reach my goals (e.g., clarify maladaptive relational patterns, use process comments or self-involving statements, reframe, reflect predominant affect, validate subjective worldview, improve parenting skills, link presenting problems to our current interaction, clarify the pathogenic beliefs that emerge when the client is distressed, role-play)? How did the client respond to these interventions? What can I do next time to differentiate myself and our process from the relational themes that have been problematic with others? How consonant are my strategies with the client's familial and cultural background?

_____

_____

_____

_____

_____

_____

_____

_____

_____

Guidelines for Treatment Planning

It seems that nothing can make a new therapist feel anxious more quickly than having to write a case formulation! Realistically, it is challenging to organize your thoughts into a coherent treatment plan, but keep trying and your efforts will be rewarded with more productive sessions. The following guidelines will help therapists write treatment plans that provide a more focused therapy and help therapists respond to treatment planning requests from HMOs and other third-party insurers.

1. Formulation of the problems   Summarize the client's presenting problems and suggest why this client is seeking treatment now. What has the client done in the past to address these problems? How does the client feel about himself or herself (e.g., worthless, superior, burdened) and how is this manifested with others and with you? Clarify the organizing themes or patterns that characterize the client's most important experiences with others. What expectations does the client have of others in crisis moments or during stressful interactions (e.g., are they experienced as critical, withholding, intrusive, appreciative)? What are the conflicted feelings states that tend to recur for this client (e.g., is the client frequently anxious, ashamed, guilty)? Are there any significant personal, familial, or cultural factors that make it difficult for this client to acknowledge having problems or to ask for help? How are these potential signs of "resistance" likely to be expressed with you in treatment? How have sociocultural factors (such as race, religion, socioeconomic status, gender, and sexual orientation) affected the client and shaped his or her subjective worldview? Provide a diagnosis with your rationale.

2. Treatment focus   Anticipate the potential course of treatment. Clarify the maladaptive relational patterns (including the pathogenic beliefs and conflicted feelings that accompany them) that recur for the client in his or her presenting problems, in current and past significant relationships, and with you. For example, thinking of the client's internal working models or cognitive schemas, ask yourself: (1) What does the client want from others (e.g., to be cared for or protected)? (2) What does the client expect from others (e.g., to be criticized or rejected)? (3) What is the client's experience of self in relationship to others (e.g., a sense of being disconnected or burdensome)? (4) What are the conflicting affective states that typically result (e.g., contempt and shame or affection and security)? Thinking in terms of these relational patterns and the client's expectations of others, identify the interpersonal coping strategies this client has typically used (for example, complying with others, withdrawing from others, trying to control others). What do these coping strategies tend to elicit from the counselor and others? Try to provide a conceptual focus for treatment by integrating your formulations of this client's problems into two or three themes that capture what you think is wrong and needs to change.

3. Developmental context   How did the client's problems originally come about? Reveal attachment history (e.g., a self-absorbed and unresponsive caregiver), parenting practices (e.g., a nurturant caregiver who was, however, unable to set limits and enforce discipline), familial roles (e.g., a scapegoated or parentified child), family systems functioning and communication patterns (e.g., the nature of the marital relationship, clarity of intergenerational boundaries), trauma (e.g., sexual abuse), or other developmental experiences that may have shaped current problems. If you do not have sufficient developmental information, formulate working hypotheses about potential developmental experiences that may have transpired. Suggest how these developmental experiences (individual, familial, cultural) could interact with current stressors, crises, and social supports to shape the client's subjective worldview and current conflicts. Try to identify the client's pathogenic beliefs about himself or herself and faulty expectations of others.

4. Therapeutic process   What has treatment focused on to date, and how have you and the client interacted? How do you tend to feel toward the client? Is this similiar or different from how others tend to feel toward the client? How does the client bring his or her conflicts into the therapeutic relationship? More specifically, consider several ways the therapeutic process you are enacting with the client may be both different from and similar to the maladaptive relational patterns identified in your treatment plan. What does the client tend to elicit from you and others? How does the client "test" (i.e., confirm or disconfirm) core beliefs and expectations with you? When the client is distressed, how does he or she tend to distort or misperceive you?

5. Goals and interventions   Where do you want to go with this client and how do you plan to get there? For example, you might want to give a sexual abuse survivor the experience of secure boundaries, or to convey to the client a sense of being valued, or to support efficacy by encouraging the client's own initiative. Focusing on two or three goals that seem realistic to you and are experienced by the client as his or her own, utilize your interpersonal and developmental formulations of the client's problems to clarify what you are trying to accomplish right now (e.g., help this client feel "seen" and understood), your intermediate goals (e.g., help this client learn to say "no" and advocate on his or her own behalf more assertively), and your long-term goals (e.g., help this client clarify his or her own internal experience, identify and pursue his or her own genuine interests and goals, and express his or her own voice more authentically). Delineate your intervention strategies at each time period and how you plan to use yourself and the relationship you provide this client to reach these goals. Articulate as specifically as you can the responses this client will need from you in order to change. For example, for clients with detached parents, you would actively help them generate possible solutions. In contrast, for clients with intrusive or controlling parents, you would minimize suggestions, homework, and directions while validating and joining enthusiastically in all expressions of the client's own initiative. Evaluate the appropriateness of the goals set and interventions used, given the client's personal, familial, and cultural context. Assess the appropriateness of referrals.

6. Impediments to change   If the client terminates prematurely or treatment is not successful, what factors could be contributimg? Identify the specific situations, relational scenarios, or affects that are likely to trigger the client's anxiety or diminish his or her sense of safety. How will the client cope with this anxiety in the client's relationship with you (e.g., by withdrawing, becoming critical, or complying)? Why have these anxieties and threats made sense in terms of the client's worldview or previous experience, and how were these coping strategies adaptive and necessary in the past? Suggest potential ways that the client's old relational scenarios could be reenacted in the therapeutic relationship and how the client's conflicts or interpersonal style may interact with your own countertransference propensities.

BIBLIOGRAPHY

Alexander, F., & French, T. M. (1980). *Psychoanalytic therapy: Principles and applications.* University of Nebraska Press.

Anchin, J., & Kiesler, D. (1982). *Handbook of interpersonal psychotherapy.* New York: Pergamon Press.

Anderson, C. M., & Steward, S. (1983). *Mastering resistance: A practical guide to family therapy.* New York: Guilford.

Balcom, D., Lee, R., & Tager, J. (195). The systemic treatment of shame in couples. *Journal of Marital and Family Therapy, 21*(1), 55–65.

Bandura, A. (1977). Self-efficacy: Toward a unifying theory of behavioral change. *Psychological Review, 84*(2), 191–215.

Bandura, A. (1982). Self-efficacy mechanisms in human agency. *American Psychologist, 37,* 122–147.

Bandura, A. (1997). *Self-efficacy: The exercise of control.* New York: Freeman.

Baumrind, D. (1967). Child care practices anteceding three patterns of preschool behavior. *Genetic Psychology Monographs, 75,* 43–88.

Baumrind, D. (1971). Current patterns of parental authority. *Developmental Psychology Monograph, 4*(Whole No. 1, Pt. 2).

Baumrind, D. (1983). Familial antecedents of social competence in young children. *Psychological Bulletin, 94*(1), 132–142.

Beavers, W. R. (1982). Healthy, midrange and severely dysfunctional families. In F. Walsh (Ed.), *Normal family processes* (pp. 45–66). New York: Guilford.

Beck, A. (1976). *Cognitive therapy and the emotional disorders.* New York: International Universities Press.

Beck, A., & Freeman, A. (1990). *Cognitive therapy of the personality disorders.* New York: Guilford.

Benjamin, L. (1995). *Interpersonal diagnosis and treatment of personality disorder* (2nd ed.). New York: Guilford.

Bergin, A., & Garfield, S. (Eds.). (1994). *Handbook of psychotherapy and behavior change* (4th ed.). New York: Wiley.

Binder, J., & Strupp, H. (1997). "Negative process": A recurrently discovered and underestimated facet of therapeutic process and outcome in the individual psychotherapy of adults. *Clinical Psychology: Science and Practice, 4,* 121–139.

Boszormenyi-Nagy, I., & Spark, G. (1973). *Invisible loyalties: Reciprocity in inter-generational family therapy.* New York: Harper & Row.

Bowen, M. (1966). The use of family theory in clinical practice. *Comprehensive Psychiatry, 7,* 345–376.

Bowen, M. (1978). *Family therapy in clinical practice.* New York: Aronson.

Bowlby, J. (1988). *A secure base.* New York: Basic Books.

Burns, D., & Auerbach A. (1996). Therapeutic empathy in cognitive-behavioral therapy: Does it really make a difference? In P. Salkovskis (Ed.), *Frontiers of Cognitive Therapy* (pp. 135–164). New York: Guilford.

Cashdan, S. (1988). *Object Relations Therapy.* New York: Norton.

Davanloo, H. (Ed.). (1980). *Short-term dynamic psychotherapy.* New York: Aronson.

Ecker, B., & Hulley, L. (1996). *Depth-oriented brief therapy.* San Francisco: Jossey-Bass.

Egan, G. (1994). *The skilled helper.* Pacific Grove, CA: Brooks/Cole.

Engel, L., & Ferguson, T. (1990). *Hidden guilt.* New York: Pocket Books.

Erikson, E. (1968). *Childhood and society* (2nd ed.). New York: Norton.

Farley, J. (1979, January). Family separation-individuation tolerance: A developmental conceptualization of the nuclear family. *Journal of Marital and Family Therapy,* 61–67.

Field, T. (1996). Attachment and separation in young children. *Annual Review of Psychology, 47,* 541–561.

Frailberg, S., Adelson, E., & Shapiro, V. (1975). Ghosts in the nursery: A psycho-analytical approach to the problems of impaired mother–infant relationships. *Journal of the American Academy of Child Psychiatry, 14,* 387–421.

Fromm, E. (1982). *Escape from freedom.* New York: Avon.

Fromm-Reichmann, F. (1960). *Principles of intensive psychotherapy.* Chicago: University of Chicago Press.

Gelso, C., & Carter, J. (1994). Components of the psychotherapy relationship: Their interaction and unfolding during treatment. *Journal of Counseling Psychology,* 41(3) 296–306.

Gill, M., & Muslin, H. (1976). Early Interpretations of transference. *Journal of the American Psychoanalytic Association, 24,* 779–794.

Gilligan, C. (1982). *In a different voice: Psychological theory and women's development.* Cambridge: Harvard University Press.

Goldenberg, I., & Goldenberg, H. (1996). *Family therapy: An overview* (4th ed.). Pacific Grove, CA: Brooks/Cole.

Greenberg, J., & Mitchell, S. (1983). *Object relations in psychoanalytic theory.* Cambridge: Harvard University Press.

Greenson, R. (1967). *The technique and practice of psychoanalysis* (Vol. 1). New York: International Universities Press.

Guy, J. (1987). *The personal life of the psychotherapist.* New York: Wiley.

Haley, J. (1967). Toward a theory of pathological systems. In G. H. Zuk & I. Boszormenyi-Nagy (Eds.), *Family therapy and disturbed families.* Palo Alto, CA: Science and Behavior Books.

Haley, J. (1980). *Leaving home; The therapy of disturbed young people.* New York: McGraw-Hill.

Hartman, E. (1978, October). Using eco-maps and genograms in family therapy. *Social Casework,* 464–476.

Henry, W., Schacht, T., & Strupp, H. (1990). Patient and therapist introject, interpersonal process, and differential psychotherapy outcome. *Journal of Consulting and Clinical Psychology, 58,* 768–774.

Horney, K. (1966). *Our inner conflicts.* New York: Norton.

Horney, K. (1970). *Neurosis and human growth.* New York: Norton.

Horowitz, M., Marmar, C., Krupnick, J., Wilner, N., Kaltreider, N., & Wallerstein, R. (1984). *Personality styles and brief psychotherapy.* New York: Basic Books.

Hovarth, A. O., & Greenberg, L. S. (1994b). *The working alliance: Theory, research, and practice.* New York: Wiley.

Jordan, J., Kaplan, A., Miller, J., Striver, P., & Surrey, J. (1991). *Women's growth in connection* (pp. 126–184). New York: Guilford.

Kafka, F. (1966). *Letter to his father.* New York: Schocken Books.

Kahn, M. (1997). *Between therapist and client.* New York: Freeman.

Karen, R. (1998). *Becoming attached.* New York: Oxford University Press.

Kell, B., & Mueller, W. (1966). *Impact and change: A study of counseling relationships.* Englewood Cliffs, NJ: Prentice-Hall.

Kelly, G. (1963). *The psychology of personal constructs.* New York: Norton.

Kiesler, D. (1988). *Therapeutic metacommunication: Therapist impact disclosure as feedback in psychotherapy.* Palo Alto, CA: Consulting Psychologists Press.

Kiesler, D., & Van Denburg, T. (1993). Therapeutic impact disclosure: A last taboo in psychoanalytic theory and practice. *Clinical Psychology & Psychotherapy, 1*(1), 3–13.

Klerman, G. L., Weissman, M., Rounsaville, B., & Chevron, E. (1984). *Interpersonal psychotherapy of depression.* New York: Basic Books.

Kohut, H. (1971). *The analysis of the self.* New York: International Universities Press.

Kohut, H. (1977). *The restoration of the self.* New York: International Universities Press.

Laing, R. D., & Esterson, A. (1970). *Sanity, madness and the family.* Middlesex, England: Penguin.

Lazarus, A. (1973). Multimodal behavior therapy: Treating the BASIC I.D. *Journal of Nervous and Mental Disease, 156,* 404–411.

Levenson, E. (1982). Language and healing. In S. Slip (Ed.), *Curative factors in dynamic psychotherapy.* New York: McGraw-Hill.

Levenson, H. (1995). *Time-limited dynamic therapy.* New York: Basic Books.

Levinson, D. (1978). *The seasons a man's life.* New York: Ballantine.

Levinson, D. (1996). *The seasons of a woman's life.* New York: Knopf.

Lewis, H. (1971). *Shame and guilt in neurosis.* New York: International Universities Press.

Luborsky, L., & DeRubeis, R. (1984). The use of psychotherapy treatment manuals: A small revolution in psychotherapy research style. *Clinical Psychology Review, 4,* 5–14.

Luborsky, L., & Marks, D. (1991). Short-term supportive-expressive psychoanalytic psychotherapy. In P. Crits-Cristoph & J. Barber (Eds.), *Handbook of short-term dynamic psychotherapy* (pp. 110–136). New York: Basic Books.

Mahler, M., Pine, F., & Bergman, A. (1975). *The psychological birth of the human infant: Symbiosis and individuation.* New York: Basic Books.

Malan, D. H. (1976). *The frontier of brief psychotherapy: An example of the convergence of research and clinical practice.* New York: Plenum.

Mann, J. (1973). *Time-limited psychotherapy*. Cambridge, MA: Harvard University Press.

Mann, J., & Goldman, R. (1982). *A casebook in time–limited psychotherapy*. New York: McGraw-Hill.

Masterson, J. (1972). *Treatment of the borderline adolescent: A developmental approach*. New York: Wiley.

Masterson, J. (1976). *Psychotherapy of the borderline adult: A developmental approach*. New York: Brunner/Mazel.

May, R. (1977). *The meaning of anxiety*. New York: Norton.

McCarthy, P. (1982). Differential effects of counselor self-referent responses and counselor status. *Journal of Counseling Psychology, 29*, 125–131.

McClure, F., & Teyber, E. (1996). *Child and adolescent therapy: A multicultural-relational approach*. Fort Worth, TX: Harcourt Brace Jovanovich.

McClure, F., & Teyber, E. (2000). *Multicultural case studies in child and adolescent therapy*. Pacific Grove, CA: Wadsworth.

Miller, A. (1949). *Death of a salesman*. New York: Penguin.

Miller, A. (1981). *Prisoners of childhood*. New York: Basic Books.

Mills, J., Bauer, G., & Miars, R. (1989). Use of transference in short-term dynamic *Psychotherapy, 26*, 338–343.

Minuchin, S. (1974). *Families and family therapy*. Cambridge: Harvard University Press.

Mitchell, S. (1988). *Relational concepts in psychoanalysis*. Cambridge: Harvard University Press.

Mitchell, S., & Black, M. (1995). *Freud and beyond*. New York: Basic Books.

Mueller, W., & Kell, B. (1972). *Coping with conflict: Supervising counselors and psychotherapists*. New York: Appleton-Century-Crofts.

Najavits, L., & Strupp, H. (1994). Differences in the effectiveness of psychodynamic therapists: A process-outcome study. *Psychotherapy, 31*, 114–123.

Olson, D. H., McCubbin, H. I., Barnes, H., Larsen, A., Muxen, M., & Wilson, M. (1983). *Families: What makes them work*. Newbury Park, CA: Sage.

Passons, W. (1975). *Gestalt approaches in counseling*. New York: Holt, Rinehart & Winston.

Pinderhughes, H. (1989). *Understanding race, ethnicity and power*. New York: Free Press.

Pipher, M. (1994). *Reviving Ophelia*. New York: Ballantine.

Robitschek, C., & McCarthy, P. (1991). Prevalence of counselor self-reliance in the therapeutic dyad. *Journal of Counseling & Development, 69*, 218–221.

Rocklin R., & Levitt, D. (1987). Those who broke the cycle: Therapy with non-abusive adults who were physically abused as children. *Psychotherapy, 24*(4), 769–778.

Rogers, C. (1951). *Client-centered therapy*. Boston: Houghton-Mifflin.

Rogers, C. (1959). A theory of therapy, personality and interpersonal relationships as developed in the client-centered framework. In S. Koch (Ed.), *Psychology: A study of science:* Vol. 3. Formulations of the person and the social context (pp. 184–256). New York: McGraw-Hill.

Safran, J., & Muran, J. (Eds.). (1998). *The therapeutic alliance in brief psychotherapy*. Washington: D.C.: American Psychological Association.

Satir, V. (1967). *Conjoint family therapy*. Palo Alto, CA: Science and Behavior Books.

Schatzman, M. (1973). *Soul murder: Persecution in the family*. New York: Random House.

Seligman, M. (1975). *Helplessness*. New York: Freeman.

"Shame." (1975). [Special Issue]. *American Behavioral Scientist, 38*(8).

Sifneos, P. (1987). *Short–term dynamic psychotherapy: Evaluation and techniques* (2nd ed.). New York: Plenum.

Solomon, M. (1973). A developmental, conceptual premise for family therapy. *Family Process, 12*(2), 179–188.

Speight, S., Myers, L., Cox, C., & Highlen, P. (1991). A redefinition of multicultural counseling. *Journal of Counseling Development, 70*(1), 29–36.

Springmann, R. (1986). Countertransference: Clarifications in Supervision. *Contemporary Psychoanalysis, 22*, 252–277.

Steinbeck, J. (1952). *East of Eden*. New York: Viking Press.

Stierlin, H. (1972). *Separating parents and adolescents*. New York: Quadrangle.

Strupp, H. H. (1980). Success and failure in time-limited psychotherapy. *Archives of General Psychiatry, 37*, 595–613, 708–716, 831–841, 947–954.

Strupp, H., & Binder, J. (1984). *Psychotherapy in a new key: A guide to time-limited dynamic psychotherapy*. New York: Basic Books.

Strupp, H., & Hadley, S. (1979). Specific versus nonspecific factors in psychotherapy: A controlled study of outcome. *Archives of General Psychiatry, 36*, 1125–1136.

Sue, D., & Sue, H. (1998). *Counseling the culturally different* (3rd ed.). New York: Wiley.

Sue, S., & Zane, N. (1987). The role of culture and cultural techniques in psychotherapy: A critique and reformulation. *American Psychologist, 42*, 37–45.

Sullivan, H. S. (1968). *The interpersonal theory of psychiatry*. New York: Norton.

Sullivan, H. S. (1970). *The psychiatric interview*. New York: Norton.

Teyber, E. (1981). Structural family relations: A review. *Family Therapy, 1*, 39–48.

Teyber, E. (1983). Effects of the parental coalition on adolescent emancipation from the family. *Journal of Marital and Family Therapy, 9*, 89–99.

Teyber, E. (1994). *Helping children cope with divorce: A practical guide for parents*. San Francisco: Jossey-Bass.

Teyber, E., & McClure, F. (2000). Therapist variables. In C. Snyder & R. Ingram (Eds.), *Handbook of psychological change; Psychotherapy processes and practices for the 21st century*. New York: Wiley.

Tyron, G., & Kane, A. (1993). Relationship of working alliance to mutual and unilateral termination. *Journal of Counseling Psychology, 40*(1), 33–36.

Wachtel, P. (1982). *Psychoanalysis and behavior therapy: Toward an integration*. New York: Basic Books.

Wachtel, P. (1993). *Therapeutic communication: Principles and effective practice*. New York: Guilford.

Walborn, F. (1996). *Process variables: Four common elements of counseling and psychotherapy*. Pacific Grove, CA: Brooks/Cole.

Weiss, J. (1993). *How psychotherapy works*. New York: Guilford.

Weiss, J., & Sampson, H. (1986). *The psychoanalytic process: Theory, clinical observation and empirical research*. New York: Guilford.

Wenar, C. (2000). *Psychopathology from infancy through adolescence: A developmental approach* (4th ed.). New York: Random House.

Wheelis, A. (1974). *How people change*. New York: Harper & Row.

Winnicott, D. W. (1965). Ego distortion in terms of true and false self. In *The maturational process and the facilitating environment*. New York: International Universities Press.

Woodward, B. (1984). *Wired*. New York: Simon & Schuster.

Yalom, I. (1975). *The theory and practice of group psychotherapy* (2nd ed.). New York: Basic Books.

Yalom, I. (1981). *Existential psychotherapy*. New York: Basic Books.

NAME INDEX

Adelson, E., 55
Alexander, F., 18
Anchin, J., 273, 274
Anderson, C.M., 86
Auerbach, 33

Balcom, D., 86
Bandura, A., 98,115
Bateson, G., 10
Bauer, G., 222
Baumrind, D., 163
Beauers, 152
Beck, A., 48, 185, 208, 210
Benjamin, L., 5, 26
Bergin, A., 251
Bergman, A., 14, 237
Binder, J., 7, 48, 56, 185, 208, 303
Black, M., 28
Block-Lewis, H., 129
Borns, 33
Boszormenyi-Nagy, I., 11, 14, 57, 162, 177
Bowen, M., 7, 11, 37, 57, 151, 152, 177
Bowlby, J., 28, 33, 55, 115

Carter, J., 33
Cashdan, S., 51, 273
Chevron, E., 26
Cox, C., 78

Davanloo, H., 25
DeRubeis, R., 48

Ecker, 48
Eagan, G., 57, 148
Engel, L., 237
Erickson, E., 7, 153
Esterson, A., 40

Fairbairn, W., 10
Farley, J., 161
Ferguson, T., 237
Fraiberg, S., 55
French, T., 18
Freeman, A., 27, 48, 195, 208, 210
Freud, S., 5, 10
Fromm, E., 210
Fromm-Reichmann, F., 6, 7, 19

Garfield, S., 251
Gaston, L., 33
Gelso, C., 33
Gill, M., 230
Gilligan, C., 56
Goldenberg, H., 10
Goldenberg, I., 10
Goldman, R., 25
Greenberg, J., 33
Greenson, R., 32, 86
Guy, J., 148

Hadley, S., 50, 141
Haley, J., 10, 37, 57, 232, 237
Hartman, E., 177

VanDenberg, T., 51

Wachtel, P., 255, 274
Walborn, F., 251
Wasserman, M., 5
Weiss, J., 18, 23, 48, 138, 210, 216, 237,
    303
Wenar, C., 14, 164, 178

Wheelis, A., 90, 115
Winnicott, D., 41, 137
Woodward, B., 165

Yalom, I., 53, 70, 115

Zuk, G., 57

SUBJECT INDEX

TO THE OWNER OF THIS BOOK:

I hope that you have found *Interpersonal Process in Psychotherapy: A Relational Approach,* Fourth Edition useful. So that this book can be improved in a future edition, would you take the time to complete this sheet and return it? Thank you.

School and address: _____

_____

Department: _____

Instructor's name: _____

1. What I like most about this book is: _____

_____

_____

2. What I like least about this book is: _____

_____

_____

3. My general reaction to this book is: _____

_____

_____

4. The name of the course in which I used this book is: _____

_____

5. Were all of the chapters of the book assigned for you to read? _____

   If not, which ones weren't? _____

6. In the space below, or on a separate sheet of paper, please write specific suggestions for improving this book and anything else you'd care to share about your experience in using this book.

_____

_____

_____

OPTIONAL:

Your name: _____ Date: _____

May we quote you, either in promotion for *Interpersonal Process in Psychotherapy: A Relational Approach,* Fourth Edition, or in future publishing ventures?

Yes: _____ No: _____

Sincerely yours,

Edward Teyber

FOLD HERE

ATTN: Counseling Editor, Eileen Murphy

BROOKS/COLE/THOMSON LEARNING
511 FOREST LODGE ROAD
PACIFIC GROVE, CA   93950-9968

FOLD HERE